Nanoscale CMOS VLSI Circuits

Design for Manufacturability

About the Authors

Sandip Kundu, Ph.D., is a professor in the Electrical and Computer Engineering Department at the University of Massachusetts at Amherst specializing in VLSI design and testing. Previously, he was a Principal Engineer at the Intel Corporation and Research Staff Member at the IBM Corporation. He is a fellow of the IEEE, has been a distinguished visitor of the IEEE Computer Society and associate editor of the IEEE Transactions on Computers and the IEEE Transactions on VLSI Systems. Dr. Kundu has published more than 130 technical papers and holds 12 patents.

Aswin Sreedhar, Ph.D., is a research associate in the Electrical and Computer Engineering Department at the University of Massachusetts at Amherst. His research interests are in statistical techniques for design for manufacturability of VLSI Systems and circuit reliability. Previously, he was a graduate intern at the Intel Corporation and AMD. He is a recipient of the best paper award for lithography based yield modeling (2009) at the DATE conference.

Nanoscale CMOS VLSI Circuits

Design for Manufacturability

Sandip Kundu

Aswin Sreedhar

McGraw Hill

New York Chicago San Francisco Lisbon
London Madrid Mexico City Milan New Delhi
San Juan Seoul Singapore Sydney Toronto

The McGraw·Hill Companies

Cataloging-in-Publication Data is on file with the Library of Congress.

McGraw-Hill books are available at special quantity discounts to use as premiums and sales promotions, or for use in corporate training programs. To contact a representative please e-mail us at bulksales@mcgraw-hill.com.

Nanoscale CMOS VLSI Circuits: Design for Manufacturability

1 2 3 4 5 6 7 8 9 0 DOC/DOC 1 9 8 7 6 5 4 3 2 1 0

ISBN: 978–0-07–163519-6
MHID: 0–07-163519-X

The pages within this book were printed on acid-free paper.

Sponsoring Editor
Michael Penn

Acquisitions Coordinator
Michael Mulcahy

Editorial Supervisor
David E. Fogarty

Project Manager
Aloysius Raj, Newgen Imaging Systems Pvt. Ltd.

Copy Editor
Matt Darnell

Proofreader
Ginny Faber

Indexer
Sandip Kundu

Production Supervisor
Pamela A. Pelton

Composition
Newgen Imaging Systems Pvt.Ltd.

Art Director, Cover
Jeff Weeks

To my late father Prof. Hari Mohan Kundu,
whose encouragement for pursuit of excellence still endures,
and to my mother Mrs. Pravati Kundu,
who has been a pillar of support and strength,
my loving wife Deblina and
my daughters Shinjini and Shohini

—Sandip

To my father Mr. Sreedhar Jagannathan
and my mother Mrs. Amirthavalli Sreedhar,
whose sacrifice and ever present support drives
my thirst for knowledge,
my loving wife Srividhya and
my brother Parikshit

—Aswin

Contents

Preface

The purpose of this book is to introduce readers to the world of design for manufacturability and reliability. It is intended to be used as a text for senior-level undergraduates and for graduate students in their initial years and also to serve as a reference for practicing design engineers. Because there are entire conferences and journals devoted to this subject, it is impossible for any compendium to be complete or fully current. Therefore, we focus more on principles and ideas than on the granular details of each topic. There are references at the end of each chapter that direct the reader to more in-depth study. In order to understand this book, readers should have some knowledge of VLSI design principles, including cell library characterization and physical layout development.

This book is a result of the research interests of both coauthors, who have published actively in the area of design for manufacturability. Professor Kundu also introduced a new course on Design for Manufacturability and Reliability at the University of Massachusetts. Much of this book's organization is based on the structure of that course, which was developed for classroom instructions. Thus, it is hoped that students will benefit greatly from this book. The text also deals extensively with costs, constraints, computational efficiencies, and methodologies. For this reason, it should also be of value to design engineers.

The material is presented in eight chapters. Chapter 1 introduces the reader to current trends in CMOS VLSI design. It offers a brief overview of new devices and of contributions from material sciences and optics that have become fundamental for the design process achieving higher performance and reduced power consumption. The basic concepts of design for manufacturability (DFM) are reviewed along with its relevance to and application in current design systems and design flows. The chapter also explores reliability concerns in nano-CMOS VLSI designs from the perspective of design for reliability (DFR), computer-aided design (CAD) flows, and design optimzations to improve product lifetime.

Chapter 2 discusses the preliminaries of semiconductor manufacturing technology. The various steps—which include oxidation, diffusion, metal deposition, and patterning—are explained. This chapter concentrates chiefly on patterning steps that involve photolithography and etching processes. Techniques are discussed for modeling the photolithography system so that the manufacturability of a given design can be effectively analyzed. The techniques are classified as either phenomenological or fully physical, and the methods are compared in terms of accuracy and computational efficiency.

The focus of Chapter 3 is variability in the process parameters for current and future CMOS devices and the effects of this variability. The main subjects addressed are variations in patterning, dopant density fluctuation, and dielectric thickness variation due to chemical-mechanical polishing and stress.

Chapter 4 explains the fundamentals of lithographic control through layout-based analysis as well as the important photolithography parameters and concepts. Lithographic variability control is illustrated by resolution enhancement techniques such as optical proximity correction, phase shift masking, and off-axis illumination. This chapter discusses components of the DRM manual, including geometric design rules, restricted design rules, and antenna rules. It also contains sections on the evolution of model-based design rules check and other changes to CAD tools for traditional physical design. The chapter concludes with a presentation of advanced lithography techniques, such as dual-pattern lithography, inverse lithography technology and source mask optimization, that are used to push the resolution limit.

Chapter 5 provides an in-depth look at various manufacturing defects that occur during semiconductor manufacturing. These defects are classified as resulting either from contaminants (particulate defects) or from the design layout itself (pattern dependent). The chapter describes how critical area is used to estimate yields for particle defects as well as how linewidth-based models are used to estimate yield for pattern-dependent defects. Metrology and failure analysis techniques—and their application to semiconductor measurement for process control—are also described.

In Chapter 6, particle defects and pattern-based defects are examined in terms of their impact on circuit operation and performance. The discussion covers the defect models and fault models used to effectively identify and predict design behavior in the presence of defects. This chapter also explores yield improvement for designs through fault avoidance and fault tolerance techniques.

The physics of reliability issues and their impacts are discussed in Chapter 7. Reliability mechanisms such as hot carrier injection, negative temperature bias instability, electromigration, and electrostatic discharge (ESD) are explained and illustrated. The mean time to failure for each of these reliability failure mechanisms are also discussed with design solution to mitigate their effects.

Finally, Chapter 8 addresses the changes to CAD tools and methodologies due to DFM and DFR approaches at different stages of the circuit realization process, including library characterization, standard cell design, and physical design. This chapter then delves into the need for statistical design approaches and model-based solutions to DFM-DFR problems. The importance of reliability-aware DFM approaches for future designs is also detailed.

The central theme of this book is that decisions made during the design process will affect the the product's manufacturability, yield, and reliability. The economic success of a product is tied to yield and manufacturability, which traditionally have been based solely on the effectiveness and productivity of the *manufacturing* house. Throughout this book, readers are shown the impact of the *design* methodology on a product's economic success.

Sandip Kundu
Aswin Sreedhar

Nanoscale CMOS VLSI Circuits

Design for Manufacturability

CHAPTER 1
Introduction

1.1 Technology Trends: Extending Moore's Law

Complementary metal oxide semiconductor (CMOS) technology has been the dominant semiconductor manufacturing technology for more than two decades. The demands for greater integration, higher performance, and lower power dissipation have been driving the scaling of CMOS devices for more than a quarter century. In 1965, Gordon Moore famously predicted that the number of transistors on a chip would double every 18 months while the price remained the same. Known as Moore's law, the prediction has held true until today. This has been made possible by advances in multiple areas of technology. Design technologies such as high-level design languages, automatic logic synthesis, computer-aided circuit simulation, and physical design have enabled increasingly larger designs to be produced within a short time. Manufacturing technologies—including mask engineering, photolithography, etching, deposition, and polishing—have improved continuously to enable higher levels of device integration.

Moore's law has been sustained by concurrent improvement across multiple technologies. There has always been a demand for high-performance circuits that are cheap to produce and consume less power. Historically, scaling of transistor feature size has offered improvement in all three of these areas. "High performance" means an increase in clock frequency with every new generation of technology. Such increases are possible only with commensurate increases in transistor drive current while parasitic capacitances remain low, thus reducing propagation delay. The transistor drive current is a function of the gate dimensions and the number of charge carriers and their mobility in the channel region. *Propagation delay* is the input-to-output signal transition time of a gate or a transistor. The propagation delay depends primarily on the intrinsic capacitances of the device: the threshold voltage and load capacitance. By making changes to the contributing parameters, an integrated circuit (IC) with higher clock frequency can be produced. With the foray of semiconductor chips into portable and handheld devices, power consumption has also become a critical design parameter. Power consumption can be divided into dynamic power and static power. *Dynamic power* is the power dissipated during transistor operation. It is dependent on the supply voltage and frequency of circuit operation. It also depends on device and interconnect parasitic capacitances, which in turn depend on the manufacturing process and materials. *Static power* is the power consumed irrespective of the device use. It is chiefly dependent on the threshold voltage of a device, which in turn is related to process parameters such as dopant density in the gate poly and the transistor channel region.

Improvements in clock frequency have been achieved primarily by creating device of smaller effective channel length and reducing the threshold voltage of the device. Power consumption and some device reliability problems have been kept under control by reducing the supply voltage. Leakage control has been attained through selectively increasing the threshold voltage of transistors in noncritical gates. This is also known as multithreshold CMOS (MTCMOS) technology.

In each generation of CMOS technology, new challenges have cropped up that required new solutions. In early days, when the layout size increased to about a thousand polygons, manual layout became impractical, so layout automation tools were required. Later, when the gate count increased beyond thousands, computer-aided logic synthesis became necessary. When the gate counts became still larger, high-level design languages were invented. At about 1.5 μm or so, interconnect delays became a concern. This led to development of interconnect resistance and capacitance (RC) extraction from circuit layout. With continued scaling, coupling capacitances created problems that required development of signal integrity tools. As the feature size became even smaller, considerations of device reliability required the introduction of electromigration rules and checks. As the transistor count increased and the frequencies became larger, power density became a critical concern—especially in mobile designs. The reduction of dynamic power demanded commensurate reductions in interconnect capacitances, which in turn required development of low-K interlayer dielectric between interconnect layers.

Collectively, these techniques have helped to move the industry from one technology node to the next. Future technologies will require innovation in (1) basic device structures, (2) materials, and (3) processing technologies.

The move into transistors feature widths close to 45 nanometers and below, unlike any other shift in technology, has led to the simultaneous investigation of many changes to design and manufacturing that attempt to better satisfy the three technology requirements just listed. As identified in the *International Technology Roadmap for Semiconductors* (ITRS) report,[1] the basic transistor patterning technology will become a critically important factor in enabling future technologies. Photolithography, the core of transistor and interconnect patterning, has been found wanting at lower feature sizes. The main concerns are materials for mask and projection optics, temperature of operation, and wavelength of light source. (These issues are detailed in Chapter 3.) Lithography techniques that help produce devices of smaller effective gate lengths are central to technology scaling. The emergence of new, structurally modified metal oxide semiconductor field-effect transistor (MOSFET) devices aim at increasing the channel region's surface area in order to improve the device's drive current is also impelling the need for better manufacturing techniques. These devices also need the help of lithography to be printed on wafer. Newer materials that can provide higher performance (through increased mobility) and better variability control (through reduced intrinsic and interconnect capacitances) also aim to help break the performance and power consumption barriers. We address all these issues in the following sections.

1.1.1 Device Improvements

Conventional CMOS scaling involves scaling of multiple aspects of a transistor; these include feature length and oxide thickness as well as dopant density and profile. As we approach atomic scales for transistors, scaling of these aspects presents a new set of challenges. For example, scaling oxide thickness increases tunneling leakage through oxide. Increased channel doping increases source-drain leakage. Increased source-drain doping increases band-to-band direct tunneling leakage to bulk. Increased source-drain doping also increases the source-drain capacitance, compromising the performance of transistors. It is widely recognized that conventional scaling of bulk CMOS will encounter these difficulties and that further scaling of transistors will require modification to the conventional MOS transistor. Several alternative devices are now under investigation or actual use. They include silicon-on-insulator (SOI) MOSFET, whose design seeks to mitigate the source-drain capacitance and transistor body effects. The FinFET (an FET with a finlike vertical rather than a planar structure) and tri-gate transistors being developed seek to increase transistor ON current without increasing the OFF current. Transistors based on carbon nanotube (CNT) technology offer another alternative to device scaling. However, it is not yet patternable using current lithography processes.

Several foundries are currently manufacturing SOI-based MOSFETS used in high-performance ICs. Although the wafer cost for SOI has come down significantly, cost and yield still remain as barriers to wider adoption of SOI devices. Furthermore, FinFET, tri-gate, and other multigate devices are in the early stages of development. Once they become practical to manufacture in high volume, such devices will impel changes to conventional circuit design and optimization methods. Carbon nanotubes have been touted as a potential replacement technology for silicon transistors, but assembly, performance, and reliability remain open issues for such devices. The following sections provide an overview of these developing device technologies.

1.1.1.1 Silicon on Insulator

The SOI process uses a silicon-insulator-silicon substrate in place of a doped silicon substrate as the bulk material in MOSFET devices. A thin layer of buried SiO_2 isolates the channel regions of transistors, which enables the transistor body to float without risk of substrate coupling noise. This, in turn, reduces capacitance to body and improves both circuit performance and power attributes. Figure 1.1 depicts an SOI-based MOSFET device. From a circuit perspective, one difference between an SOI and a bulk transistor is that the body of an SOI transistor forms an independent fourth terminal of the device. This terminal is typically left unconnected, but it can be fixed at a potential to control the threshold voltage. The main advantage of using SOI over bulk CMOS is the increase in performance due to lower junction capacitance, lower body effect, and improved saturation current. Other advantages include better control of device leakage over the entire die, reduced susceptibility to soft error, and low temperature sensitivity. A partially depleted SOI process is often used to deliver both high performance and reduced power consumption, where "partially depleted" means that the channel inversion region of the SOI device does not completely consume the body region.[2,3] Partially depleted SOI uses the same materials, tools, processes, and parameter specifications as the older bulk technology. It also uses the same design technology and computer-aided design (CAD) tools, and changes in CAD tools for circuit simulation and physical design are easily accommodated within current framework. This fact facilitates quick adoption of this technology.

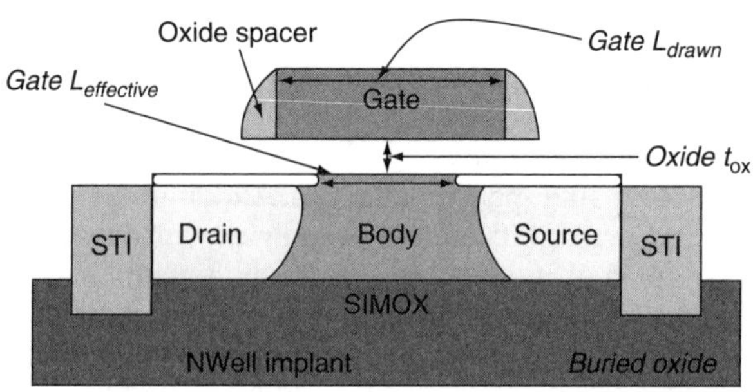

FIGURE 1.1 Partially depleted SOI MOSFET.

1.1.1.2 Multigate Devices

A group of multigate devices that includes FinFETs, DG-FETs ("DG" denotes "double-gate"), and tri-gates has been touted to replace the traditional MOSFET. Multigate devices are MOSFETs that have multiple gate terminals. The gate terminals can be joined together to form a single connection that performs the same operation as a MOSFET, or they can be controlled independently to offer circuit designers greater flexibility. In this latter configuration the devices are simply termed *independent gate* FET devices. Multigate transistors are being manufactured solely to create ever-smaller transistors that can provide higher performance with the same or smaller chip area. These devices can be classified based on the direction of gate alignment. Horizontally aligned multigate devices are called *planar* gate devices, which can be double-gate or multibridge transistors with common or independent gate controls. Both FinFETs and tri-gates are *vertical* gates, which (for manufacturing reasons) must all have the same height. This constraint forces all transistors to have the same width. Consequently, current approaches to device sizing and circuit optimization techniques must be tweaked to accommodate discrete transistor sizes. Planar double-gate devices form a natural extension to SOI technology, and they can be manufactured using any of three conventional techniques: (1) the layer-on-layer process, (2) wafer-to-wafer bonding, and (3) the suspended channel process. The channel region of the device is sandwiched between two independently fabricated gate terminals with oxide stacks. Figure 1.2 shows a planar double gate fabricated using the layer-on-layer technique.

A FinFET is a nonplanar vertical double-gate transistor whose conducting channel is formed by a thin polysilicon "fin" structure that wraps around the body of the device.[4,5] The dimensions of the fin dictate the channel length of the device. Figure 1.3(a) illustrates the structure of a FinFET. As shown, the gate region is formed over the silicon that connects the drain and source regions. Transistors with effective gate length of 25 nm have been reported.[5] Tri-gate transistors were devised by Intel[6] as its trademark new device for future technologies that attempt to extend Moore's law for higher performance and lower leakage. They are quite similar to FinFETs

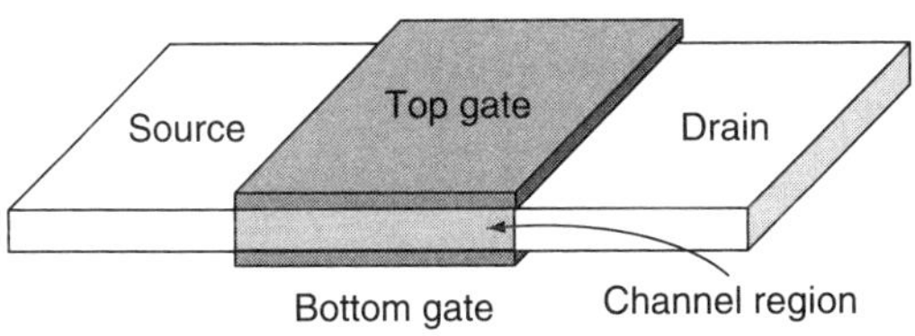

FIGURE 1.2 A planar double gate.

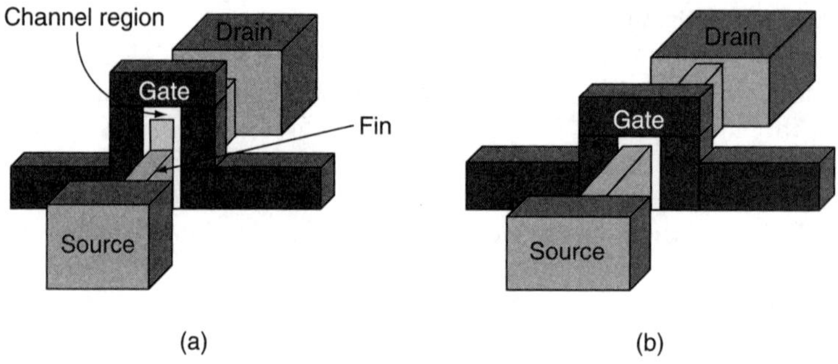

FIGURE 1.3 Structure of (a) FinFET and (b) Tri-gate.

and are also considered as nonplanar vertical multigate devices (see Figure 1.3(b)). These devices provide increased surface area for the channel region, thus creating higher drive currents by wrapping a single gate in place of multiple gate structures.[7,8]

The new multigate devices described here provide greater control over gate threshold voltage and also increase the surface area of the channel for improved drive current, thereby producing faster chips.

1.1.1.3 Nanodevices

Nanodevices are created using materials other than silicon. These materials are used to realize nonconventional devices capable of mimicking the operation of a MOSFET. Not only are nanodevices an order of magnitude smaller than conventional MOSFETs produced today, they are also unique in terms of materials and the manufacturing technology used. MOS transistors operate on the basis of movement of charge carriers across the channel region, whereas the operation of nanodevices is based on quantum mechanical principles. Nanodevices can be classified, in terms of their working mechanism, as molecular and solid-state devices. *Molecular* devices use a single molecule (or a few molecules) as the switching device, and they can be as small as 1 nm.[9] Examples include switches using catenanes[10] or rotaxanes[11] as well as DNA-strand-based devices.[12] Molecular computing systems are highly sensitive to electrical and thermal fluctuations. They also require large-scale changes to current design practices in order to accommodate significantly higher failure rates and power constraints.

Solid-state nanodevices that have been investigated with an eye toward forming logic circuits of reduced density include (1) carbon nanotubes, (2) quantum dots, (3) single-electron transistors, (4) resonant tunneling devices, and (5) nanowires. Without delving into details of these devices, we list the underlying mechanism of conduction in each. Carbon nanotubes and silicon nanowires are further along in terms of manufacturing developments. These devices use "ballistic transport" (i.e., the movement of electrons and holes are unhindered by obstructions) of charge carriers as the charge-conducting mechanism. Quantum dots interact with each other based on Coulomb forces but without actual movement of electrons or holes.[13] Resonant tunneling diodes exhibit negative differential resistance characteristics when a potential is applied across the device, so they can be used to build ultrahigh-speed circuitry.[14,15] Finally, single-electron transistors are three-terminal devices whose operation is based on the "Coulomb blockade," a quantum effect whereby the gate voltage determines the number of electrons in a region.[16] Nonconventional manufacturing techniques (i.e., not based on lithography) are being used to reduce the cost of fabricating these devices, but large-scale manufacturing is not yet practical.

1.1.2 Contributions from Material Science

Except for SOI, all the new devices described so far are still in their nascent stages and have not yet been manufactured in large quantities. In order to obtain consistent results while scaling transistors, process improvements were made in the materials domain. New materials that target specific process stages and device regions have been suggested either to improve performance or reduce leakage and so allow extension of CMOS scaling. Examples include strained silicon materials, low-*K* and high-*K* dielectrics, metal gates, and copper conductors. A discussion of the purpose and properties of these developments is presented next.

1.1.2.1 Low-K and High-K Dielectrics

Dielectrics (oxides) form an integral part of CMOS IC manufacturing today. A liner composed of silicon dioxide (SiO_2) or silicon oxynitride (SiON) has traditionally been at the heart of transistor operation, providing high impedance control for the conduction path between source and drain of a transistor. Similarly, SiO_2 is used as a barrier layer between active regions of devices and also between layers of metal interconnects. The drive current (and hence the speed) of a transistor is directly proportional to the gate oxide capacitance. The gate oxide capacitance depends on the oxide thickness t_{ox} and the dielectric constant ε_{ox} of the material used as oxide:

$$I_D = \mu C_{ox} \frac{W}{L} V; \qquad C_{ox} = \frac{\varepsilon_{ox}}{t_{ox}} \tag{1.1}$$

In order to increase the oxide capacitance, the thickness of the oxide is scaled in tandem with transistor scaling until the oxide thickness reaches only a few layers of molecules. Oxide thickness is measured both in optical/microscopy and electrical terms. Because of various field effects, thickness as measured in terms of capacitance tends to be slightly higher than thickness observed via microscopy. At a thickness of about 20 Å, tunneling leakage through gate oxide becomes a serious problem (see Figure 1.4).[17] Clearly, the tunneling current for SiO_2 is many orders of magnitude higher than other gate oxides at this thickness.

Thicker oxides are required to reduce tunneling leakage. However, thicker oxides reduce both gate capacitance and transistor drive current. This necessitates a high-*K* oxide material to maintain higher gate capacitance with thicker oxides. Hafnium oxide ($\varepsilon = 25$) has reportedly been used as a high-*K* gate dielectric. With the introduction of high-*K* gate oxides, threshold voltages tend to increase significantly, which is addressed by changing the gate electrode materials. Unlike conventional SiO_2 gates, high-*K* gate dielectrics need metal gates and a complex gate stack structure. Required properties of high-*K* materials include (but are not limited to) high dielectric constant, low leakage current density, small flat-band voltage shifts, low concentration of bulk traps, and reliability comparable to that of current SiO_2 dielectrics. Table 1.1[18] lists some high-*K* dielectrics along with their dielectric constants and compatibility with silicon substrate. The crystal structure and stability of the Si substrate together determine possible defect levels for a given type of oxide.

As the transistor count increases, the interconnect length increases even faster. In today's chips, interconnect capacitance dominates gate

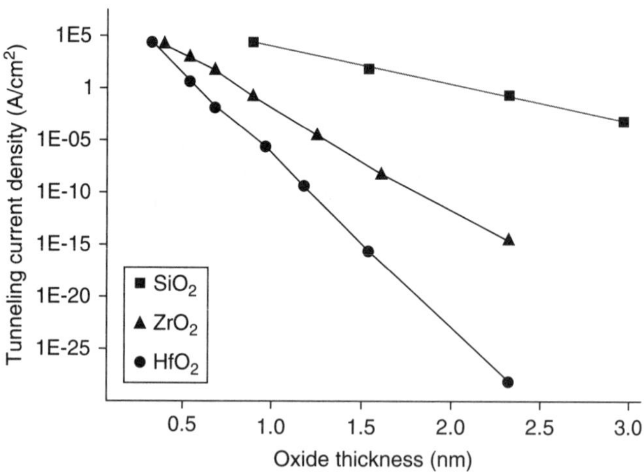

Figure 1.4 Gate oxide tunneling currents for three oxide types at different oxide thicknesses.

Material	ε	Crystal structure	Stable on silicon?
SiO_2	3.9	Amorphous	Yes
Si_3N_4	7.8	Amorphous	Yes
Y_2O_2	15	Cubic	Yes
TiO_2	80	Tetragonal	No
HfO_2	25	Monoclinic, tetragonal	Yes
Ta_2O_5	26	Orthorhombic	No
Al_2O_3	9	Amorphous	Yes

TABLE 1.1 High-K oxide materials with dielectric constants and silicon substrate compatibility

capacitance and is the greatest source of active power dissipation. Because power has emerged as the most significant barrier to transistor usage, reducing power dissipation has become a shared goal for both process and design engineers. At manufacturing level, this is addressed by reducing the dielectric constant for the dielectric material between metal layers, which directly reduces the interconnect capacitance and contributes to power reduction. Organic materials and porous SiO_2 have been explored as possible alternatives. The interlayer dielectric (ILD) must meet a thermal specification to transport heat effectively; today, $K \leq 2.5$ is used.

1.1.2.2 Strained Silicon

The movement of charge carriers, such as electrons in an *n*-channel device and holes in a *p*-channel device, cause current to flow from the source to the drain of the transistor. Under the influence of an electric field, the speed at which the carriers move is called *mobility*. It is defined as $\mu = v/E$ where v is the velocity of charge carriers and E is the applied electric field. The strength of the drain current (I_D) is proportional to the mobility (μ_n, μ_p) of the carriers. This mobility is a function of temperature and is also a function of crystal stress. The latter property is used by modern processes to improve mobility by straining the atoms in the channel. *Straining* refers to the technique through which the interatomic distance between the atoms in the channel is increased or decreased. This causes an increase in the mean-free path of the charge carriers present in the channel. For nMOS devices, a tensile stress improves electron mobility; for pMOS devices, a compressive stress improves hole mobility. The source and drain regions of the nMOS transistor are doped with silicon-germanium atoms to induce a tensile stress on the channel region.

A layer of compressive or tensile nitride liner can be applied over the gate region. Typically, this is done at a higher temperature. Differential coefficients of thermal expansion produce strain upon cooling. Higher stress may contribute to crystal defects and reliability issues. Higher mobility may also contribute to increased transistor leakage. These factors limit the amount of stress that can be applied to the device. See Sec. 3.6 for more details on strain engineering.

1.1.3 Deep Subwavelength Lithography

Manufacturing small MOSFET devices and interconnect wires today requires printing of polygons that can have feature widths of less than a quarter wavelength of the light source. Photolithography is at the heart of the semiconductor manufacturing process; it involves multiple steps that lead to the formation of device and interconnect patterns on the wafer. Without photolithography it would not have been possible to assemble billions of transistors on a single substrate. A simple photolithography setup involves an illumination system consisting of a UV light source; a mask that carries the design patterns; the projection system, which comprises a set of lenses; and the wafer. With everyday lighting equipment, an object whose width is smaller than the wavelength of the light being used to project it will not be projected with good resolution and contrast. The resolution of the patterns being printed is defined as the minimum resolvable feature on the wafer for a given illumination source and projection system parameters. Thus the *resolution R* depends on the numerical aperture (NA) of the lens system and the wavelength λ of the light source. The equation that describes this relation is known as *Rayleigh's equation* and is given as follows:

$$R = k_1 \frac{\lambda}{\mathrm{NA}} \tag{1.2}$$

The minimum resolvable linewidth on a particular mask is also referred to as the critical dimension (CD) of the mask. The numerical aperture of a lens system is the largest diffraction angle that a lens system can capture and use for image formation. Mathematically, it the sine of the maximum angle incident on the lens multiplied by the refractive index (n) of the medium:

$$\mathrm{NA} = n \sin\theta \tag{1.3}$$

Because air is the medium in optical systems, the limit value of numerical aperture is 1. Manufacturing limitations are such that the NA limit has not been achieved. But new inventions using water as medium have increased the NA above 1, since the refractive index of water is higher that that of air. Numerical aperture will continue to play an important role because the higher the NA, the better the resolution of the system.

Another important parameter that controls the robustness of patterns printed on wafer is the depth of focus (DOF), which is defined as the maximum vertical displacement of the image plane such that an image is printable within the resolution limit. This is the total range of focus that can be allowed if the resulting printed image on the wafer is to be kept within manufacturing specifications. The maximum vertical displacement is given by

$$\mathrm{DOF} = k_2 \frac{\lambda}{\mathrm{NA}^2} \tag{1.4}$$

Since this focus tolerance depends inversely on the square of the numerical aperture, there is a fundamental limit on extremely high NA processes. Improvement in resolution means a reduction in R in Eq. (1.2). One way to improve resolution is to use a light source with wavelength less than or equal to the required minimum feature width of the mask. Figure 1.5[19] shows the historical trend in wavelength of the illumination system. For a light source to be used in lithography, it should be of single frequency with nearly no out-of-band radiation, coherent in phase, and flicker-free with minimum dispersion. In addition, the lens system must be available to focus light at that frequency. Ordinary glass tends to be opaque to UV rays and is unsuitable for lithography[20].

Another method for improving resolution is to reduce the k_1 factor, which depends on processing technology parameters. This factor can be written as

$$k_1 = R_{\text{half-pitch}} \frac{\mathrm{NA}}{\lambda} \tag{1.5}$$

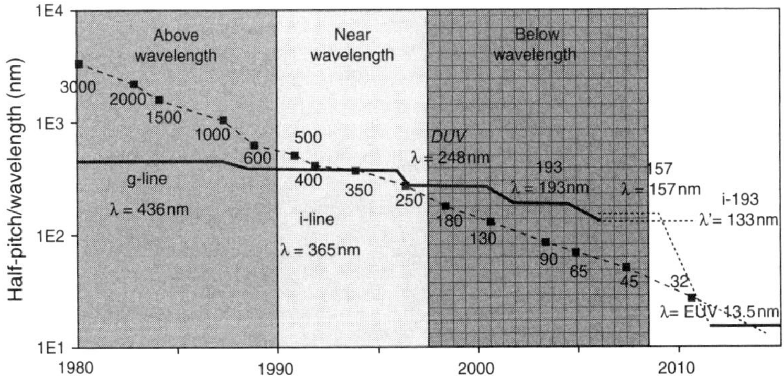

FIGURE 1.5 Progress trend of illumination light source across technology generations.

Here R is defined by the patterning rules of the technology. It is typically referred to as the *minimum half-pitch* used in the technology node or process. Factors that determine k_1 are the imaging system's numerical aperture (NA), wavelength (λ), and half-pitch (R). As seen in Figure 1.6, the k_1 factor has been progressively reduced by technology scaling to produce smaller features on the mask. With current technology using 193-nm light source for printing 45-nm features, the theoretical limit for the k_1 factor can be obtained from Eq. (1.5) as $k_1 = 0.25$. The theoretical limit is calculated by assuming a value of 1 for the numerical aperture. In practice, however, attaining a k_1 anywhere near 0.25 with the current single-exposure systems requires the use of high-index fluids having NA close to or greater than 1. Another approach is to use light sources of smaller wavelengths. These two options are still being investigated, and neither has been shown to perform reliable image transfer.[21]

Double-pattern lithography has been seen as a viable technique to improve the k_1 factor below 0.25. This is accomplished by increasing the pitch size while holding constant the minimum resolvable dimension of patterns. More details on double patterning are provided in Sec. 4.5.1.

1.1.3.1 Mask Manipulation Techniques

Resolution enhancement techniques (RETs) are methods used to improve the resolution of the lithography system by manipulating various parameters of the system. Most RETs aim to manipulate the patterns on the mask. The resolution of a feature being printed depends on neighboring features and the spacing between them.

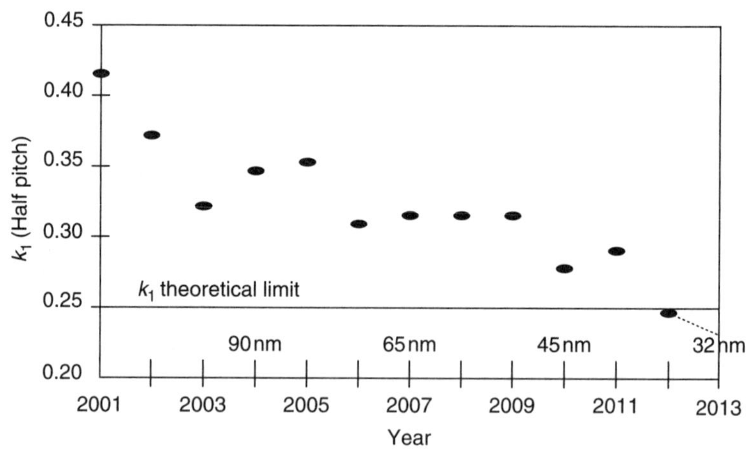

FIGURE 1.6 Trend in k_1 reduction for printability improvement.

The principle of diffraction governs the interaction of light waves that pass through the mask patterns on the mask while being projected onto the wafer. As shown in Figure 1.7, RET modifications to the mask improve the resolution of the features being printed.

Resolution enhancement techniques include optical proximity correction, phase shift masking, off-axis illumination, and multiple exposure systems. Optical proximity correction (OPC) changes the shape of the feature by adding extra jogs and serifs to improve the resolution (see Sec. 4.3.2). Phase shift masking (PSM) utilizes the superposition principle of light waves to improve resolution by creating phase changes in spaces between the features; see Sec. 4.3.3 for more details. Off-axis illumination (OAI) is based on the principle that, if the light rays are incident at an angle on the mask, then higher-order diffraction patterns can be made to pass through the lens and thereby improve resolution. This method and the type of lenses it uses are described in Sec. 4.3.4.

Another promising technique that improves the resolution of patterns being printed is multiple exposure systems. In one

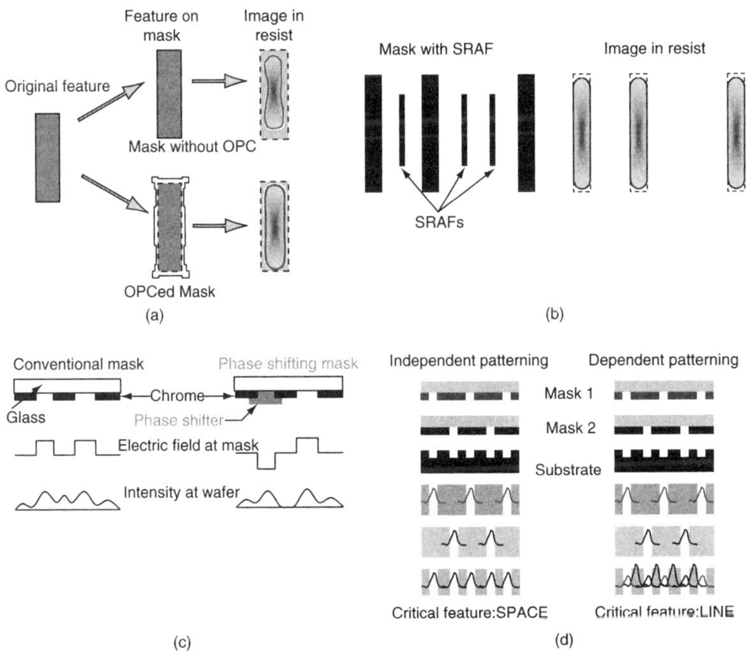

FIGURE 1.7 RET mask manipulation to improve pattern transfer: (a) optical proximity correction; (b) SRAF insertion; (c) phase shift masking; (d) double patterning.

embodiment of such a system, the pattern is carried by two separate masks that are exposed in separate steps, leading to the final image on wafer. This is known as double-pattern lithography, and it is being used to print critical masks by decomposing them into two masks. This method increases the spacing between metal lines and hence reduces the minimum resolvable feature that can be printed on the wafer; see Sec. 4.5.1 for details about this technique. Ongoing research seeks to use triple and quadruple patterning techniques to further push the resolution barrier. One negative consequence of such patterning systems is that fabrication throughput and the overall process yield may decrease, leading to an increase in product cost.

1.1.3.2 Increasing Numerical Aperture

As Eq. (1.2) indicates, increasing the system's numerical aperture will improve the resolution. The numerical aperture is given by Eq. (1.3), where n is the refractive index of the medium between the projection system and the wafer and θ denotes the maximum angle of incidence for a ray passing through the projection lens. Air, with refractive index 1, is typically used as the medium. Figure 1.8[1] illustrates goals for future techniques that aim to improve NA. Immersion lithography (Figure 1.9) uses a liquid medium to increase the refractive index n. Water, with refractive index of 1.3, is currently being used as the immersion fluid, although other high-index fluids have been suggested as possible replacements. Immersion can lead to process issues such as spot defects from water molecules and error in handling wafers. Additional suggestions for improving numerical aperture include the use of high-index lenses, lenses with increased curvature, and high-index resist materials.

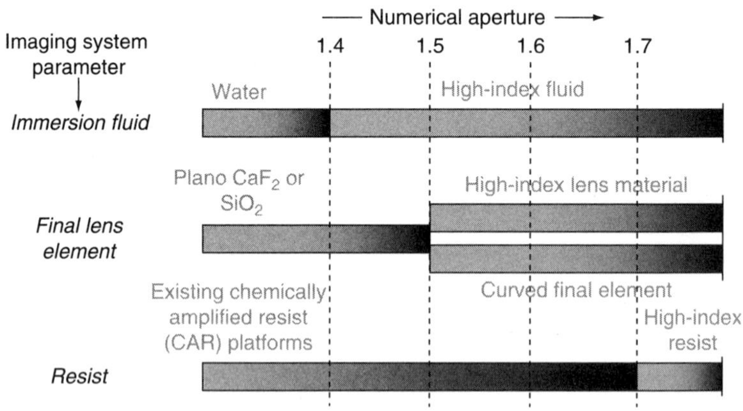

Figure 1.8 Future trends in techniques to improve numerical aperture.

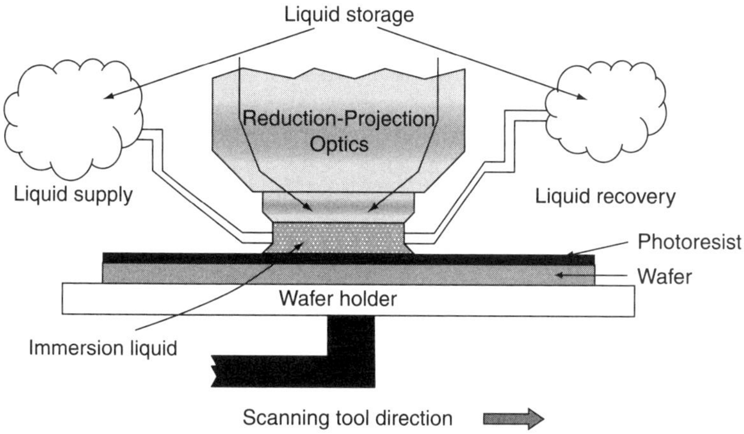

FIGURE 1.9 Immersion lithography technology.

1.2 Design for Manufacturability

Design for manufacturability (DFM) in the current context refers to the new design techniques, tools, and methodologies that ensure printability of patterns, control the parametric variation, and enhance yield. A broad definition of DFM encompasses various methodologies from the starting point of design specification to the product launch of an IC, which include circuit design, design optimization, mask engineering, manufacturing metrology—to name just a few technologies that aim to manufacture chips with repeatability, high yield, and high cost effectiveness. The two most important metrics by which all DFM methodologies are assessed are the cost of the entire process and the cumulative chip yield loss due to irregularities at various manufacturing steps.

With continued scaling, patterning is conducted almost at the resolution limit, while the transistor count is growing exponentially. When lithographic patterning is pushed to its limit, a single defect may invalidate a chip consisting of millions of transistors. This underscores the importance of DFM. Concerns about manufacturability have become so pervasive that DFM considerations—once dealt with entirely by design rules and mask engineering—are moving upstream in the design process. The next few sections will examine the economic value of DFM, current parameter uncertainties, traditional DFM approaches, and the requirement for model-based DFM techniques.

1.2.1 Value and Economics of DFM

In Sec. 1.1 it was noted that scaling from one technology node to the next often required changes in design methodologies, CAD tools, and process technology. Design for manufacturability is one of those necessities that became a design concern with the advent of subwavelength lithography. Advances in this technology have translated into more intrusive changes in design methodology. Long ago, optical diffraction effects were handled through design rules check (DRC), and any remaining issues were handled in mask preparation. Since the introduction of subwavelength lithography, rule-based DRC has been supplanted by model-based DRC: simple rules have been replaced with quick and approximate optical simulation of layout topology. With an increase in the field of optical influence (aka optical diameter), the DRC models became more complex. The true value of DFM is not yet understood by many designers still using older process technologies. However, as designs move to 45 nm and below, DFM steps are becoming more critical in the design process. In 130-nm to 65-nm technologies, DFM issues were mostly handled by postprocessing of the layout (i.e., resolution enhancement techniques). But postprocessing alone cannot ensure manufacturability when deep subwavelength lithography is involved. If one-pass postprocessing proves insufficient, then iteration between physical layout generation and RET steps becomes necessary. When all is said and done, DFM methodologies increase the number of tools that are run through the design as well as the number of iteration cycles, thus increasing the design's time to tape out (TTTO). Designers are increasingly concerned about the fact that improving their designs using DFM tools requires more in-depth knowledge of the process techniques. Designers already juggle multiple design targets, which include area, performance, power, signal integrity, reliability, and TTTO. Of course, adding new design objectives will affect existing ones. Thus, the effectiveness of any DFM methodology must be judged in terms of its impact on other design objectives. In addition to the imaging system parameters already discussed and to the associated processes, DFM also pertains to gate CD and interconnect CD variations, random dopant fluctuations, mobility impacts, and other irregularities that are subjects of TCAD (technology CAD) studies. Once the design netlist rolls over to the DFM step (see Figure 1.10), the outcome may be (1) changes to the physical design netlist such that the desired parameters are within specifications; (2) information feedback to designers regarding areas of design where such DFM changes cannot be incorporated automatically; or (3) The parametric impact of DFM on the design process, including static timing analysis as well as signal integrity and reliability. With outcome (2), where a one-pass DFM step does not succeed, the remaining issues may be addressed by automatic layout tools or may require manual intervention. This is an area in which tools and methodologies are still evolving.

The first outcome results in a modified version of the design, incorporating anticipated postsilicon effects. The techniques used

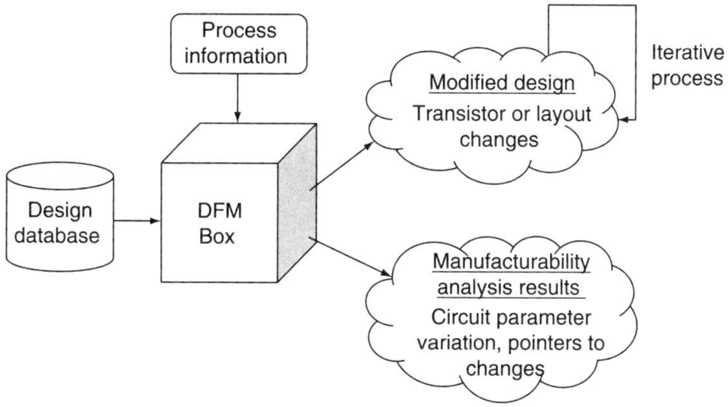

FIGURE 1.10 Purpose of design for manufacturability (DFM) methodologies.

include, for example, circuit hardening, resolution enhancement, and dummy fills. The third outcome involves analyzing the given design and providing design-specific parameter variability ranges. Instead of providing yield benefits with DFM-based suggestions, these results help designers perform more effective optimization with the promise that postsilicon circuit variability will be minimized and will fall within specifications.

Other than the value of DFM perceived by the designer, DFM also bears an important economic aspect. The economics of DFM aims to establish a cost-benefit metric for each DFM methodology by assessing its return on investment. Most such methodologies seek to improve the overall design yield by taking different approaches to optimizing parameters of the design and manufacturing process. As described by Nowak and Radojcic,[22,23] the economics of DFM can be classified into three areas of potential profit or loss. (1) DFM investment cost; (2) design re-spin costs; and (3) DFM profit (see Figure 1.11).[23] The costs of investing in DFM tools and methodology are incurred during the phases of product concept, design, optimization, and tape out. The benefit of this investment is realized after tape out, with improvements in the yield and reliability of the chip. This means that the yield curve of a process that includes DFM-based methodologies is sharper.

Quantification of DFM benefits requires silicon feedback. Given the high cost of such direct observation techniques as microscopy, the benefits of DFM are usually assessed via indirect silicon feedback such as manufacturing yield and parametric yield. Indirect measurements are contaminated with multiple parameters, so decorrelating these parameters requires carefully constructed test structures. Design for manufacture is important not only for continual product yield improvement but also for enabling future technology nodes.

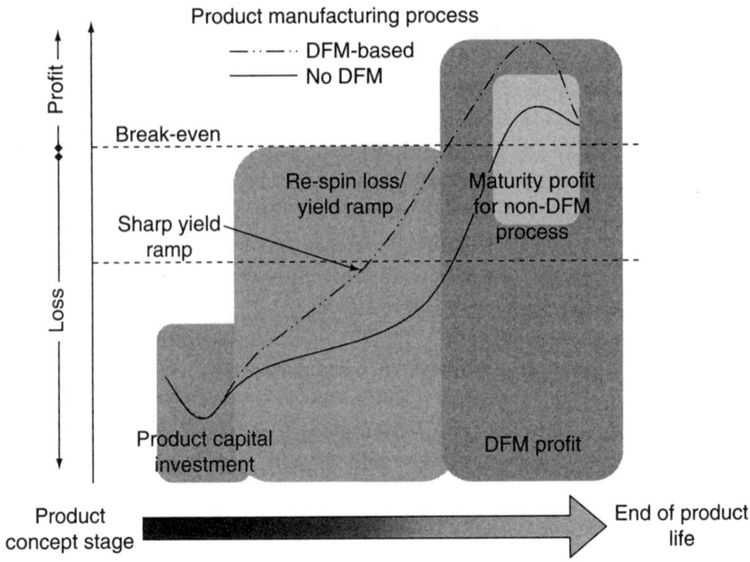

FIGURE 1.11 The value and economics of design for manufacturability (DFM).

1.2.2 Variabilities

Parametric variation has emerged as a major design concern. For *correct-by-construction* methodology, circuit models need to be accurate and model parameters need to be correct; otherwise, the behavior of the design cannot be reliabily predicted before construction. In reality, a design may vary from model parameters owing to variations in manufacturing process parameters. Current designs may consist of billions of transistors, so when these variations become large there is always the possibility of circuit failure, which can significantly reduce yield. Also, current design practice is to assume that the underlying hardware continues to be correct during the product lifetime. However, the relentless push for smaller devices and interconnects has moved the technology closer to a point where this design paradigm is no longer valid.[24–27] For example, with the advent of 90-nm technology, negative bias temperature instability (NBTI) became a major reliability concern,[28] since a pMOS device degrades continuously with voltage and temperature stress. For nMOS devices fabricated using 45-nm technology, positive bias temperature instability (PBTI) is likewise becoming a concern.[29] Windows XP failure data compiled by Microsoft Research also points to increased occurrences of hardware failures.[30] According to ITRS, these problems are expected to worsen in future technologies,[31] as designs are more likely to experience failure due to what designers call PVT issues—that is, process corner, voltage, and temperature issues.

Table 1.2 categorizes these variations from both a source and impact point of view. Columns 1 and 2 form the first source and effect relationship for variations in semiconductor manufacturing processes. As mentioned previously, variations in the manufacturing process lead to variations in the properties of the device and interconnect. Manufacturing variations can be categorized as irregularities in equipment and processing, such as in lithography, and chemical processing. Other sources of variations include mask imperfections caused during mask manufacturing, mask mishandling, tilting, and alignment issues. Additional sources of variation are improper focal position and exposure dose of the imaging system and variation in photoresist thickness. Sources of device and interconnect variation include such process steps as dopant implant onto the source, drain, or channel regions of devices on the wafer and planarization of metal lines and dielectric features during chemical-mechanical polishing (CMP).

The effects of such manufacturing variations are observed through changes in the circuit parameters. The most important among them are the parameters of the active devices, notably transistors and diodes. Variations in circuit parameters have engendered several modeling and analysis techniques that attempt to predict parameter behavior after fabrication. Among physical features, circuit

Manufacturing process	Circuit parameters	Circuit operation	CAD analysis
Mask imperfections	Channel length	Temperature	Timing analysis
Alignment, tilting	Channel width	Supply voltage	RC extraction
Focus, dosage	Threshold voltage	Aging, PBTI/NBTI	I-V curves
Resist thickness, etch	Overlap capacitance	Coupling capacitance	Cell modeling
Doping	Interconnects	Multiple input switching	Process files
Chemical mechanical polishing			Circuit simulations

TABLE 1.2 Variations in IC Manufacturing and Design: Sources and Impacts

performance is more sensitive to postlithography channel length and interconnect width than any other. Consequently, they are known as critical dimensions (CD). Poly-CD variation leads to change in effective channel length. Circuit delay tends to increase linearly with increasing channel length, whereas leakage current tends to increase exponentially with decreasing channel length. As shown in Figure 1.12, a 10 percent variation in gate CD induces large variation in threshold voltage (V_T) and delay. Because the V_T of devices can fall below the minimum allowable for leakage control, poly-CD control has become a critical aspect of overall process control. Interconnect CD variation leads to changes in path delay, coupling capacitance effects, increased susceptibility to electromigration, and spot defects.

Although the manufacturing variations have always existed and the manufacturing tolerances have generally improved with successive generations of technology, the impact on circuit parameters has been otherwise. This is reflected in terms of wider variation in circuit performance and power dissipation due to leakage.

Finally, manufacturing sources of variation—which include mask imperfections, wafer handling, alignment, and tilting—lead to errors in overlay and dielectric thickness. Focus, dose, and resist thickness variation are factors that lead to CD variations on wafer. Many modeling methods apply statistical techniques to predict the effect of

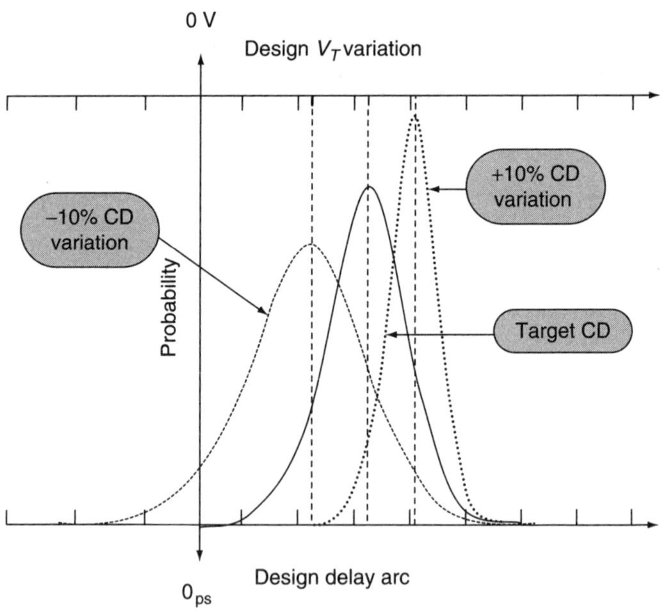

FIGURE 1.12 Design V_T and delay variation due to change in critical dimensions (CD).

these variations on electrical parameters, layout printability, and die yield. The etching process is used to remove parts of the material not covered by protective layers. It can lead to pattern fidelity issues because wet, chemical, and plasma etch processes cannot be error-free. The most important effects of etching problems are line edge roughness (LER), which refers to the horizontal deviation of the feature boundary, and line width roughness (LWR), which refers to random deviation in width along the length of the polygon. One effect of LER on transistor is changes in threshold voltage V_T, as shown in Figure 1.13.[32] Fluctuation in dopan density also induces such variation in V_T. Figure 1.14[33] illustrates three devices that have an equal number of dopant atoms in the channel but have different V_T values. Chemical-mechanical polishing is used to planarize the wafer after deposition

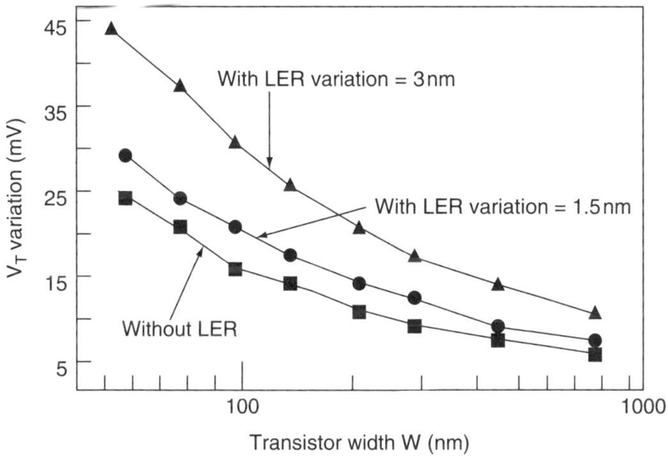

FIGURE 1.13 V_T variation due to LER (produced with 45-nm gates using predictive technology models).

FIGURE 1.14 V_T variation due to random dopant fluctuation (RDF).

of the metal and dielectric layer material. Pattern density on the wafer causes CMP to create surface roughness, defined as vertical deviation of the actual surface from an ideal one. Such changes in the surface lead to focus changes during subsequent lithography steps, contributing to further CD variation (see Figure 1.15).

Circuit operation can be affected by several sources other than variation in manufacturing process and circuit parameters. For example, environmental factors, which include supply voltage and temperature variation, affect the amount of current that flows through a device. Temperature has an effect on circuit reliability (i.e., aging). Circuit reliability effects, such as electromigration, NBTI, and hot carrier degradation, change interconnect and gate delays over time. These effects are related to interconnect width and thickness, which in turn depend on the effectiveness of patterning and CMP (respectively). Thus, a link can be seen between physical design, patterning structures in the surrounding regions, and the circuit aging process.

At each step of the circuit realization process, CAD tools are used to predict circuit performance. As the realization gets closer to the transistor and physical levels, the model parameters become progressively more accurate to better predict circuit performance. Initial performance prediction models do not consider variation. In a typical design environment, interconnect RC extraction may be based on nominal process parameters, while transistor models may take parametric variation into account. Subsequent to manufacturing, if silicon fails to meet performance expectations, such unlisted variations have been identified as sources of errors. Considering all the possible sources of variations is an expensive proposition in terms of design optimization and timely convergence of design. Thus, a company must be constantly evaluating new techniques for its DFM arsenal.

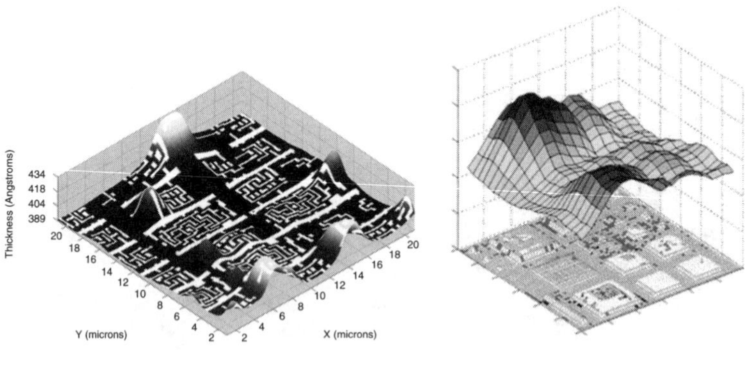

FIGURE 1.15 Design-dependent chip surface undulations after CMP. (*Courtesy of Cadence Design Systems.*)

1.2.3 The Need for a Model-Based DFM Approach

Design for manufacturing has been in use since the late 1990s. Traditional DFM relied on design rules and guidelines for the polygons and shapes present in an IC layout. Rules suggested by tools were based on interaction between two adjacent metal line or two adjacent poly line features. If a design layout passed all specified design rules and abided by all suggested guidelines, then it was set to produce a high yield. All traditional DFM methodologies were applied at the full chip level, with corner-based functional analysis and parametric yield-loss analysis predominating.

With the advent of subwavelength lithography, design rules check alone is not sufficient to ensure high yield. This fact has been chiefly attributed to the printability problems introduced by subwavelength lithography. Printing of features whose width is less than half the wavelength of the light source creates diffraction-induced pattern fidelity issues. The interaction between polygons has been found to extend well beyond adjacent features. This region of influence on neighboring features is called *optical diameter*. As the number of polygons increase, it is impossible to bring about rule checks for each type of polygon-polygon interaction. The number of DRC rules has increased exponentially to a point where it has become virtually impossible to produce an optically compliant layout by rule based DRC alone. Since the introduction of subwavelength lithography, rule-based DRC has been supplanted by model-based DRC, wherein simple rules are replaced with quick and approximate optical simulation of layout topology. As the optical diameter increased, these models became more complex. Because of this complexity, model-based DFM methodologies typically limit themselves to smaller regions of the circuit. These models have evolved over time to incorporate multiple effects, including diffraction, CMP-induced wafer surface modulations, random dopant fluctuations, and LER.

As fabrication moves into the 32-nm technology node, layout modifications based on phase shift masking and double patterning will have to consider interactions of second and third alternative neighbors. Another new aspect for DFM methodologies is the need for model-based techniques to predict DFM impact on timing, power, signal integrity, and reliability issues. There is also a need to provide early-design-stage feedback so that correct circuit operation within the process variation window is assured. Standard cell methodologies today incorporate model-based postlithography analyses that yield highly compact, printable, and functional cells on silicon.

1.3 Design for Reliability

Transistors and interconnects are known to fail during their life time under circuit operations. Some of the known failure mechanisms include gate-oxide shorts, interconnect voids or blobs caused by electromigration, V_T shift during the lifetime of transistor operation that is due to negative and positive bias temperature instability, and other mechanical, chemical, or environmental factors associated with manufacturing. When such failures are modeled correctly, product lifetime can be improved by design changes that involve device and interconnect sizing and well as floorplanning to reduce thermal hotspots. Collectively, this process is known as design for reliability (DFR). Although DFR is distinct from DFM, the two may be integrated from the perspective of design methodology because the correction mechanisms are similar.

Design for reliability comprises the techniques, tools, and methodologies employed to analyze, model, and predict the reliability of a given device or circuit. Reliability parameters are known to evolve over process maturity. Nonetheless, it is important to establish a relation between circuit parameters and product reliability so that clear targets can be set during the DFR process. The reliability models use information about failure mechanisms and how they relate to circuit design parameters in order to model a product's reliability; the models aim to predict the mean time to failure (MTTF) of a device. The MTTF is a function of manufacturing parameters and also of the parameters associated with circuit operation, such as the device's supply voltage and temperature.

Design for reliability is an exercise in circuit and layout sizing, floorplanning, and implementing redundancies to address failures. Redundancies could be added at the circuit, information, time, and/or software levels.[30,31] Dual-rail encoding and error-correcting codes are examples of information redundancy. Spare circuits and modules are examples of circuit redundancy. Modern memories often incorporate spare rows and columns to improve yield; in addition, spare processor cores, execution units, and interconnects have been used in commercial circuits. Multisampling latches enable time redundancy, and software redundancy includes redundant multithreads (RMT); many of these features are now found in commercial systems.

1.4 Summary

An effective DFM-DFR methodology provides early feedback to the design during its nascent stage. In this chapter we introduced the reader to current trends in the design of nanoscale CMOS and very large-scale integration (VLSI) circuits, explaining the various changes that have been incorporated toward the end of achieving the two principal goals of higher performance and lower power consumption. We also provided a brief overview of new device structures in the 22-nm technology node that have been touted as replacements for traditional MOSFET devices. We discussed the role of material science and optics in improving device operation, printability, and design reliability. Also discussed were the applicability of DFM in the presence of process and design parameter variability as well as the process of integrating design and manufacture. We examined the need for newer, model-based DFM methodologies given the use of subwavelength lithography and higher density of layout patterns. Finally, we mentioned some important reliability concerns in nanoscale CMOS VLSI design and described the DFR-based CAD tools that can help increase the anticipated lifetime of designs. In short, we have described the trends in technology and the rising importance of DFM and DFR.

References

1. "Lithography," in *International Technology Roadmap for Semiconductors Report*, http://www.itrs.net (2007).
2. K. Bernstein and N. J. Rohrer, *SOI Circuit Design Concepts*, Springer, New York, 2000.
3. K. K. Young, "Analysis of Conduction in Fully Depleted SOI MOSFETs," *IEEE Transactions on Electron Devices* **36**(3): 504–506 (1989).
4. D. Hisamoto, T. Kaga, Y. Kawamoto, and E. Takeda, "A Fully Depleted Lean-Channel Transistor (DELTA)—A Novel Vertical Ultra-Thin SOI MOSFET," in *Technical Digest of IEEE International Electron Device Meeting (IEDM)*, IEEE, Washington, DC, 1989, pp. 833–836.
5. X. Huang, W. C. Lee, C. Kuo, D. Hisamoto, L. Chang, J. Kedzierski, E. Anderson, et al., "Sub-50nm FinFET:PMOS," in *Technical Digest of IEEE International Electron Device Meeting (IEDM)*, IEEE, Washington, DC, 1999, pp. 67–70.
6. R. S. Chau, "Integrated CMOS Tri-Gate Transistors: Paving the Way to Future Technology Generations," *Technology @ Intel Magazine*, 2006.
7. F. L. Yang, D. H. Lee, H. Y. Chen, C. Y. Chang, S. D. Liu, C. C. Huang, T. X. Chung, et al., "5 nm-gate nanowire FinFET", in *Technical Digest of IEEE Symposium on VLSI Technology*, IEEE, Dallas, TX, 2004, pp. 196–197.
8. B. Doyle, B. Boyanov, S. Datta, M. Doczy, J. Hareland, B. Jin, J. Kavalieros, et al., "Tri-Gate Fully-Depleted CMOS Transistors: Fabrication, Design and Layout," in *Technical Digest of IEEE Symposium on VLSI Technology*, IEEE, Yokohoma, 2003, pp. 133–134.
9. K. Sandeep, R. Shukla, and I. Bahar, *Nano, Quantum and Molecular Computing. Implications to High Level Design and Validation*, Springer, New York, 2004.
10. C. P. Collier, G. Mattersteig, E. W. Wong, Y. Luo, K. Beverly, J. Sampaio, F. M. Raymo, et al., "A Catenana-Based Solid State Electronically Reconfigurable Switch," *Science* **289**(5482): 1172–1175, 2000.
11. W. R. Dichtel, J. R. Heath, and J. F. Stoddart, "Designing Bistable [2]Rotaxanes for Molecular Electronic Devices," *Philosophical Transactions of the Royal Society A: Mathematical Physical and Engineering Sciences* **365**: 1607–1625, 2007.
12. N. B. Zhitenev, H. Meng, and Z. Bao, "Conductance of Small Molecular Junctions," *Physical Review Letter* **88**(22): 226801–226804, 2002.
13. C. S. Lent, P. D. Tougaw, and W. Porod, "Quantum Cellular Automata: The Physics of Computing with Arrays of Quantum Dot Molecules," in *Proceedings of Physics and Computation*, IEEE, Dallas, TX, 1994, pp. 5–13.
14. J. P. Sun, G. I. Haddad, P. Mazumder, and J. N. Schulman, "Resonant Tunneling Diodes: Models and Properties," *Proceedings of the IEEE* **86**(4): 641–660, 1998.
15. K. Ismail, B. S. Meyerson, and P. J. Wang, "Electron Resonant Tunneling in Si/SiGe Double Barrier Diodes," *Applied Physics Letters* **59**(8): 973–975, 1991.
16. K. K. Likharev, "Single-Electron Devices and Their Applications," *Proceedings of the IEEE* **87**(4): 606–632, 1999.

17. F. Saconi, J. M. Jancu, M. Povolotskyi, and A. Di Carlo, "Full-Band Tunneling in High-κ Oxide MOS Structures," *IEEE Transactions on Electron Devices* **54**(12): 3168–3176, 2007.
18. S. K. Ray, R. Mahapatra, and S. Maikap, "High-k Gate Oxide for Silicon Heterostructure MOSFET Devices," *Journal of Material Science: Material in Electronics* **17**(9): 689–710, 2006.
19. R. Doering, "Future Prospects for Moore's Law," in *High Performance Embedding Computing Workshop*, MIT Press, Lexington, MA, 2004.
20. Harry J. Levinson (ed.), *Principles of Lithography*, SPIE Press, Bellingham, WA, 2005.
21. Chris A. Mack (ed.), *Fundamental Principles of Optical Lithography*, Wiley, West Sussex, U.K., 2007.
22. M. Nowak, "Bridging the ROI Gap between Design and Manufacturing," *SNUG*, Santa Clara, CA, 2006.
23. M. Nowak and R. Radojcic, "Are There Economic Benefits in DFM?" in *Proceedings of Design Automation Conference*, IEEE, Anaheim, CA, 2005, pp. 767–768.
24. S. Y. Borkar, "Designing Reliable Systems from Unreliable Components: The Challenges of Transistor Variability and Degradation," *IEEE Micro* **25**(6): 10–16, 2005.
25. J. M. Carulli and T.J. Anderson, "Test Connections—Tying Application to Process," in *Proceedings of International Test Conference*, IEEE, Austin, TX, 2005, pp. 679–686.
26. P. Gelsinger, "Into the Core…," *Stanford EE Computer Systems Colloquium*, http://www.stanford.edu/class/ee380/Abstracts/060607.html (2006).
27. J. Van Horn, "Towards Achieving Relentless Reliability Gains in a Server Marketplace of Teraflops, Laptops, Kilowatts, & 'Cost, Cost, Cost'… (Making Peace between a Black Art and the Bottom Line)," in *Proceedings of the International Test Conference*, IEEE, Austin, TX, 2005, pp. 671–678.
28. M. Agostinelli, S. Pae, W. Yang, C. Prasad, D. Kencke, S. Ramey, E. Snyder, et al., "Random Charge Effects for PMOS NBTI in Ultra-Small Gate Area Devices," in *Proceedings of International Reliability Physics Symposium*, IEEE, San Jose, CA, 2005, pp. 529–532.
29. M. Denais et al., "Interface Trap Generation and Hole Trapping under NBTI and PBTI in Advanced CMOS Technology with a 2-nm Gate Oxide," *IEEE Transactions on Device and Materials Reliability* **4**(4): 715–722, 2004.
30. B. Murphy, "Automating Software Failure Reporting," *ACM Queue* **2**(8): 42–48, 2004.
31. G.Groseneken, R. Degraeve, B. Kaczer, and P. Rousel, "Recent Trends in Reliability Assessment of Advanced CMOS Technology," *Proceedings of IEEE 2005 International Microelectronics Test Structure* 18: 81–88, 2005.
32. W. Zhao and Y. Cao, "New Generation of Predictive Technology Model for Sub-45nm Design Exploration," *IEEE Transactions on Electron Devices* **53**(11): 2816–2823, 2006.
33. A. Asenov, A. R. Brown, J. H. Davies, S. Kaya, and G. Slavcheva, "Simulation of Intrinsic Parameter Fluctuations in Decananometer and Nanometer-Scale MOSFETs," *IEEE Transactions on Electron Devices* **50**(9): 1837–1852, 2003.

CHAPTER 2

Semiconductor Manufacturing

2.1 Introduction

The art of manufacturing an integrated circuit involves various stages of physical and chemical processing of the semiconductor wafer (i.e., substrate). The major processing steps are oxidation, patterning, etching, doping, and deposition. An integrated circuit is obtained by repetitive processing of the wafer through these steps in a given sequence. Silicon is now the dominant material used in high-volume semiconductor manufacturing. However, the basic steps discussed in this chapter are applicable to other types of compound integrated circuits that use germanium, gallium arsenide, or indium phosphide substrates. The two basic goals of semiconductor manufacturing today are:

1. Creating three-dimensional semiconductor device and interconnect structures using semiconducting, conducting (metals) and insulating (oxides) materials.
2. Pattern processing and doping to transfer design objectives related to device connectivity and parametric properties to structures thus created.

Silicon dioxide is used as gate dielectric and also as insulator between metal layers. The *oxidation* step is used to create gate dielectric while the chemical vapor deposition (CVD) step is typically used to create inter metal layer insulator. In oxidation, silicon dioxide is obtained by heating the silicon wafer to a temperature of 1000 to 1200°C in the presence of oxygen. The thickness of oxide is controlled by the parameters of the oxidation process. CVD requires a silicon containing precursor such as silane (SiH_4) which is oxidized to produce amorphous SiO_2. *Patterning* is the most important step in semiconductor fabrication. It involves using photolithography to transfer abstract geometries onto a coated silicon substrate. Prior to patterning, the silicon substrate is coated with a photo-sensitive material, or photoresist. Patterns representing abstract design information are imprinted on a mask used during the patterning process. The simplest mask is the chrome-on-glass (COG) type; this is made of chrome-covered quartz substrates, and patterns are etched to form opaque and transparent (i.e., chromeless) regions. Basic COG masks are used only in higher, noncritical metal layers today, because they do not work well for regions that require high contrast and fine resolution. See Sec. 4.3.3 for more details on contrast and resolution of patterns and their impact on different mask types.

Etching is the process of removing certain regions of the substrate or oxide layer not covered by a protective layer (e.g., photoresist or nitride). There are both liquid and gaseous forms of chemical etching, and the choice of chemical depends on the material being etched and the protective layer used. Dry plasma etching is the predominant method used in IC fabrication because of its precision and its ability to avoid etching underneath the protective layer. Etching also plays a vital part in pattern formation on the wafer. Section 2.2.2 provides the details on different methods of etching.

Doping is the process of introducing impurities such as boron, phosphorus, or antimony into the semiconductor material so as to control the type of majority carrier in the wafer. Diffusion, ion implantation, and rapid thermal annealing are the techniques commonly used to introduce impurities into regions of the semiconductor. Doping is typically done successively to obtain a proper doping profile, and high temperature is maintained throughout the process.

Deposition is the process by which any material is laid onto the substrate during the manufacturing process; these materials include metals, polysilicon, silicon nitride, and silicon dioxide. Deposited metals such as aluminum and copper are used as conductors in today's integrated circuits. Evaporation, chemical vapor deposition (CVD), and sputtering are all techniques that can be used to deposit material onto a substrate.

Photolithography and etching processes control the shape of patterns formed on the wafer. Because IC defects are increasingly tied to design patterns, in this chapter we delve into the details of photolithographic patterning and the etching process. The other processes involved in manufacturing an integrated circuit are beyond the scope of this text.

2.2 Patterning Process

The pattern formation process involves transfering mask patterns onto the wafer followed by etching to create the required contours. Photolithography makes up the overall process of mask fabrication and the transfer of patterns from the mask to wafer, and etching is the

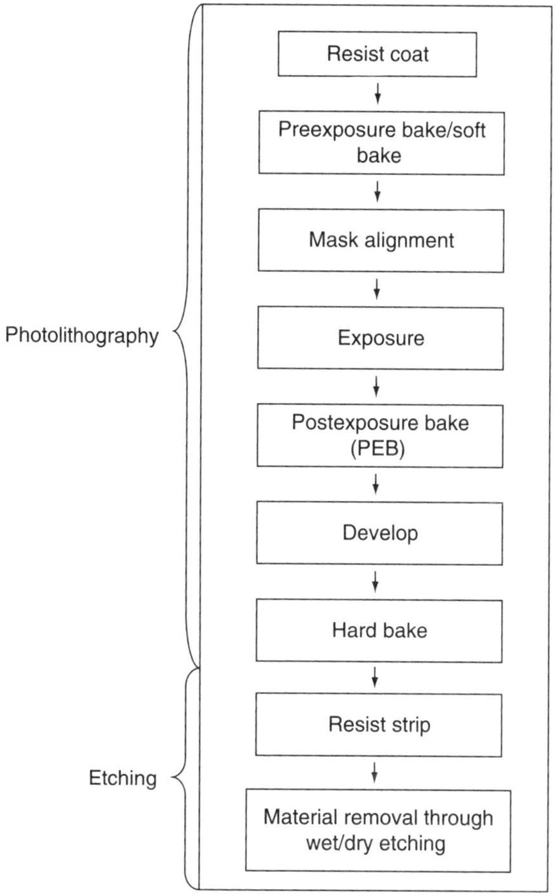

FIGURE 2.1 Steps involved in the patterning process.

process of defining the patterns formed after each photolithography step. Figure 2.1 lists all the steps involved in the patterning process.

2.2.1 Photolithography

Lithography, which was invented near the end of the eighteenth century, is the process of printing patterns on a planar surface such as a smooth stone or metal plate. The term *lithos* means stone, and *grapho* means to write or print. Photolithography is a form of lithography in which light sources are used to transfer the patterns that are present on the mask (the smooth surface) onto the wafer (the plate). Semiconductor manufacturing uses the photolithography technique for printing abstract design patterns. A brief description of each step in the photolithography process follows.

2.2.1.1 Resist Coat

First, the wafer surface is cleaned to ensure good adhesion. Then the wafer is coated with a photosensitive material called *photoresist*. Because the photoresist does not adhere well to the silicon dioxide (or silicon nitride) on the wafer, an extra adherent layer is often applied before the photoresist. Figure 2.2 shows the resist coat process and the resulting uniform layer of resist. Photoresist is typically applied in liquid form. The wafer is placed on a vacuum chuck that spins at a high speed while the photoresist is poured onto the wafer. Figure 2.3 shows a photoresist application table where the wafer is spun while the resist polymer is sprayed from the top using a nozzle. The resist can be dispensed through a nozzle in one of two ways. *Static* dispensing keeps the nozzle positioned at the center of the wafer, whereas *dynamic* dispensing moves the nozzle around the wafer at a fixed radius from the center. Although the static method is simpler, dynamic dispensing ensures better uniformity of the resist. As the chuck spins at about 1500 rpm for a 300-mm wafer, the wafer is subject to a centrifugal force that creates a thin uniform layer of photoresist. A thin photoresist layer increases resolution of the patterns printed, but thinness is not preferred when an anti-etching protective layer is needed. Wafers are typically coated with thick layers of photoresist material—nearly twice the minimum width of the feature being printed. The resist thickness depends on the size of the wafer and the viscosity of the photoresist material. It is also inversely proportional to the square root of the speed at which the wafer is spun. Highly viscous liquids form a thicker resist layer. Viscosity of the resist can be controlled by changing the speed at which the chuck spins.

2.2.1.2 Preexposure (Soft) Bake

Soft or preexposure baking is a drying step used to increase the adhesion between the photoresist and the wafer and to remove any

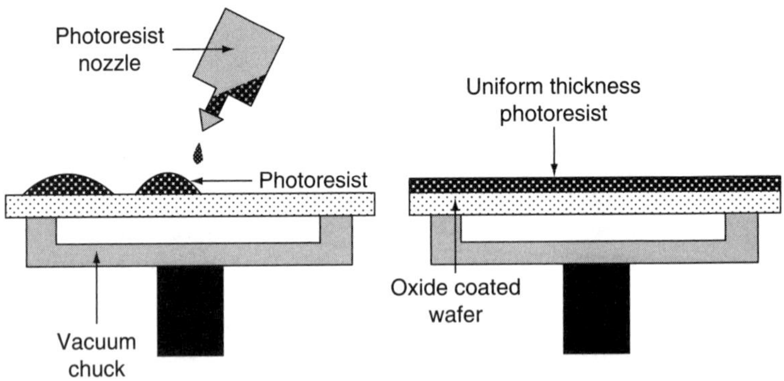

Figure 2.2 Photoresist coating process.

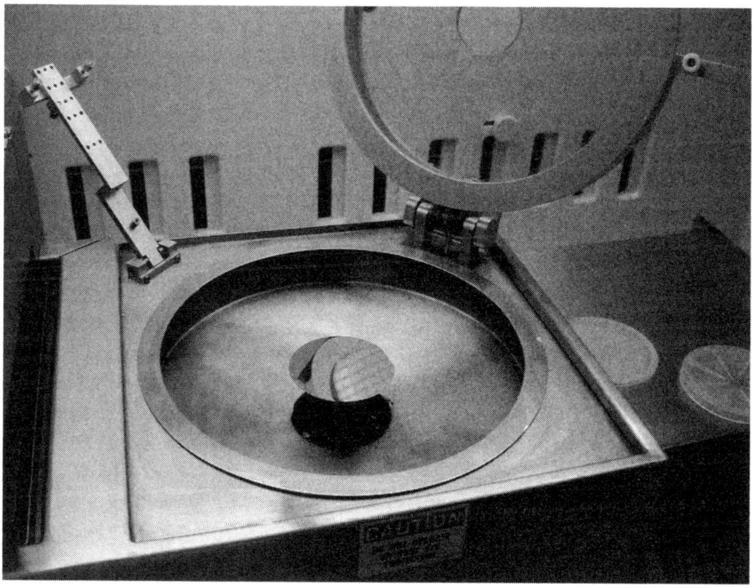

FIGURE 2.3 A bare silicon wafer mounted on a vacuum chuck and ready to be coated with photoresist (*Courtesy of Center for Hierarchical Manufacturing, University of Massachusetts, Amherst.*)

solvent present in the photoresist. Soft baking is performed before the resist-coated wafer is sent to the exposure system. Baking is accomplished by placing the wafer on a heated pad whose temperature is maintained at near 90°C. The wafer is placed on the pad for specific duration in the presence of air or nitrogen. Optimal soft bake conditions must be maintained in order to remove the solvents and reduce the decomposition of the resist. Once the soft bake is completed, the wafer is ready for mask alignment and patterning.

2.2.1.3 Mask Alignment

A typical fabrication step can require as many as thirty exposure steps. Before each such step, the mask is aligned with the previous stage to avoid errors. In some technologies, even the first mask is aligned with an underlying axis.[1,2] Mask alignment is performed with the aid of special alignment marks printed on the mask. An alignment mark is transferred onto the wafer as a part of the IC pattern. The alignment mark on mask n is carefully overlaid on the wafer's alignment pattern created by the mask $n-1$ exposure step. Simple marks such as boxes and crosses are used to minimize alignment errors, otherwise known as *overlay* errors. Mask pattern geometries have become extremely small because of technology scaling, so today's VLSI designs have extremely tight alignment tolerances. Computer-controlled alignments are made to obtain the required precision for all exposure steps.

2.2.1.4 Exposure

The exposure stage involves the actual transfer of mask patterns onto the photoresist-coated wafer. Photoresist is a material that undergoes a chemical reaction in response to incident light. Positive and negative photoresist undergo different chemical reactions: a positive photoresist starts to become soluble in the presence of light, but the opposite happens with a negative photoresist. Patterns on the mask are either transparent or opaque. Light passes through transparent regions and falls on the photoresist while the opaque regions block light from passing through. Over the past four decades, exposure systems have changed to ensure higher resolution, fewer defects, and smaller exposure times. Different exposure systems and their impact on resolution and contrast are discussed in the next section. One other issue that sometimes arises after exposure is the effect of reflected light waves on the sides of the photoresist. As light waves pass through the photoresist, some of them reflect off the base of the resist layer and form patterns—called *standing waves*—on the sidewalls. Bottom antireflection coating (BARC) is applied during the photoresist coat stage in order to reduce the reflectivity of the resist surface.

2.2.1.5 Postexposure Bake (PEB)

After exposure, the wafer is baked at high temperature (60 to 100°C) to facilitate photoresist diffusion during development. The wafer's PEB is done for a longer period than for the soft bake process. For conventional resists, PEB can reduce the standing waves caused by increased surface reflectivity. For chemically amplified resists, the PEB stage aids the diffusion process in exposed regions of the photoresist. The postexposure bake causes the resist to create photoacid generators (PAGs) that increase its solubility. Just as in the soft bake stage, care must be taken to ensure correct operating environment and temperature in order to reduce decomposition of the resist.

2.2.1.6 Development

Immediately after exposure, a developer solution is sprayed onto the wafer to aid in the process of removing the regions exposed by light (positive resist). Developers are typically water-based solutions. Because the characteristics of the resist-developer interaction determine the shape of the resultant photoresist, development is a crucial aspect of photolithography. Spinning, spraying, and puddle development are techniques that are used to develop the exposed resist. The flow of the developer solution determines the speed and effectiveness of the resist diffusion.

2.2.1.7 Hard Bake

A final baking occurs after development in order to solidify the resist for subsequent fabrication stages. Baking at a high temperature (150 to 200°C) ensures cross linking of the resist polymer, which is required for thermal stability. The hard bake also removes solvents, liquids, and gases to optimize the resulting surface's adherent characteristics.

2.2.2 Etching Techniques

Etching refers to the process of removing regions of material not covered by photoresist or any other protective material after the development process. The final shape of remaining material depends on the rate of removal of unprotected material and the direction of etching. The material-removal rate is called the *etch rate,* and it depends on the type of etch process. Etching is said to be *isotropic* if the etching proceeds evenly in all directions, whereas *anisotropic* etching removes material by moving in only one direction. Figure 2.4 illustrates the two etching processes. Etching techniques may also be classified as wet etching or dry etching.

2.2.2.1 Wet Etching Techniques

Wet etching employs the use of chemicals in liquid form to remove unprotected barrier material. Wet etching typically involves one or more chemical reactions that eventually diffuse the material to be removed without affecting the other materials or the etchant solution. The material to be removed by the etchant material is often called the barrier layer. Those portions of the material in this layer that are not covered by a protective layer are etched away by the etchant.

As shown in Figure 2.5, wet chemical etching involves three stages: (1) diffusion of etchant solution or immersion of the wafer in etchant solution, (2) formation of barrier material oxide (i.e., oxide of uncovered material), and (3) removal of the oxide. After the first step, the next two steps continue iteratively until all the unprotected barrier material is removed. The etchant solution used in this process is

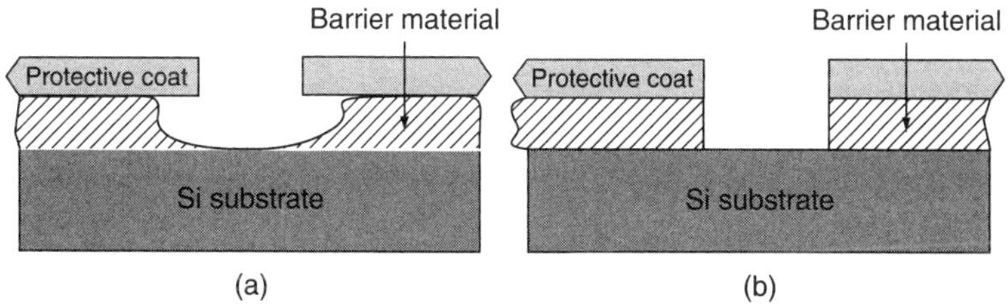

FIGURE 2.4 Etching profiles obtained for (a) isotropic etching and (b) anisotropic etching.

Material to be etched	Etchant solutions
Silicon dioxide	Hydrofluoric acid, ammonium fluoride
Silicon substrate	Nitric acid, hydrofluoric acid
Aluminum	Nitric acid, acetic acid, phosphoric acid

TABLE 2.1 Common Barrier Layers and Their Respective Etchant Solutions

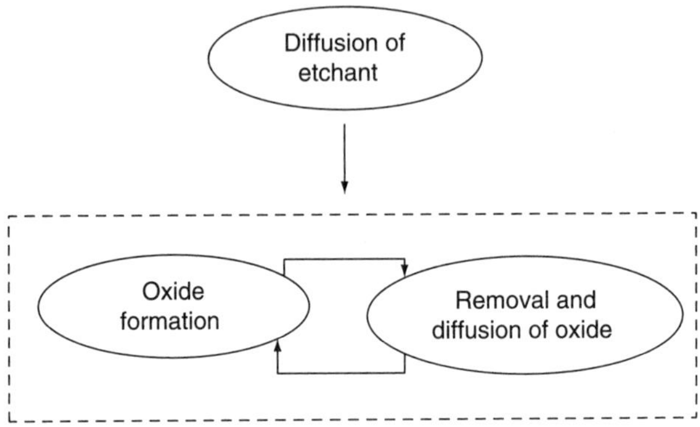

FIGURE 2.5 Stages in wet chemical etching.

chosen based on the material required to be removed. Table 2.1[3] lists a few common barrier layers and their corresponding etchant solutions.

The two parameters that control etching are etch rate and sensitivity. *Etch rate* refers to rate at which the barrier layer is removed. This rate typically depends on the chemical composition of the etchant solution and the oxide being formed. The etch rate is slower for oxides formed with dry oxygen than for those formed in the presence of water vapor. The *sensitivity* of the etching process determines the amount of material removed in the barrier, protective, and substrate layers. Etchants of greater sensitivity are required for unprotected layers.

Wet chemical etching is an isotropic etch process, which means that the material is removed at an equal rate in all directions. This results in undercutting of the barrier material, as seen in Figure 2.4. *Undercutting* is the removal of material under the protective layer. This extra reduction is measured during test chip manufacturing and creates an etch bias for which etch masks must compensate.

2.2.2.2 Dry Etching Techniques

The dry etching technique employs gaseous chemicals or ions to etch out unprotected regions of material. Dry etching is based on either chemical reactions (as with plasma etching) or physical momentum (as with ion milling).

Plasma etching is a pure dry chemical etching process that is performed in the presence of an excitation source installed within a vacuum chamber. The excitation field is set up to excite atoms of the gas that diffuse into the material to be etched (see Figure 2.6). These atoms combine with the barrier material to form compound materials through chemical reactions. The compound materials facilitate the dissipation of by-products after removal of material. Like wet etching, plasma etching is an isotropic process. Table 2.2[3] lists some gases used for plasma etching.

Ion milling is a dry etching technique that uses gases such as argon (Ar^+) to bombard the wafer. This technique is considered a momentum-based method because the etching aims to knock atoms off the wafer surface. The electric field present around the wafer accelerates the

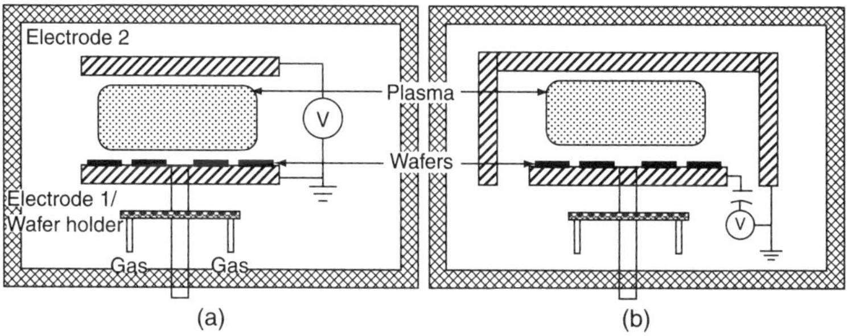

FIGURE 2.6 Dry etching techniques: (a) plasma etching; (b) reactive ion etching using asymmetric fields.

Material to be etched	Etchant solutions
Silicon dioxide	CF_4, C_2F_6, C_3F_8, CHF_3
Polysilicon	CF_4, CCl_4, SF_6
Aluminum, copper	CCl_4, Cl_2, BCl_3
Silicon nitride	CF_4, C_2F_6, C_3F_8, CHF_3

TABLE 2.2 Commonly Used Gases in Plasma and Reactive Ion Etching Processes

ions to the surface in order to physically knock atoms off the barrier material. Anisotropic etching can be ensured by targeting the ions that are perpendicular to the wafer surface. However, the ion milling process is limited by poor selectivity.

Reactive ion etching (RIE) combines chemical- and momentum-based methods. In RIE, plasma systems ionize the gases used in the vacuum chamber, which are then accelerated to the wafer by means of an asymmetric field applied to the wafer (see Figure 2.6(b)). Because it combines the two types of dry etching, RIE is better able to achieve the required anisotropy and sensitivity.

Photolithography and etching are the fundamental stages at which patterns are transferred and defined. In fact, defects are becoming increasingly dependent on the effectiveness of the pattern transfer process. It is therefore important to model the patterning process in order to produce mask patterns that are resilient to any errors in the process. The following sections will discuss the details of modeling and simulation with a lithographic system.

2.3 Optical Pattern Formation

A simple lithographic exposure system is shown in Figure 2.7.[4] The system consists of four basic elements: an illumination system, the mask or reticle, the imaging lens system, and the resist-coated wafer substrate. Light from the source is directed on to the reticle, which contains both transparent and opaque regions. Light coming through the transparent regions of the reticle then pass though the lens system and fall onto the wafer. The objective of the exposure system is to deliver light to the reticle from the coherent source with appropriate intensity, directionality, and spatial characteristics to translate the mask patterns accurately into light and dark regions on the wafer.

Patterns to be formed on the wafer are etched onto the reticle. Lines (dark patterns) with spaces (white patterns) and contact or via holes (the square regions), as shown in Figure 2.8, are the types of patterns typically drawn in a layout. The transmittance of light through the mask is determined by the pattern. In a binary image mask (BIM), the dark area of patterns have zero transmittance (and so light cannot pass through) and the other areas have maximum transmittance. The light passes through these areas of maximum transmittance and then passes through the lens system. The purpose of the lens system is to shrink the image to the required dimension and to project correctly (i.e., without any loss of information) all patterns on the reticle to the wafer. The illumination system creates a pattern that resembles the mask image, and the projection system projects this image onto the wafer to form light and dark regions on the photoresist. The resist patterns are developed and etched in preparation for subsequent processing stages.

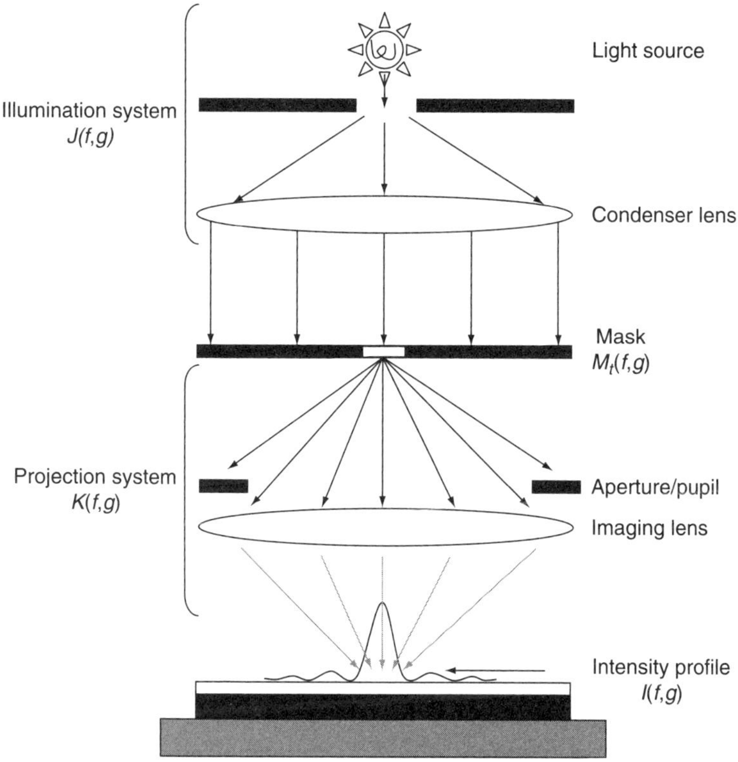

FIGURE 2.7 A simple semiconductor photolithography system.

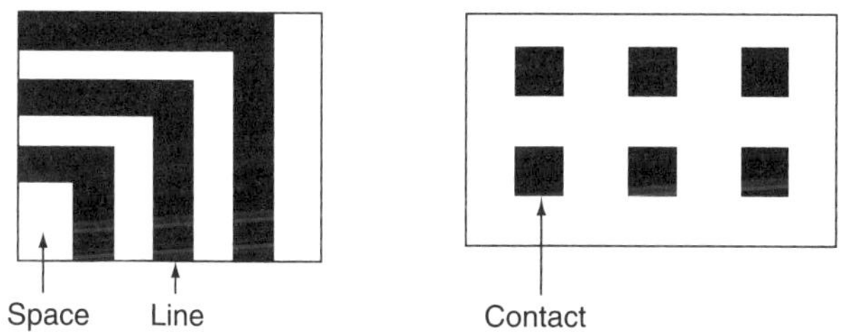

FIGURE 2.8 Typical mask features in nanoscale CMOS layouts.

2.3.1 Illumination

Choosing the proper light source is a function of the type of patterns being printed, the resolution required for the system, and the properties of the lens system. The power and wavelength of the light source are the fundamental criteria on which the choice of a light source is made. Wavelength is directly related to the smallest feature that can be printed; power is directly related to the exposure time and hence throughput. Smaller wavelengths can yield better performance, and higher throughput leads to reduced unit cost and hence greater profits.

Different light sources with varying power and wavelength have been used over past generations of semiconductor manufacturing technology. Early illumination systems used high-pressure mercury arc lamps with wavelengths between 300 and 450 nm. These light sources were used for technology nodes with feature widths greater than 1 μm. Later, sodium lamps were used as illumination sources in photolithography. The g-line and i-line sources of (respectively) 436-nm and 365-nm wavelengths are shown in Figure 2.9[5] along with their corresponding intensities. To improve the resolution of images transferred using sodium lamps, chemically amplified resists (CARs) were introduced to aid in pattern formation. However, sodium lamps were not able—even in the presence of better lens systems and chemically amplified resists—to produce the high-resolution images required for printing features smaller than 250 nm.[6]

It became necessary to find other illumination sources with adequate intensity that could properly image features below 300 nm. Lasers seemed to be an ideal choice, but the available continuous single-mode lasers exhibited too much self-interference to be used in lithography. However, excimer lasers were capable of ensuring dose uniformity by providing interference-free radiation at wavelengths of 248 nm and 193 nm. Source directional uniformity throughout the wafer is ensured by using the Köhler illumination technique, in which the light source is placed at the front focal plane of the condenser (see Figure 2.10). When the source of an imaging system is placed at the

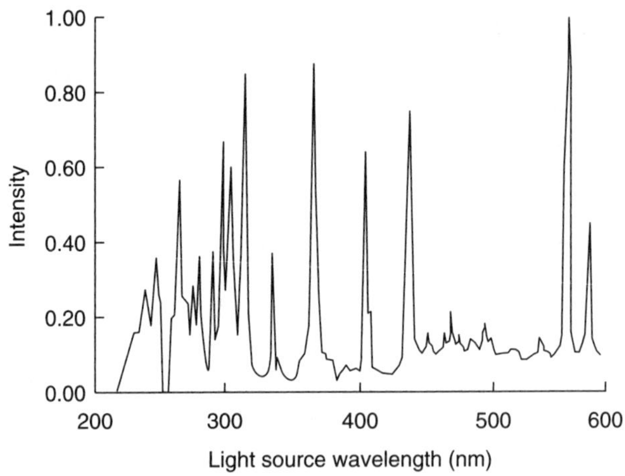

FIGURE 2.9 Spectral content of mercury lamps.

front focal plane of a lens in this manner, the light rays refracted through the lens tend to be focused at infinity (i.e., they are parallel to each other and to the optical axis).[7] Nonuniformity of source intensity is averaged out so that each point on the mask receives the same intensity.

In lithography, the shape of a conventional light source is circular. Other types of illumination shapes have been used to print certain types of features. Figure 2.11 displays some of the commonly used

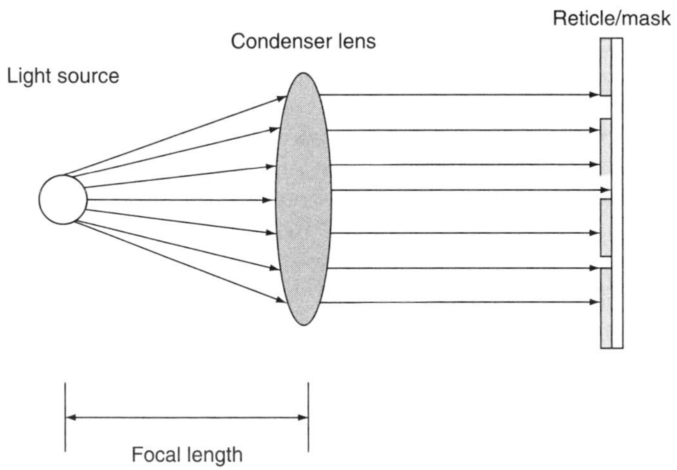

FIGURE 2.10 Köhler illumination: placing the light source at the front focal plane of the condenser lens ensures uniform light directionality.

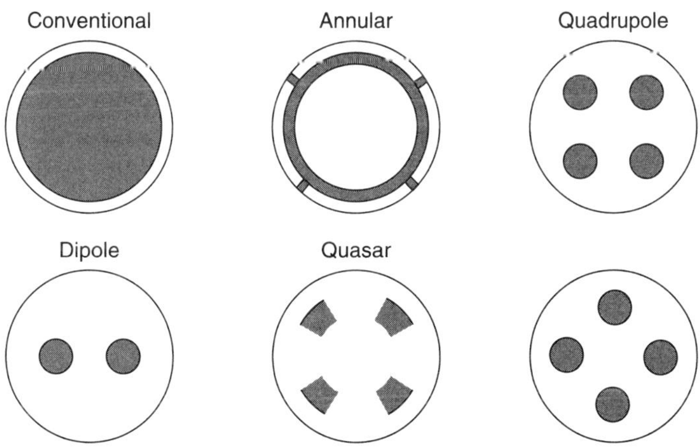

FIGURE 2.11 Types of illumination sources; the outer circle forms a support region of unit radius.

illumination schemes in lithography today. Annular light sources emit light that produce rays at specific focal regions. The quadrupole and dipole sources are used to expose features based on their particular orientation. See Chapter 4 for details regarding the use of such sources and their ability to control lithographic abnormalities.

2.3.2 Diffraction

Diffraction is the foremost phenomenon in projection imaging. The word "diffraction" originated in the 1600s from the published work of Grimaldi, who defined it as a general characteristic of waves that occurs whenever a portion of the light wavefront is obstructed in some way. This phenomenon is also referred to as the "deviation" of light from a rectilinear path. The Dutch scientist Christian Huygens proposed, in his 1690 work entitled *Treatise on Light*, that each point on a light wavefront be considered as a source of secondary spherical wavelets: at each point, a new wavefront can be constructed by superposition of all the waves at that point. Figure 2.12[8,9] is a simplified illustration of the so-called Huygens principle that shows the formation of spherical wavelets and the phenomenon of diffraction through bending of light. However, the Huygens principle did not take the wavelength (λ) of light into consideration, and neither could it explain the phenomenon of different phases of the wavelets.

In 1882, Gustav Kirchhoff formulated an equation for the diffraction pattern based on the condition that the wavelets must satisfy Helmholtz's equation and the conservation of energy.[10,11] Kirchhoff proposed that, if the distance from the mask to the image

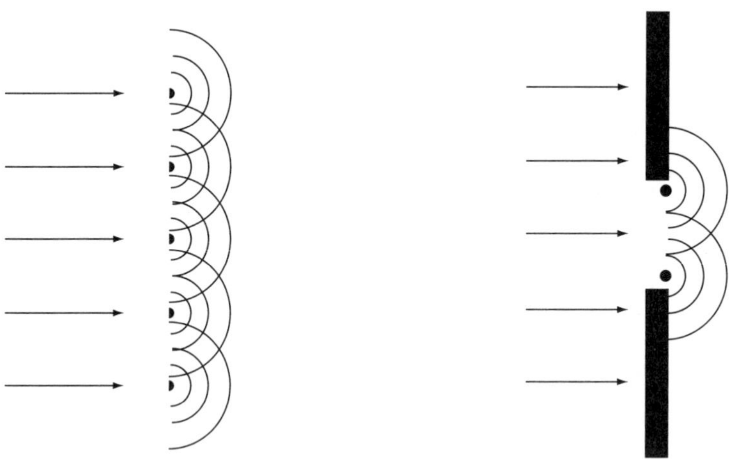

FIGURE 2.12 Propagation of plane waves and illustration of diffraction through a slit.

plane is less than the wavelength of the light source (which is typically the case when feature sizes exceed 2λ), then the diffracted image can be assumed to be just the shadow of the mask pattern. The diffraction is actually omitted, and the output pattern is close to a step function (see Figure 2.13).[11]

Augustin-Jean Fresnel simplified matters in 1818 by describing the diffraction pattern in terms of the phase of the waves and the sum of the amplitudes of all the spherical waves at a given point to form the diffraction pattern. This approximation, later coined the Huygens-Fresnel principle, is valid only when the distance between the source and the image plane is greater than the wavelength of the light source used. As shown in Figure 2.13, varying the wavelength of the light source, the aperture width, or the distance to the image plane yields different interpretations of diffraction theory. In optical lithography systems, the distance between source and image plane is assumed to be greater than the light source wavelength; hence, the *Fraunhofer diffraction* approximation is used to estimate the diffraction pattern's intensity (see Eq. [2.7]).

Consider an imaging experiment involving a coherent illumination source of wavelength λ and a mask consisting of a slit of width w (see Figure 2. 14). The image plane is placed at a distance $R >> w$. Consider a harmonic wave whose E-field on the X axis is given by the wave equation

$$E(x,r) = E_0 \sin(kr - \omega t) \tag{2.1}$$

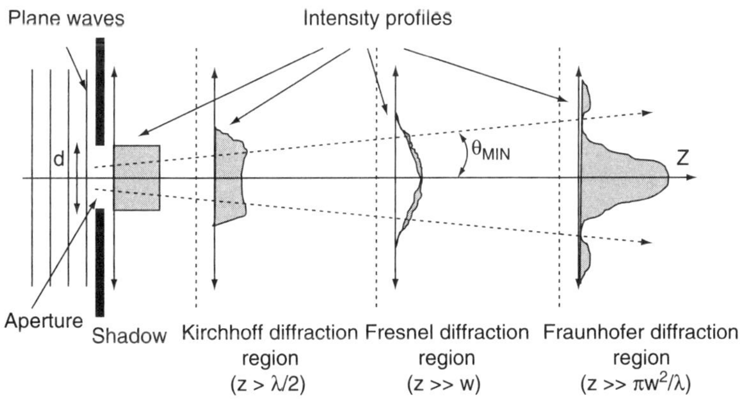

FIGURE 2.13 Diffraction pattern profiles at different regions based on the distance between source and image planes.

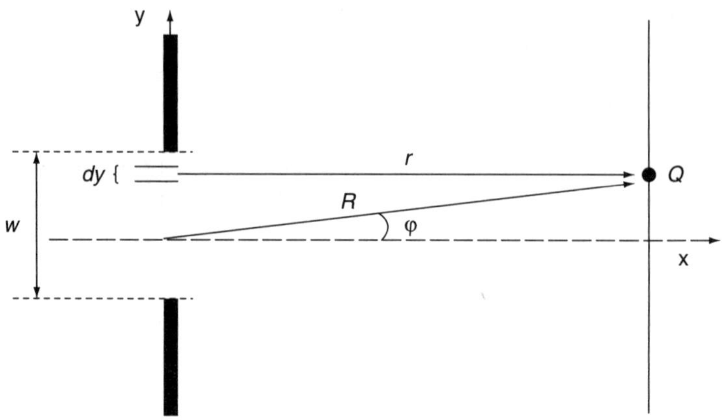

FIGURE 2.14 Single-slit experiment to estimate the electric field at a point Q.

where k is the wave number and ω is the angular frequency. The field dE at any point Q at a distance r due to an infinitesimally small region dy of the slit depends on the source strength per unit length s_L and the distance R from mask to image plane:

$$dE = \frac{s_L}{R}\sin(kr - \omega t)\,dy \tag{2.2}$$

where r is the distance between the infinitesimal slit and the image plane. The term r can be expanded by using the Maclaurin series in terms of R, y, and φ (i.e., the angle made by the line from the slit's center to the imaging point).[9] The first-order approximation yields

$$r = R - y\sin\varphi + \cdots \tag{2.3}$$

Now considering the entire vertical slit width, the electric field is given by the integral,

$$E = \frac{s_L}{R}\int_w \sin\left[k(R - y\sin\varphi) - \omega t\right]dy \tag{2.4}$$

which leads to

$$E = \frac{s_L w}{R}\frac{\sin\left[(kw/2)\sin\varphi\right]}{(kw/2)\sin\varphi}\sin(kR - \omega t) \tag{2.5}$$

Now the electric field can be written as

$$E = \frac{s_L}{R}\left(\frac{\sin\zeta}{\zeta}\right)\sin(kR - \omega t);\ \zeta = (kw/2)\sin\varphi \tag{2.6}$$

The *intensity* of a diffraction pattern is defined as the amount of light that passes through the slit before the lens system determines pattern shape. Also known as *irradiance,* intensity is the time-averaged value of the square of the electric field:

$$I = \langle E^2 \rangle_t \tag{2.7}$$

Given $\langle \sin(kR - \omega t) \rangle_t^2 = 1/2$, the irradiance resulting from the diffraction of a coherent light source through a single slit is expressed as

$$I = \frac{1}{2}\left(\frac{s_L w}{R}\right)^2 \left(\frac{\sin\zeta}{\zeta}\right)^2 \tag{2.8}$$

Equation (2.8) is a typical sinc^2 function, one that is symmetric about the Y axis. As shown in Figure 2.15,[9,12] the irradiance amplitude falls rapidly as we move away from the center of the slit. The width of the central maximum depends on the slit width w and the source wavelength λ.

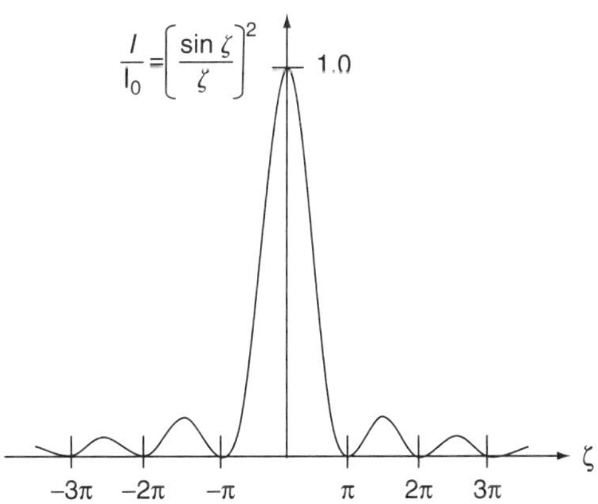

FIGURE 2.15 Intensity distribution resulting from Fraunhofer diffraction of a single slit.

Young's single-slit diffraction experiment is a well-known but simple experiment that illustrates the diffraction effect. Figure 2.16 depicts the single slit used in the experiment and the resulting diffraction pattern in the image plane. This single-slit experiment can be used to envision the diffraction of patterns on a mask. Each specific feature width generates a diffraction pattern that rapidly deteriorates from its center. The diffraction patterns from each mask source will interact with each other, reflecting the superposition principle. When two waves are of the same phase, constructive interference occurs and the amplitudes are additive; conversely, out-of-phase waves cause destructive interference and so their amplitudes are subtractive.

In order to obtain a mathematical equation for the diffraction of the mask pattern, let $m_t(x, y)$ be the mask E-field transmission function in the spatial domain. The value of the mask transmission function at each point on the mask is determined by the presence or absence of a transparent pattern. For a chrome-on-glass mask, $m_t(x, y)$ has a value of 1 for regions that are devoid of chrome (i.e., transparent) and a value of 0 for regions where chrome was coated on the quartz glass surface (see Figure 2.17). The image plane represented by the xy plane is the plane before the objective lens. In the spatial frequency domain, the coordinates f and g represent the plane and are proportional to the wave number k and inversely proportional to the distance R and wavelength λ. If $E_j(x, y)$ denotes the electric field of the light source,

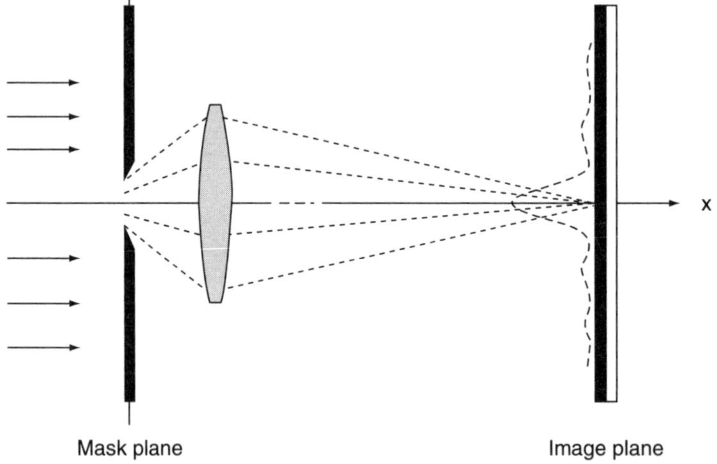

FIGURE 2.16 Formation of the Fraunhofer diffraction pattern from a single slit.

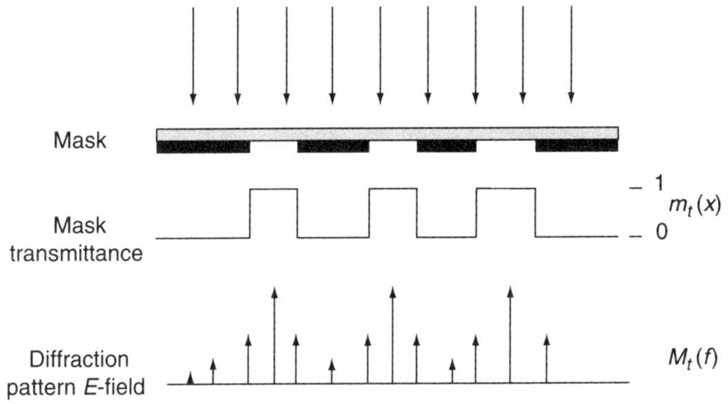

FIGURE 2.17 Mask pattern, its transmission function, and the diffraction pattern *E*-field.

then the electric field of the diffraction pattern for the mask transmission function $m_t(x, y)$ is given by the Fraunhofer integral:

$$M_t(f,g) = \iint m_t(x,y) \cdot E_j(x,y) e^{-2\pi i(fx+gy)}\, dxdy \tag{2.9}$$

This equation is nothing more than the Fourier transform. In essence, the diffraction pattern created by the illumination system by the passage of light through the mask is the Fourier transform of the patterns on the mask.

2.3.3 Imaging Lens

The diffracted light now travels through the lens system. Because it is of finite size, the objective lens allows only a limited portion of the diffracted light to pass through. Lenses are circular in general, and the region of light that is allowed to pass through can be simply quantified by the lens diameter. This region can also be viewed as a circular aperture filtering out diffracted light outside its diameter. The simple ray diagram of Figure 2.18 shows that the diffraction limit is given by the largest diffraction angle that can be captured by the lens to be used for image formation. This angle is represented by $\theta_{\max}$ in the figure. The numerical aperture (NA) of the lens system is a characteristic parameter that determines the quality of the diffracted image on the wafer. It is defined as the sine of the largest diffraction angle, $\theta_{\max}$:

$$\text{NA} = n \sin \theta_{\max} \tag{2.10}$$

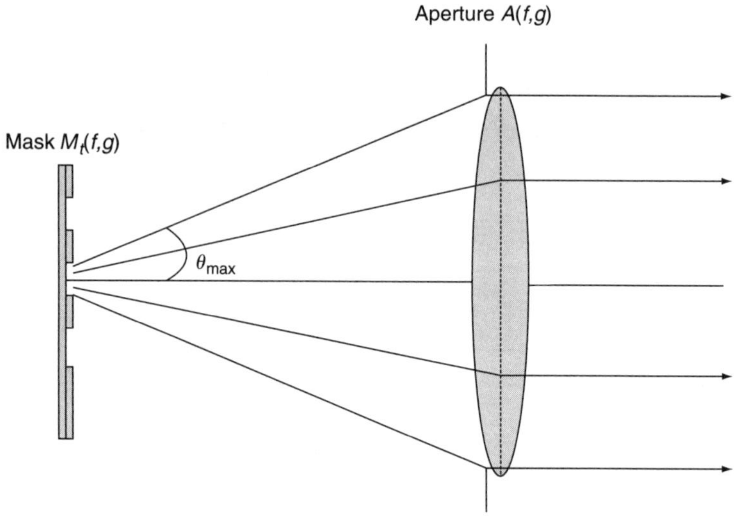

FIGURE 2.18 Numerical aperture (NA) of an imaging system.

The mathematical limit to the numerical aperture is 1 when air is the medium between the objective lens and the wafer. Because of many accuracy problems in lens manufacturing, this NA limit has not been achieved. However, newer technologies use water (with $n > 1$) as the medium to increase the numerical aperture and hence the amount of light being imaged onto the wafer.

A simple function for the aperture can be written as,

$$A(f,g) = \begin{cases} 1, \sqrt{f^2 + g^2} < \dfrac{\text{NA}}{\lambda} \\ 0, \sqrt{f^2 + g^2} > \dfrac{\text{NA}}{\lambda} \end{cases} \tag{2.11}$$

where f and g are coordinates used to represent the aperture function in the frequency domain. It can be seen that the function described by Eq. (2.11) is a filter that allows all rays inclined at angles below $\theta_{\max}$ through the lens.

There is always a need for lithography to print increasingly smaller features. Smaller feature sizes lead to more devices being printed onto the wafer. The smallest feature that can be printed on a wafer with acceptable quality and control defines the resolution of the imaging system. The resolution R of the imaging system was proposed by Lord Rayleigh as[13]

$$R = k_1 \frac{\lambda}{\text{NA}} \tag{2.12}$$

where k_1 is the Rayleigh factor used as proportionality constant. A high k_1 factor value ensures better pattern fidelity. With changes in resolution requirements, k_1 has been reduced, leading to increased instances of fidelity and contrast issues. Equation (2.12) shows that the resolution of the system is directly proportional to the wavelength of the light source. This direct correlation is a key enabler in the quest to find newer light sources that can print smaller features.

The properties of the lens system determine its resolution, and another important factor is the depth of focus (aka depth of field). *Depth of focus* (DOF) is the maximum vertical displacement of the image plane such that the image is printable within the resolution limit. This is the total range of focus points that can be allowed and still keep the resulting wafer image within manufacturing specifications. The depth of focus is given by

$$\mathrm{DOF} = k_2 \frac{\lambda}{\mathrm{NA}^2} \tag{2.13}$$

The focus is inversely proportional to the square of the numerical aperture. Thus, lower R and greater DOF requirements conflict.

2.3.4 Exposure System

The functions of an exposure system consist of exposing the features on the mask and channeling the light from the mask to the wafer. Exposure system configuration and techniques change with the source wavelength and the chemical properties of the resist.

The imaging system now used in photolithography for semiconductor manufacturing is *projection optics*, where the mask and the wafer are placed far apart from one another and a lens system is used to project the features (see Figure 2.19).[14] Other types of printing techniques include proximity printing and contact printing. In *contact printing*, the photomask and resist-coated wafer are actually in contact with each other. Contact printing was popular in 1960s, when device dimensions exceeded 2 μm. Patterns on the mask were transferred onto the wafer using light sources with wavelengths between 300 nm and 500 nm. The theoretical resolution limit for contact printing is given by

$$R = 3\sqrt{\frac{\lambda z}{8}} \tag{2.14}$$

where λ is the source wavelength and z the resist thickness. Because of practical difficulties and the inability to push the resolution limit, contact printing could not be used for later technology generations. In *proximity printing*, as the name suggests, the photomask and the wafer are placed near each other but are not in contact. The theoretical resolution limit for proximity printing is given by[15]

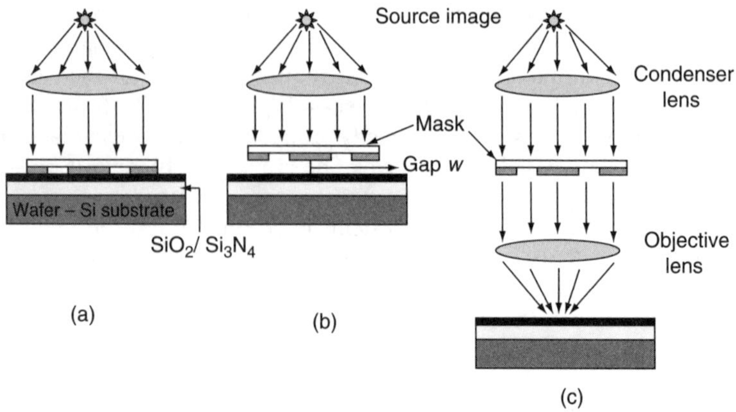

FIGURE 2.19 Exposure systems used in photolithography: (a) contact printing; (b) proximity printing, with gap w between wafer and mask; (c) projection printing.

$$R = 3\sqrt{\frac{\lambda}{4}\left(w + \frac{z}{2}\right)} \tag{2.15}$$

where w is the gap between the wafer and the photomask. The highest resolution that could be achieved with a 450-nm light source and a very small gap of 10 μm was 3 μm. Diffraction issues due to the gap limited the use of proximity printing, and projection printing has been employed ever since.

The two techniques used to expose the mask in projection lithography are the scanner approach and the stepper approach (see Figure 2.20). The scanning technique projects a slit of light onto the wafer while the mask and the wafer are being moved across it. The exposure dose depends on the slit width, resist thickness, and speed of mask and wafer movement. The stepper projects a rectangular region, called a *field*, for one exposure step at a time. The field region is a function of the mask size, exposure dose, and required throughput. This stepper technique can be used to perform reduction imaging, described in the next section.

2.3.5 Aerial Image and Reduction Imaging

As the diffraction pattern passes through the lens system, some of the pattern is filtered out by the aperture. We know that the diffraction pattern is the Fourier transform of features in the photomask, so an inverse Fourier transform of this function should produce features that resemble the original mask. This inverse Fourier transform operation is performed by the lens system. A carefully designed lens system can produce a nearly ideal pattern on a wafer. In the frequency

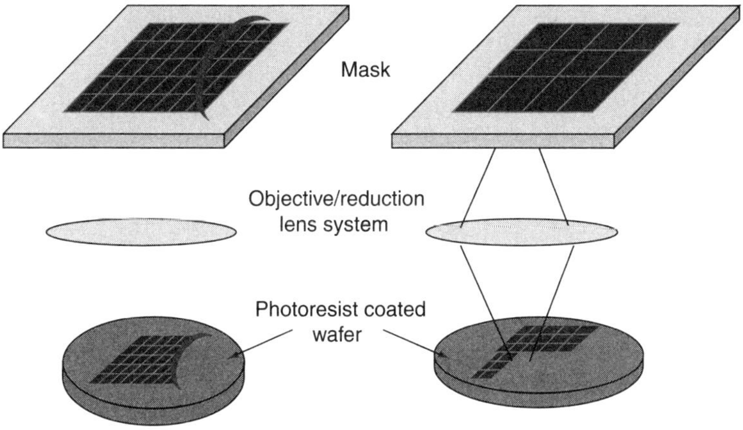

FIGURE 2.20 Set-up for reticle exposure: scanner (left) and reduction stepper (right).

domain, the filter function can be multiplied by the diffraction pattern to obtain the electric field of the diffracted image above the resist. The electric field of the final projected aerial image (AI) in spatial domain is given as

$$E_{\text{AI}}(x, y) = F^{-1}\{M_t(f, g)A(f, g)\} \tag{2.16}$$

where $F^{-1}\{\cdot\}$ denotes the inverse Fourier transform function. The intensity distribution of this image is known as the *aerial image* of the photomask pattern. The intensity distribution is simply the time-averaged square of the electric field.

The control of feature width is critical in the photomask process, because any error in the shapes will be replicated on the resist. So-called 1X exposure systems have features on the photomask that are of the same dimension as the required wafer dimensions. With technology scaling and as light sources of smaller wavelength were found, reduction projection systems came into use. In *reduction imaging,* the features on the mask undergo demagnification during the photolithography process. Reduction imaging employes a series of lenses with focal lengths matched to create the required demagnification (see Figure 2. 21). In effect, the reduction lens system has a different numerical aperture at each end of the system. The critical dimension (CD) of the printed feature based on reduction imaging is given by

$$\Delta\text{CD}_{\text{wafer}} = \frac{\Delta\text{CD}_{\text{mask}}}{M_D} \tag{2.17}$$

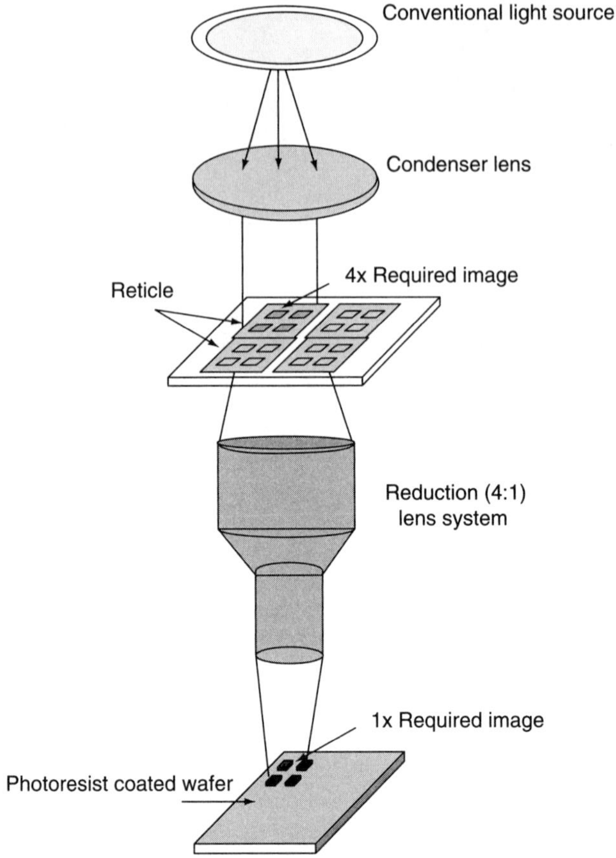

FIGURE 2.21 Imaging system with reduction optics; demagnification factor $M_D = 4$.

Here M_D is the demagnification factor, which is typically 4 or 5 in modern projection systems. Of course, errors in the mask are demagnified by the same factor M_D. But for systems with features smaller than the light source wavelength, there is a mask error enhancement factor (MEEF) that must be taken into account when considering wafer side errors:

$$\Delta \mathrm{CD}_{\mathrm{wafer}} = \mathrm{MEEF} * \frac{\Delta \mathrm{CD}_{\mathrm{mask}}}{M_D} \tag{2.18}$$

The importance of reduction is the ease of mask production. Because it is much easier to produce a mask of 4X feature size than one of 1X feature size, reduction imaging plays a significant role in mask production for all layers. Although this helpful feature of reduction imaging has been exploited to project ever smaller images, future technology generations may require a greater degree of reduction if mask preparation becomes a bottleneck. The 2007 ITRS technical report[16] suggests that higher magnification might be considered in order to reduce mask preparation problems. Increase in magnification requires changes to the projection imaging system and also increases the burden on the controlling MEEF factor for mask features.

2.3.6 Resist Pattern Formation

The aerial image created by an exposure system propagates onto the resist to form a latent image. The resist is exposed, and diffusion reactions begin within the exposed regions. Light rays may also reflect off the resist wafer boundary to form standing waves. The postexposure bake process and antireflection coats mitigate the standing wave effect. Also, as shown in Figure 2.22(a),[15] the relative intensity of the light traveling through the resist deteriorates with depth. This creates a variation between the width of the resist profile at the top and the bottom (see Figure 2.22[c]). An acceptable variation between the intensity at the top and the bottom is given by the resist sidewall angle (θ) specification. The focal position of the lens system on the resist layer is decided based on the required sidewall angle and the absorption coefficient of the resist material. An intensity gradient that is too high will cause tapering of the resist feature, leading to variation in feature width after the etch stage.

The *exposure dose* (ED) is the strength of the light source used by the exposure system. The dose of the system is the chief contributing factor in resist pattern formation. The dose of an exposure step is decided based on the thickness of the resist, the absorption coefficient, and the imaging system parameters. Higher absorption of light is controlled by using a special contrast enhancement layer (CEL) over the photoresist.[17] Depending on the dissolution property, the photoresist can be classified as either positive or negative. The *contrast* of a resist refers to the ease of distinguishing between exposed and unexposed areas of the photoresist. The rate of change in resist thickness varies linearly with the logarithm of exposure dose, which enhances the contrast of the resist. The contrast of a positive photoresist is given by

$$\gamma_{+} = \frac{1}{\ln \mathrm{ED}_{+l} - \ln \mathrm{ED}_{+h}} \tag{2.19}$$

(see Figure 2.23), where ED_{+l} is the exposure dose below which there is no resist development, and ED_{+h} is the exposure dose above which the resist is completely consumed (i.e., exposed region). A similar

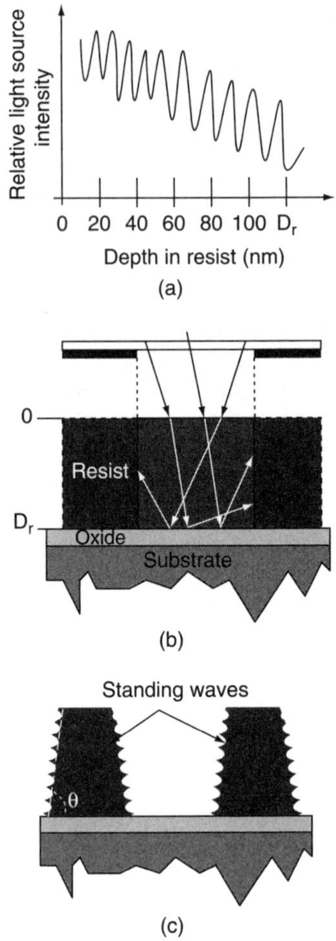

FIGURE 2.22 Resist pattern formation: (a) relative intensity of illuminating light source inside the resist, where Dr is the resist thickness on the wafer; (b) light rays passing through chrome-free regions and reflecting off the resist-wafer interface; (c) the reflected rays create standing waves in the resulting resist profile.

expression can be used for the negative photoresist. The diffused areas of the resist are treated with a developer followed by PEB. Finally, the etching process creates the required profile, which is then ready for the next stage of processing. It is important to note that a mask's resolution depends also on the width of the final resist profile after etching. Because of such issues as standing wave formation, the edge of the pattern formed after etching process is not regular. The irregularity at pattern edges is called line edge roughness (LER), which leads to line width roughness (LWR)—that is, variation with LER in the width of the pattern. In essence, LWR is variation in resolution of the feature. More details on etching-induced variation in resolution are provided in Sec. 3.2.2.

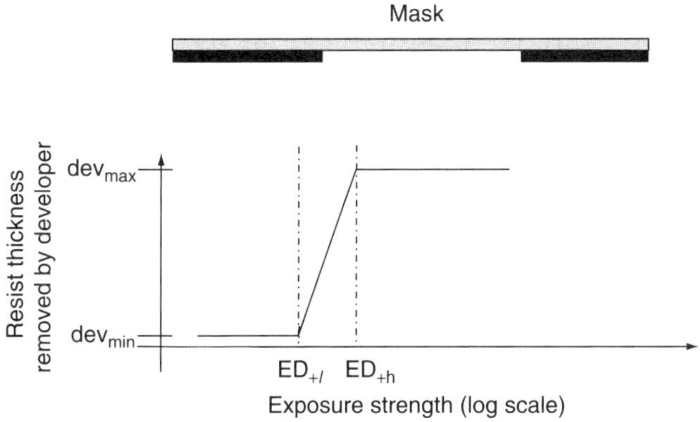

FIGURE 2.23 Calculating the contrast of an exposure system.

2.3.7 Partial Coherence

The concepts of spatial and temporal coherence must be considered when analyzing the phase relationship between signals. *Spatial coherence* is the phase correlation between light sources at different points in space at the same moment of time. *Temporal coherence* is the correlation between signals at different moments in time. Spatial coherence implies temporal coherence, but the opposite is not true. To make things clear, all light sources discussed in this book are temporally coherent. Future references to coherent, incoherent, or partially coherent sources refer only to spatial coherence.

An ideal point source, whose light rays are highly coherent and parallel to each other, form a plane wavefront when incident on the mask; this produces the intensity profile shown in Figure 2.24(a). But if the mask is illuminated by oblique light rays, as shown in Figure 2.24(b), then the intensity profile is shifted; and if the mask is illuminated from different directions, as in Figure 2.24(c), the resulting profile is a broadened version of the initial intensity profile seen in Figure 2.24(a). This technique of *partially coherent imaging* is widely used today to improve intensity profiles on wafer.

Diffraction patterns produced by light rays with a large angle of incidence may not fall within the aperture thus leading to loss of information during the inverse Fourier transform. In this case, a feature at some point in the mask cannot be replicated exactly at the same position on the wafer. *Kohler illumination* is used to prevent this loss of information by (1) making an image of the source fall at the entrance pupil of the objective lens and (2) placing the mask at the exit pupil of the condenser lens. This setup can be seen in Figure 2.25. Since the mask is placed at the exit pupil of the condenser lens, all the rays fall on the mask at incidence angles of less than 90°; hence all

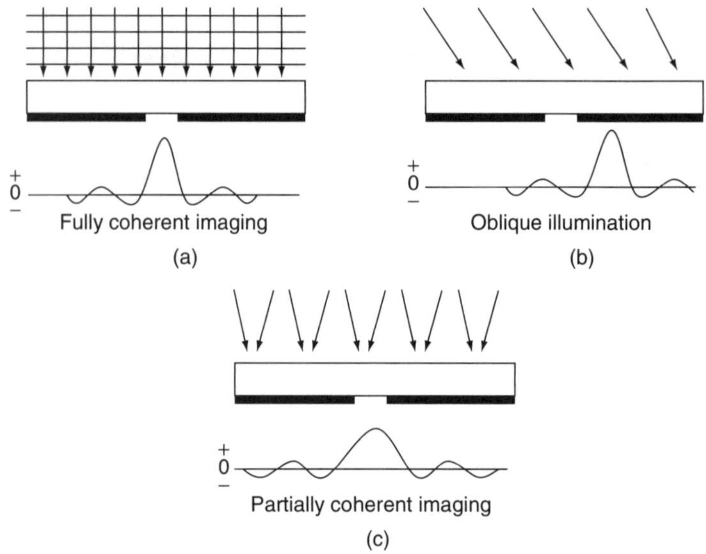

FIGURE 2.24 Intensity profiles for different imaging techniques; panel (c) shows the broadening of an intensity profile in response to incident rays from different angles.

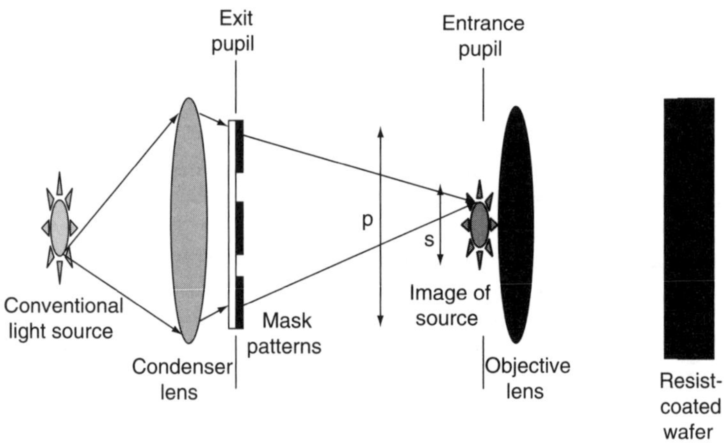

FIGURE 2.25 Kohler illumination used in the context of partial coherent imaging.

diffraction information is retrieved. Any change in the incidence angle of light rays will affect the system's resolution. Such variation is quantified by a factor known as the *partial coherence factor* (σ):

$$\sigma = \frac{n \sin\theta}{\text{NA}} \tag{2.20}$$

The change in resolution of the imaging system as a function of this coherence is given by[4]

$$R = k_1 \frac{\lambda}{\text{NA}(1+\sigma)} \tag{2.21}$$

Another conventional way of representing the partial coherence factor is as the ratio of the diameter of the source image at the aperture plane to the diameter of the aperture itself:

$$\sigma = \frac{s}{p} \Rightarrow \frac{\text{Source diameter}}{\text{Pupil diameter}} \tag{2.22}$$

2.4 Lithography Modeling

Several techniques that model photolithography can be classified into two main categories: physics-based and phenomenological models. *Physics-based models* accurately model the underlying optics and also the chemical reactions that take place within the resist. These models are highly complex, and they are difficult to calibrate and verify because their parameters have complex, nonlinear relationships. *Phenomenological models* (aka parameter-centric models) do not consider the chemical reactions as such; instead, they model image formation and estimate edge location based on available empirical resist models. Physics-based models consider many physical and chemical effects, so they are much slower than their parameter-centric counterparts. PROLITH and Solid-C are well-known commercial tools based on underlying physics-based models. Parameter-centric models are fast but compromise on accuracy. But these models are simple enough to be integrated with design flows in order to verify printability of a design.

The following sections describe the steps involved in aerial image simulation and chemically amplified resist modeling. Both physics-based and phenomenological lithography modeling methods use the aerial image simulation technique; the difference lies in the resist diffusion and development modeling.

2.4.1 Phenomenological Modeling

Phenomenological modeling involves the following stages:

1. Modeling the optics behind aerial image formation on top of the resist; this modeling incorporates the behavior of different illumination schemes, aperture types, and defocus effects.
2. Modeling the formation of latent PEB images with the aid of a first-order resist diffusion model.
3. Determining the feature edge by using a intensity threshold that is based on empirical resist models.

2.4.1.1 Hopkins Approach to Partially Coherent Imaging

Aerial image is defined as the aperture-filtered intensity distribution of the photomask pattern observed in a plane above the photoresist surface. Ernst Abbe proposed the extended source method for partially coherent imaging system. H. H. Hopkins, in his paper on diffraction,[8] devised an extension of this work for estimating the intensity profile of patterns on the photomask. Since partial coherence shifts the diffraction pattern at each point relative to the aperture, the converse observation can be used to obtain the intensity profile. We know that the electric field is given by the following Fourier transform representation:

$$E(x,k) = \iint A(k+k')M_t(k)e^{2\pi i(k+k')x}\,dk \tag{2.23}$$

where $E(k)$ is the frequency-domain representation of $E(r)$ in the spatial domain. In order to simplify the equation, we consider only the x dimension. It is important to note that, since all illumination sources shown in Figure 2.11 have a circular outer ring, their support region is a circle. Therefore, the two-dimensional intensity profile of a feature is referenced according to its radial distance (r) from the center of the feature. The intensity profile is the square of the electric field; its value, which is obtained by integration over all source points, is given by

$$I(r) = \iint \mathrm{TCC}(k,k'').M_t(k)M_t^*(k'').e^{2\pi i(k-k'')x}\,dk\,dk'' \tag{2.24}$$

where $M_t^*(k)$ is the complex conjugate of $M_t(k)$. Hopkins used an intermediate step to integrate over the source before the mask function is considered. Because only the aperture function is affected by the source, the two functions can be merged to form the transmission cross coefficient (TCC):[10,18]

$$\mathrm{TCC}(k,k'') = \iint S(k)A(k+k')A^*(k+k'')d^2k \tag{2.25}$$

where $S(k)$ is the source shape function. The intensity value obtained in Eq. (2.24) is also known as the pre-PEB latent image.

2.4.1.2 Resist Diffusion

The postexposure baking is performed on the wafer after it has been exposed. For a positive photoresist, exposed regions begin to diffuse during the PEB process. A simple convolution of the diffusion function with the pre-PEB latent image intensity is performed to estimate the post-PEB image in resist. A basic Gaussian diffusion model can be used (among various others that have been suggested). In the frequency domain, the diffusion function can be given by

$$D(k,k') = \exp\left\{-2\pi^2 d^2 (k^2 + k'^2)\right\} \tag{2.26}$$

Considering diffusion of photoresist, the change in Hopkins' TCC function ($\mathrm{TCC}_{w/.diff}$) can be expressed as

$$\mathrm{TCC}_{w/.diff}(k,k'') = \iint S(k) D(k,k') A(k+k') A^*(k+k'') d^2k \tag{2.27}$$

Hence the post-PEB latent image intensity can be written as

$$I(x;k,k'') = \iint \mathrm{TCC}_{w/.diff}(k,k'') M_t(k) M_t^*(k'') e^{2\pi i (k-k'')x} dk\, dk'' \tag{2.28}$$

2.4.1.3 Simplified Resist Model

Simplified models can be used to predict the photoresist response from the aerial image. Examples of such models include the aerial image threshold model, the variable threshold model, and the lumped parameter model. These models do not capture the mechanistic response of the photoresist, but they do describe the resist with a minimal number of parameters. The simplified models have no physical meaning and are devised primarily for the purpose of rapidly obtaining a resist response. A naïve resist model has been used in this chapter to establish the location of feature edges. This model, termed the *threshold bias* model, has been shown to be reasonably accurate in predicting projected critical dimensions. The model assumes that, by applying to the printed contour a constant bias that is equal to intensity threshold value, the exact location of a feature edge—and hence its CD—can be computed. See Figure 2.26.

The intensity threshold value is typically obtained by fitting resist profile data. An empirical resist model (such as the variable threshold model) is used to obtain the intensity threshold value.[11,15]

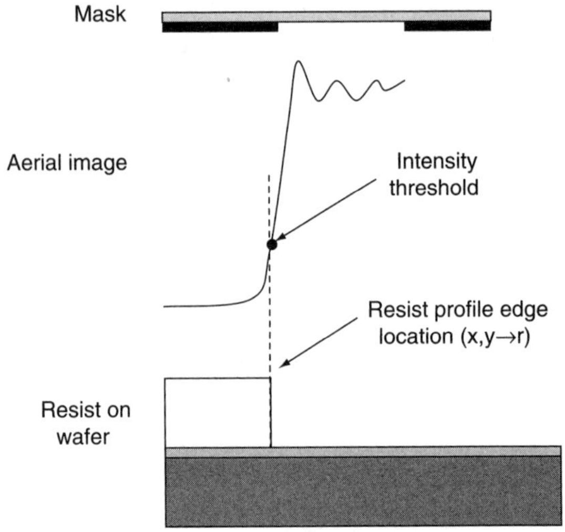

FIGURE 2.26 Using aerial image intensity threshold to estimate the edge location.

2.4.1.4 Sum-of-Coherent-Systems Approach

Hopkins formulated an approach for estimating the aerial image intensity distribution of photomask features. A technique was proposed to simplify this formulation by "decomposing" the imaging system. This technique, called the *sum-of-coherent-systems* (SOCS) approach, performs eigenvalue decomposition of the transmission cross coefficient function obtained in the original Hopkins formulation:[10,11]

$$\mathrm{TCC}_{w/.diff}(k,k'') = \iint S(k)D(k,k')A(k+k')A^{*}(k+k'')d^{2}k \tag{2.29}$$

In partially coherent illumination scheme, lights rays incident from multiple angles do not interfere with one another and so their effects on the diffraction pattern of the mask feature are independent. The decomposition is described by a set of coherent sources whose interaction is highly incoherent. The aerial image intensity is decomposed into the sum of the intensities due to different point sources. The exposure system function, which consists of the source and the aperture functions, is decomposed into eigenfunctions called *kernels*. These optical system kernels are arranged based on their eigenvalues and are used to estimate the latent image intensity. Thus, the TCC is decomposed into a discrete set of eigenvectors with their respective eigenvalues:

$$\text{TCC} = \sum_{u} \zeta_m \varphi_m(k) \cdot \varphi_m^*(k) \tag{2.30}$$

where the ζ are eigenvalues and the $\varphi(k)$ are frequency components of eigenvectors, or imaging system kernels. The aerial image intensity is given as

$$I(r) = \sum_{u} \zeta_m \cdot (\varphi_m(k,k') ** M_t(k,k'))^2 \tag{2.31}$$

where the double asterisk denotes convolution in the time domain.

The number u of such vectors is determined by the error requirement. Yet because the relation between each eigenfunction and the intensity value is nonlinear, a higher u does not necessarily lead to lesser error in intensity prediction. In short, the SOCS approach is a decomposition-based technique for calculating the aerial image intensity profile of partially coherent imaging.

2.4.2 Fully Physical Resist Modeling

Fully physical resist models are built from physical and mechanistic chemical reactions of the photoresist—from the onset of application on the resist to its removal at the etching station. Each step in the lithography process is described by a complex, physical model. The advantage of such models is the complete correlation between simulation results and actual experimental data. Another important advantage is that, when the technology parameters change, the model as such requires only tuning and not a complete revamp, as would be required with simplified models. Disadvantages include lack of speed and scalability.

There are two different types of resists. Typical resists used with i-line and g-line illumination sources are called *conventional* resists. Higher resolutions have been achieved by using deep ultraviolet (DUV) illumination sources with smaller wavelengths. Higher sensitivities and greater throughput are achieved using *chemically amplified resists.*[17] The CAR type of resist "amplifies" the exposure dose response by undergoing chemical reactions that change the dissolution properties of the resist during the PEB stage. Models used today for CAR resists are simple modifications of the conventional resist models.[26]

Resist composition depends on the tone, technology generation, illumination wavelength, and exposure system parameters. The photoresist functions by converting the spatial light energy distribution into a solubility distribution of the resist during development. Prior to exposure, the inherent chemical constituents of the photoresist are polymer resin, dissolution inhibitor, photoacid generator (PAG), and base quencher. Physical models concentrate chiefly on three stages of the resist response when modeling the photoresist: exposure, resist solubility, and development.

The first stage in the chemical process is the exposure response, which mostly consists of absorption. This absorption can be defined empirically by Lambert's law as

$$\frac{dI}{dz} = -\alpha I \tag{2.32}$$

The intensity profile within the resist during exposure is characterized as a function of the resist thickness z and the absorption coefficient α:[20,21]

$$I(z) = I_0(z) \cdot \exp\{-\alpha_{\text{eff}} \cdot z\} \tag{2.33}$$

The second stage is resist solubility, or diffusion. For CAR resists, diffusion can be characterized in three steps. First, obtain the relative concentration of PAG molecules;[22,23] next, use this value to obtain the concentration of unreacted photoacids. The third step is to model the polymer resin deblocked by the unreacted acid during PEB. This process is described well by a reaction-diffusion model. The rate of acid deblocking is a function of the diffusivity of the acid, the relative concentration of the base quencher, and PEB time.[24]

The last stage involves determining the development rate. Various models[19,25,26,27] have been proposed to obtain the bulk and relative development rates of the photoresist dipped in an aqueous developer solution. All physical resist models provide the maximum and minimum rates of resist development, which are used to estimate the shape of the resist profile.

2.5 Summary

In this chapter we looked into the various stages involved in fabricating an integrated circuit. We chiefly concentrated on two important processes that control the formation of patterns on wafer: photolithography and etching. The photolithography process was explained to help the reader understand the steps required to form the pattern on the wafer. Details of the optical imaging system's components and of the parameters that control the final pattern were discussed. We described lithography modeling in some depth because it has become an important constituent of many model-based DFM methodologies. Two types of modeling were analyzed, phenomenological and physics-based ("fully physical") modeling. We also demonstrated how the aerial image is formed above the wafer and how photoresist models respond to different values of light intensity. Understanding the fundamentals of pattern formation above and on the resist, as explained in this chapter, is required for modeling process variations and tying them to device and interconnect parameters.

References

1. M. Madou, *Fundamentals of Microfabrication,* CRC Press, Boca Raton, FL, 1997.
2. R. C. Jaeger, *Introduction to Microelectronic Fabrication,* Prentice Hall, Englewood Cliffs, NJ, 2002.
3. W. R. Runyon and K. E. Bean, *Semiconductor Integrated Circuit Processing Technology,* Addison-Wesley, Reading, MA, 1990.
4. E. C. Kintner, "Method for the Calculation of Partially Coherent Imagery," *Applied Optics* **17**: 2747–2753, 1978.
5. M. E. Dailey et al., "The automatic microscope," *MicroscopyU,* http://www.microscopyu.com/articles/livecellimaging/automaticmicroscope.html.
6. A. K. Wong, *Optical Imaging in Projection Microlithography,* SPIE Press, Bellingham, WA, 2005.
7. J. W. Goodman, *Introduction to Fourier Optics,* McGraw-Hill, New York, 1968.
8. H. H. Hopkins, "On the Diffraction Theory of Optical Images," *Proceedings of the Royal Society of London, Series A* **217**: 408–432, 1953.
9. E. Hecht, *Optics,* Addison-Wesley, Reading, MA, 2001.
10. M. Born and E. Wolf, *Principles of Optics,* Pergamon Press, Oxford, U.K., 1980.
11. C. A. Mack, *Fundamental Principles of Optical Lithography: The Science of Microfabrication,* Wiley, New York, 2008.
12. G. B. Airy, "On the Diffraction of an Object-Glass with Circular Aperture," *Transactions of Cambridge Philosophical Society* **5**(3): 283–291, 1835.
13. Lord Rayleigh, "Investigations in Optics, with Special Reference to the Spectroscope," *Philosophical Magazine* **8**: 261–274, 1879.
14. S. M. Sze (ed.), *VLSI Technology,* McGraw-Hill, New York, 1983.
15. A. K. Wong, *Resolution Enhancement Techniques in Optical Lithography,* SPIE Press, Bellingham, WA, 2001.
16. "Lithography," in *International Technology Roadmap for Semiconductors Report,* http://www.itrs.net (2007).
17. H. Ito, "Chemical Amplification Resists: History and Development within IBM," *IBM Journal of Research and Development* **341**(1/2): 69–80, 1997.
18. H. H. Hopkins, "The Concept of Partial Coherence in Optics," *Proceedings of the Royal Society of London, Series A* **208**: 263–277, 1951.
19. M. D. Smith, J. D. Byers, and C. A. Mack, "A Comparison between the Process Windows Calculated with Full and Simplified Resist Models," *Proceedings of SPIE* **4691**: 1199–1210, 2002.
20. N. N. Matsuzawa, S. Mori, E. Yano, S. Okazaki, A. Ishitani, and D. A. Dixon, "Theoretical Calculations of Photoabsorption of Molecules in the Vacuum Ultraviolet Region," *Proceedings of SPIE* **3999**: 375–384, 2000.
21. N. N. Matsuzawa, H. Oizumi, S. Mori, S. Irie, S. Shirayone, E. Yano, S. Okazaki, et al., "Theoretical Calculation of Photoabsorption of Various Polymers in the Extreme Ultraviolet Region," *Japan Journal of Applied Physics* **38**: 7109–7113, 1999.
22. K. Shimomure, Y. Okuda, H. Okazaki, Y. Kinoshita, and G. Pawlowski, "Effect of Photoacid Generators on the Formation of Residues in an Organic BARC Process," *Proceedings of SPIE* **3678**: 380–387, 1999.
23. M. K. Templeton, C. R. Szmanda, and A. Zampini, "On the Dissolution Kinetics of Positive Photoresists: The Secondary Structure Model," *Proceedings of SPIE* 771: 136–147, 1987.
24. F. H. Dill, "Optical Lithography," *IEEE Transactions on Electron Devices* **22**(7): 440–444, 1975.
25. R. Hershel and C. A. Mack, "Lumped Parameter Model for Optical Lithography," in R. K. Watts and N. G. Einspruch (eds.), *Lithography for VLSI,* Academic Press, New York, 1987, pp. 19–55.

26. T. A. Brunner and R. A. Ferguson, "Approximate Models for Resist Processing Effects," *Proceedings of SPIE* **2726**: 198–207, 1996.
27. J. Byers, M. D. Smith, and C. A. Mack, "3D Lumped Parameter Model for Lithographic Simulations," *Proceedings of SPIE* **4691**: 125–137, 2002.

CHAPTER 3

Process and Device Variability: Analysis and Modeling

3.1 Introduction

The most important concern today for design and process engineers alike is the increasing impact of parameter variation in semiconductor manufacturing. The percentage of parameter variations have increased drastically from 10 percent in 250-nm technology node to around 50 percent in 45-nm technology.[1] There is always a certain amount of variation in any manufacturing process. The degree of variability that can be tolerated is often provided with the product specification, and any variation exceeding it will lead to a low-yield process. Parameter variations can be classified into different categories based on process purpose, region of correlation, and behavior.

The basic steps in semiconductor manufacturing involve geometric patterning to create devices such as transistors, diodes, and capacitors and then connecting those devices using wires (metal interconnects). Photolithography is central to patterning that creates devices and wires. Creating a device involves poly or metal gate patterning, oxidation to create gate oxide, a development process, and introducing source and drain impurities via diffusion or ion implantation. Lithography is also used to pattern interconnect metals. Variations in patterning process are chiefly due to problems with projection lithography. As the feature width of patterns printed in the wafer have become less than a quarter of the wavelength of the light source, diffraction-induced printability variations have become highly prevalent. Other than the inherent resolution and contrast problems, further variations are caused by defocus and lens aberrations in the imaging system. These variations affect the patterns being printed, including gate and interconnect features.

Polysilicon patterning is used to create gate terminals of a MOSFET. The image transfer of gate patterns on the wafer is an important step in the creation of self-aligned gate structures. Variations in such patterns include changes in the gate length L_G and/or gate width W_G, two dimensions that determine the area over which the inversion region is formed when the transistor is in its ON state. Variations in L_G and W_G lead to changes in electrical characteristics of transistor operation. For example, a transistor's drain (drive) current I_D—which is the most important parameter characterizing its operation—is inversely proportional to the gate length L_G and directly proportional to the gate width W_G. A reduction in L_G increases I_D causing the transistor to switch faster. When such reductions are systematic across a chip the result can be improved performance, although such improvements usually come at the expense of increased leakage through transistors. At lower channel lengths, transistor threshold voltage tends to become lower, a phenomenon known as *V_T roll-off* (see Figure 3.1).[2] In contrast, leakage current increases exponentially with reduction in threshold voltage. Figure 3.2 shows the various leakage currents for a reduced V_T n-channel MOSFET. The gate width is defined by the region over which the polysilicon lies on the active region. When the patterning of active region under the poly is not perfect, gate widths may vary. This effect of improper diffusion formation, which leads to W_G variation, known as *diffusion rounding* (see Figure 3.24[b]). Change in W_G leads to proportional variation in drain current and hence in the transistor's operation.

The drain current is a function of amount of charge in the inversion layer of a transistor in the channel region, which in turn is a function of the gate capacitance. Transistor size and the gate oxide thickness are the principal determinants of gate capacitance. With scaling, the size of the transistor gate length shrinks; this calls for a commensurate

FIGURE 3.1 nMOS V_T roll-off characteristics.

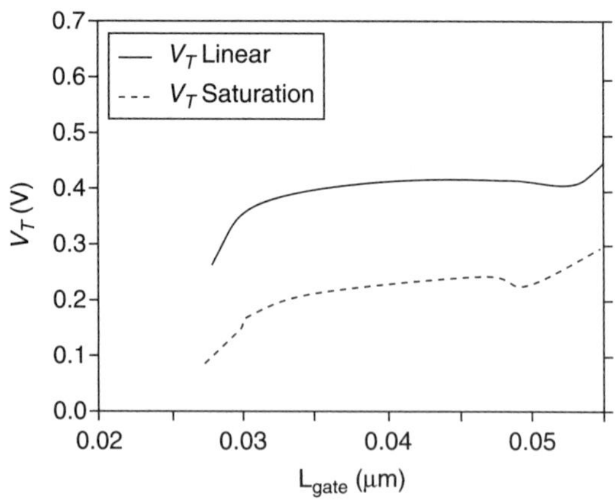

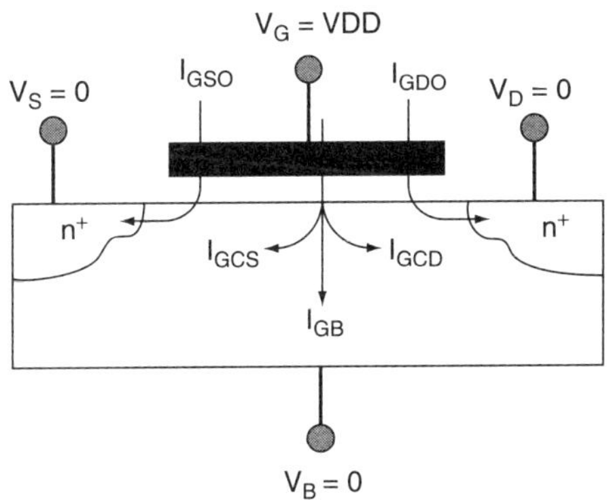

FIGURE 3.2 Various leakage currents in n-channel MOSFET with reduced V_T.

reduction in gate oxide thickness t_{ox} to improve transistor performance. Variation in the thickness of the oxide layer may lead to transistor breakdown and failure. Gate dielectric breakdowns are a known cause of circuit reliability problems.

Transistor threshold voltage V_T depends on the concentration of impurities in the channel region. Impurities are introduced through an ion implantation process, which cannot guarantee a specific number or location of dopant atoms in a nanoscale structure. Consequently, as the transistor size scales and the transistor channel area shrinks, the number of dopant atoms scales similarly. With fewer dopant atoms, even a small variation in their number leads to threshold voltage variation in transistors. New impurities are introduced into the source-drain regions of the transistor in order to increase the mobility of charge carriers through stress. The extent of this stress depends on the amount of active area, shallow trench isolation (STI), and other regions within the standard cell. Variations in the stress alters the mobility of electrons and also the extent of device leakage. A summary of past and projected process control tolerances is given in Table 3.1.[3]

Creating interconnect metal lines involves deposition and planarization. Aluminum has long been used as the material for metal interconnect wires, although modern processes use copper for interconnects. Aluminum is deposited on the wafer by processes such as sputtering, chemical vapor deposition (CVD), and epitaxy, whereas copper is deposited using a dual-damascene process. In the dual-damascene process, a single metal deposition step is used to form the vias and the interconnect metal lines. Both trenches and vias are first formed in a single dielectric layer by two separate lithography steps; then the via and trench recesses are filled in by a single metal-deposition step. After this filling, the excess metal that is deposited outside the trench is removed by a chemical-mechanical polishing (CMP) process. An electroplating process is used to deposit copper interconnects on a seed metal liner.

	Year of production								
	2001	2002	2003	2004	2005	2006	2007	2008	2009
Tech. node (commercial, not ITRS)	130		90		65		45		30
DRAM 1/2 pitch (nm)	130	115	100	90	80	70	65	57	50
MPU 1/2 pitch (nm)	150	130	107	90	80	70	65	57	50
Printed gate	90	75	65	53	45	40	35	32	28
Physical gate (postetch)	65	53	45	37	32	28	25	22	20
t_{ox} Thickness control, EOT (% 3σ)	<+4%	<+4%	<+4%	<+4%	<+4%	<+4%	<+4%	<+4%	<+4%
L_{gate} 3σ var. (nm) WiW, W2W, L2L	6.31	5.3	4.46	3.75	3.15	2.81	2.5	2.2	2
L_{gate} 3σ var. as % of Physical gate	10%	10%	10%	10%	10%	10%	10%	10%	10%
Total max allowable litho 3σ	5.51	4.33	3.99	3.35	2.82	2.51	2.24	1.97	1.79
Total max allowable etch 3σ including resist trim & gate etch	3.64	3.06	1.99	1.88	1.41	1.26	1.12	0.98	0.89
CD bias: dense & isolated lines	≤15%	≤15%	≤15%	≤15%	≤15%	≤15%	≤15%	≤15%	≤15%

TABLE 3.1 Roadmap for Overall Process Control

The deposited metal must be planarized not only to achieve uniformity of capacitance but also to establish the surface planarity required for creating subsequent upper metal layers. The planarity of the deposited metal after CMP depends on the patterns present beneath the metal and also on the underlying metal layers. Isolated and dense patterns are polished at different rates, leading to surface nonplanarity and variation in metal thickness. Such metal thickness variation can lead to variation in interconnect capacitances, contributing to variation in circuit performance and circuit noise tolerance.

Correlation factors can be used to categorize process parameter variation as lot-to-lot, wafer-to-wafer, die-to-die, or intradie variation. Process parameter variation is least when correlation is high. Process variation may be temporal or spatial. Variations within a wafer are called *spatial variations,* whereas variations across wafers are called *temporal variations.* Lot-to-lot and wafer-to-wafer are types of temporal variation or correlation. There are many sources of spatial variation, including overlay errors, reticle errors, lens errors, and focus errors. Additionally, handling errors and particulates may contribute to spatial variations. These variations initially form a considerable portion of the total process yield but are reduced as a process matures. Die-to-die variations have a spatial correlation that depends on the wafer size. Typical die-to-die (interdie) variations include those that are due to chemical-mechanical polishing and oxide thickness. Intradie variations occur within a die. In the past, intradie variations were not considered to be significant. However, with continued scaling of transistor feature size, intradie variations have become more important. In fact, some studies show that intradie variation today may be as prevalent as interdie variation.[4] Intradie variations include patterning variation, random dopant fluctuation (RDF), and CMP-induced thickness variation—to mention just a few. Variations within a die are of design concern because they influence the operation of a transistor. Patterning variations and RDF are both sources of parametric variations in design.

Variations may also be classified as being either systematic or random. This distinction is vital for analyzing the underlying causes of variation and for addressing them through specific changes in design or manufacturing. *Random variation,* as the name suggests, is not attributable to a specific cause or parameter. Such variation is often assumed to be gaussian, with mean μ and standard deviation σ. A gaussian distribution usually follows from the law of large numbers. Examples of random variation include dopant atom fluctuations in a MOS transistor, supply voltage variations in a circuit, spot defects in a wafer, and line edge roughness due to lithography. *Systematic variations* are functions of design or manufacturing parameters; hence they can be modeled by a function or a set of functions. Examples of systematic process variation include V_T skew between nMOS and pMOS devices due to systematic doping variation. Systematic variations may also arise from design attributes. The pattern density of a mask is known to cause systematic variation in oxide thickness leading to variation in interconnect capacitance. Proximity effects are known to cause distortion of shapes that lead to device and interconnect parameter variations. Such variations can often be modeled and predicted before or after manufacturing. It is also important to note that variations within a design (die) can be both systematic and random in nature. Such variations can be modeled only by factoring the two types of variations and then modeling them individually. The factoring process is complex and usually statistical in nature.[4]

Sources of intradie process variability in semiconductor manufacturing can be related to the illumination source wavelength and to the minimum feature size of patterns being printed on the wafer. It has been observed that, for "above wavelength" lithographic processes, the variations are chiefly random in nature. Such random errors may result from handling, particulates, overlay error, dopant fluctuations, and so on.

The wavelength of lithographic light sources has historically scaled in tandem with technology. For example, 0.35-μm transistors used a light source of wavelength 365 nm. But as shown in Figure 1.5, from 180-nm technology onward the wavelength of lithographic source light has remained constant at 193 nm. A continuation of the past progression toward shorter wavelengths for optical lithography has been thwarted by several factors. Although a157-nm light source exists, it did not gain traction because lithography at this wavelength requires the exposure system to be purged of oxygen and water, which strongly absorb radiation at 157 nm. Therefore, the cost of moving to a 157-nm system outweighed the benefits. Another problem with wavelengths at 157 nm or its designated successor at (e.g., 13.4-nm extreme ultraviolet of EUV light source) involves the photomask itself. The photomask must have high transmissivity of light because otherwise it absorbs energy, causing it to heat up and suffer from thermal expansion. A typical exposure system reduces the mask by a factor of 4 or 5. Thus, if we contemplate manufacturing a chip 1 cm on the side, the mask must be (4 or) 5 cm on the side. For the current generation of masks based on fused silica, a 1°C rise corresponds to a 25-nm expansion of the mask. As a result, for small-resolution targets the temperature rise of the photomask needs to be much less. Achieving this requires high reflectivity in the dark areas of the mask and high transmissivity in the exposed areas. Given the lack of highly transparent or reflective materials at 13.4 nm, EUV-based fabrication is still far on the horizon. Similarly, lenses need to exhibit a certain refractive index and transmissivity, which presents another challenge for projection systems below

193 nm. Parabolic reflectors have also been studied for optical projections systems that can obviate the need for lenses, although such a reflector must have a low absorption coefficient. This remains an open area of research for light sources at lower wavelength.

Consequently, devices of feature width 180 nm and below (i.e., minimum feature < λ) have been printed using the argon fluoride (ArF) 193-nm light source. These devices have dimensions that fall outside the diffraction limit (i.e., minimum feature ≥ λ) of a nominal projection or illumination system. The diffraction limit issue makes it difficult to print features with high contrast and within a reasonable width tolerance, which leads to an increase in design-related variability for such technology nodes. For technology nodes below 180 nm, diffraction-related issues have become much more prominent than particle-related errors (see Figure 3.3).[5] Design-dependent lithographic process variability is of major concern in today's semiconductor manufacturing.

Linewidth variation between intent and imprint is unavoidable unless a better flare-free light source with smaller wavelength is found. Parameter variations have been rising since the advent of subwavelength lithography; this can be seen in Figure 3.4, whose trend plot is derived from Nassif[6] and Borkar et al.[7] New design and process methods to control these variations are required for a high-yield process in future technology generations. In this chapter, the most important process parameter variations will be discussed in detail. Each section will examine the manifestation of these variations and their impact on design performance and yield. Analysis and estimation techniques for such variations will also be discussed.

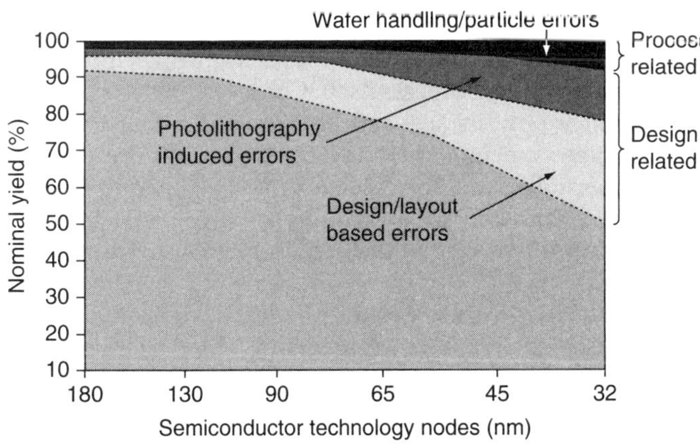

FIGURE 3.3 Process variability trends.

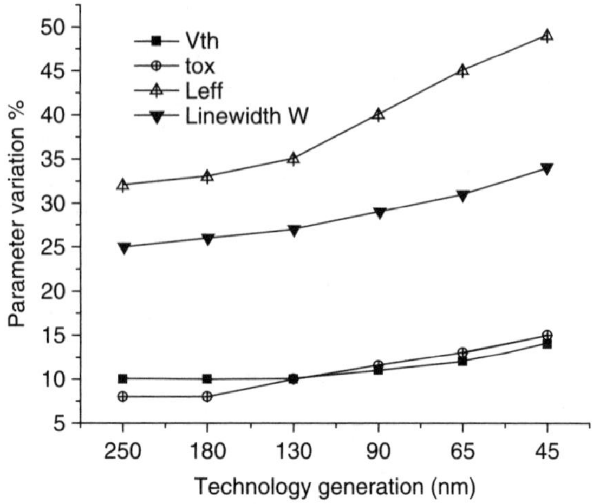

FIGURE 3.4 Rise in parameter variation with each technology generation.

3.2 Gate Length Variation

The operation of a transistor according to given specifications is dependent on the polysilicon gate pattern being printed on wafer. Variation in gate length can be caused by patterning effects or by etching-induced edge effects. We delve into each of these issues in the sections that follow.

3.2.1 Patterning Variations Due to Photolithography

Image transfer from mask to silicon is influenced by optical diffraction, which affects the intensity of light on the image plane and also the resist etch process. Depending on whether the photoresist is positive or negative, the dimensions of the printed pattern will vary for the same optical exposure. These dimensions are further influenced by the chemistry of the etch process itself. Consequently, a printed line—whether metal or polysilicon—may be characterized by a trapezoidal model. Parameters include the trapezoid base width w, the sidewall angle θ of the profile, and the resist height h. See Figure 3.5(a).[8] The critical dimension (CD) of a mask is defined as the minimum resolvable feature that can be printed on the resist within required specifications.

The width of a feature cannot be clearly defined by a single parameter based solely on the cross section of the resist. Photoresist profiles after development are fitted to a trapezoidal feature model using multiple parameters of the profile. Whenever one talks about the CD of a mask, the first measure that is recognized about the feature is its width on the resist; this is the base width w from

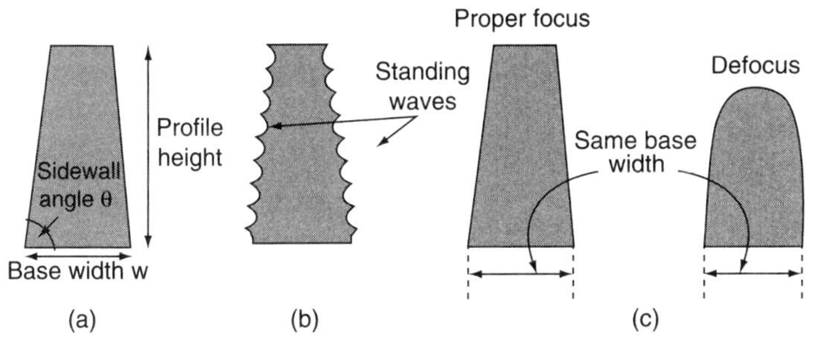

FIGURE 3.5 Assessing critical dimension (CD): (a) parameters controlling CD; (b) standing wave formation due to reflections from the resist-substrate interface; (c) varying resist profile with equivalent width at different focus.

Figure 3.5(a). Yet a change in focus can yield completely different resist profiles even with the same base width, as shown in Figure 3.5(c). Hence, when measuring CD, it is also important to consider the profile height and the effect of sidewall angle. The CD of a mask can be described by any one of these three parameters, provided the other two are within required specifications. Any one of the three parameters can be used to define CD and in postdevelopment metrology.

3.2.1.1 Proximity Effects

The required width of patterns in today's VLSI designs are well below the minimum resolvable feature width of the 193-nm light source; as a result, linewidth variation is a major issue. Linewidth varies as a result of proximity effects in optical lithography, and these proximity effects result from optical diffraction. Linewidth variation of gate-poly features may result from the diffraction patterns of adjacent lines.

As we saw in the previous chapter, diffraction patterns are centered at the midpoint of a feature and decay rapidly in intensity as they spread over a finite region. The peripheral intensity patterns overlap with diffraction patterns from neighboring lines, which leads to optical interference. The resulting intensity profile is based on superposition of the phase and amplitude of diffraction patterns from neighboring lines. Constructive interference occurs when the diffraction patterns are in phase; conversely, destructive interference occurs when they are out of phase with each other. See Figure 3.6.

Consider the following one-dimensional plane wave represented by the equation

$$\frac{\partial^2 \psi}{\partial x^2} = \frac{1}{v^2}\frac{\partial^2 \psi}{\partial t^2} \tag{3.1}$$

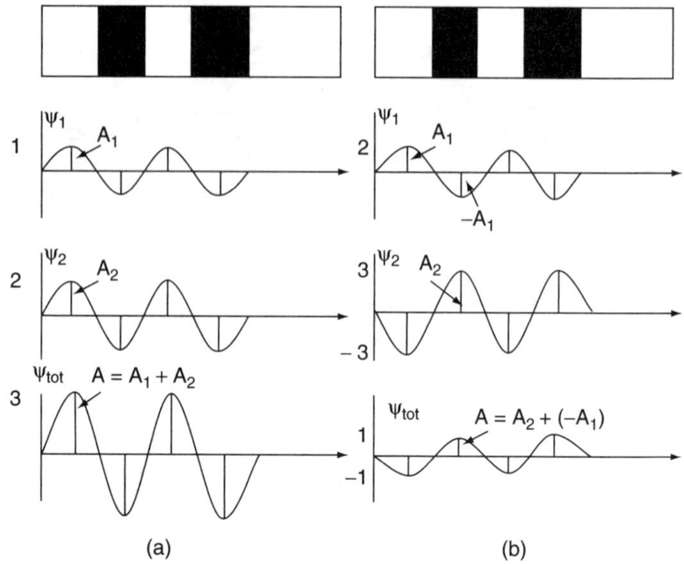

FIGURE 3.6 Superposition principle: (a) constructive interference (in-phase waves); (b) destructive interference (out-of-phase waves).

where $\psi(x, t)$ represents the wave and v the wave's velocity. Now, according to the principle of superposition, if two waves ψ_1 and ψ_2 propagate in the same direction, then their amplitudes are additive. The new wave ψ_{tot} is given by

$$\begin{gathered} \frac{\partial^2 \psi_1}{\partial x^2} = \frac{1}{v^2}\frac{\partial^2 \psi_1}{\partial t^2} \text{(wave 1)} \quad \frac{\partial^2 \psi_2}{\partial x^2} = \frac{1}{v^2}\frac{\partial^2 \psi_2}{\partial t^2} \text{(wave 2)} \\ \frac{\partial^2 (\psi_1 + \psi_2)}{\partial x^2} = \frac{1}{v^2}\frac{\partial^2 (\psi_1 + \psi_2)}{\partial t^2} \Rightarrow \frac{\partial^2 \psi_{tot}}{\partial x^2} = \frac{1}{v^2}\frac{\partial^2 \psi_{tot}}{\partial t^2} \end{gathered} \tag{3.2}$$

Figure 3.7 illustrates the interference between two-dimensional intensity patterns. One effect of such interference is the widening or shrinking of resist patterns; this is known as the *proximity effect*. Proximity effects depend on the feature width, distance between adjacent patterns (spacing), type of mask used, and other illumination system parameters. Proximity effects often occur in lower metal interconnect layers and poly-gate patterns, which can lead to parametric variation in the design process.

Proximity effects widen or constrict metal lines based on the spacing of adjacent features (see Figure 3.8). *Spacing* is defined as the distance between any two successive edges of adjacent features, and the *pitch* is defined as the distance between two successive similar edges (e.g., right edge to right edge) of two adjacent features in a

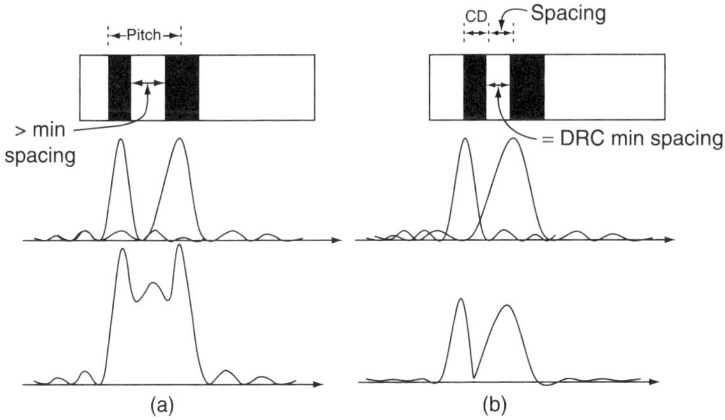

FIGURE 3.7 Interference: (a) constructive; (b) destructive.

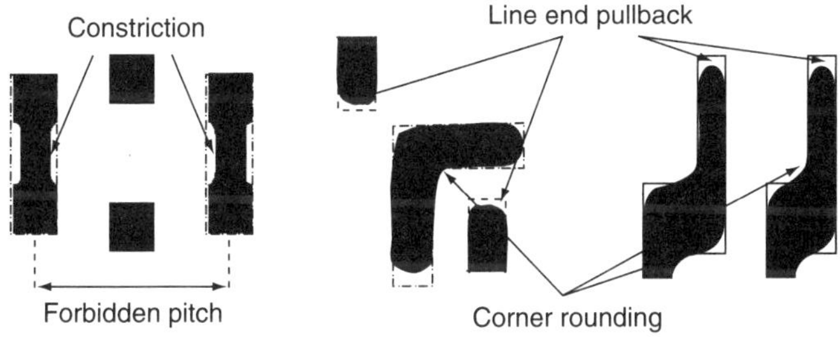

FIGURE 3.8 Proximity effects on metal lines.

layout. Socha et al.[9] studied proximity effects on linewidth by varying the spacing between features while keeping the linewidth of the mask constant. It was observed that, at certain pitches, the metal line width shrinks dramatically from the original width; see Figure 3.9.[10] The spacing at which this narrowing of width occurs defines the pitches that should be barred from the design process, and each becomes a *forbidden pitch* for that metal width.[9,10] For 45-nm technology and below, the large number of forbidden pitches imposes multiple constraints on layout topology. Geometric layout rules do not allow adjacent metal lines to be placed at forbidden pitches. Even within tolerable pitches, proximity effects lead to linewidth variation caused by neighboring lines; this variation is known as *across-chip linewidth variation* (ACLV).[11] Other proximity effects on metal lines include line end shortening and corner rounding, as shown in Figure 3.8.

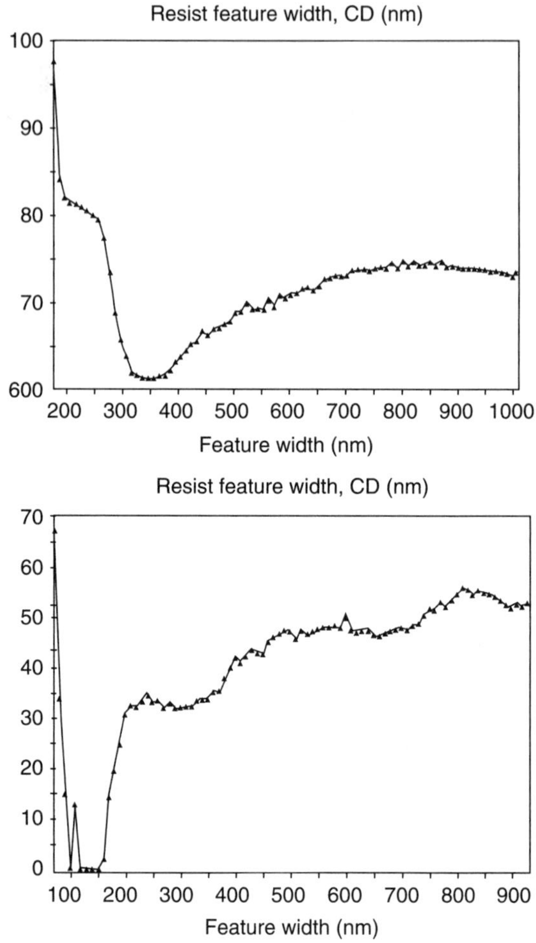

FIGURE 3.9 Forbidden pitches simulated by using a one-dimensional mask in Prolith: top, 65-nm technology; bottom, 45-nm technology.

3.2.1.2 Defocus

Defocus is defined as the difference between the focal position in the resist on wafer and its position at target focus. Focus of the exposure system depends on the light source and the reduction lens system used as well as the thickness of the resist on the wafer. Defocus leads to blurring of the image being printed on the wafer. Variation in focus causes linewidth variation due to improper pattern formation on the wafer. The impact of defocus on linewidth variation for patterns at different pitches is systematic and can be modeled. The printed linewidth for pitch variation and defocus is shown using a Bossung plot (see Figure 3.10).[12] The Bossung plot has a "smile" and a "frown" feature that is associated with the change in linewidth.[11,12] Highly dense features tend to have increased

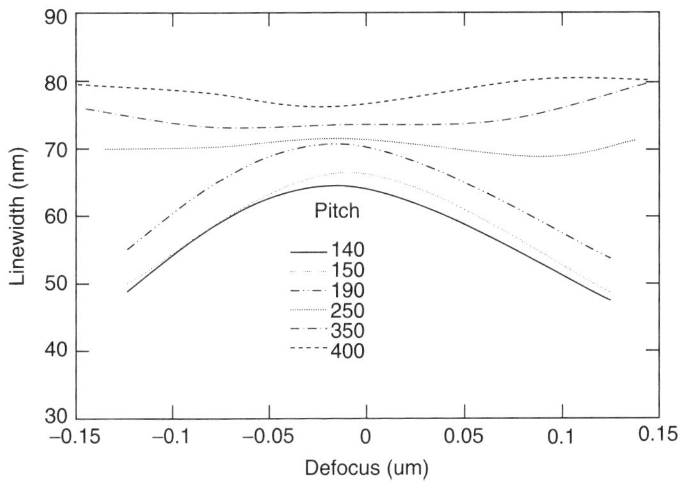

FIGURE 3.10 Linewidth change due to through-pitch variation with defocus for various pitch configurations (Bossung plot).

linewidth, with defocus causing a "smile"; isolated features tend to have decreased linewidth value, with defocus causing a "frown." *Dense patterns* are those where features are closely spaced over a region of the layout, and *isolated patterns* are those where features are sparsely populated. Defocus affects dense and isolated lines differently. As shown in Figure 3.10, the rate of change in linewidth for dense features is higher than the rate of change for isolated features.

Resist thickness variation due to improper CMP planarization leads to defocus across the lens field, as shown in Figure 3.11.[13] Chemical-mechanical polishing is layout dependent, and planarization errors across varying density regions can be modeled and predicted (see Sec. 3.5 for detailed treatment of this subject). The high-energy light sources used by exposure systems cause the lens to heat during patterning; this changes the refractive index of the lens material, which is another cause of defocus. Lithographic input parameters such as exposure dose and system focus can vary systematically and also have associated random components. Typical random components of defocus, including wafer misalignment and wafer tilt, are difficult to model. The linewidth variation that is due to such sources is modeled statistically.

3.2.1.3 Lens Aberration

Aberration is defined as a departure from ideal behavior. Thus, aberration in the lens system is any deviation from the ideal operation of a lens. Understanding lens aberration is facilitated by a review of

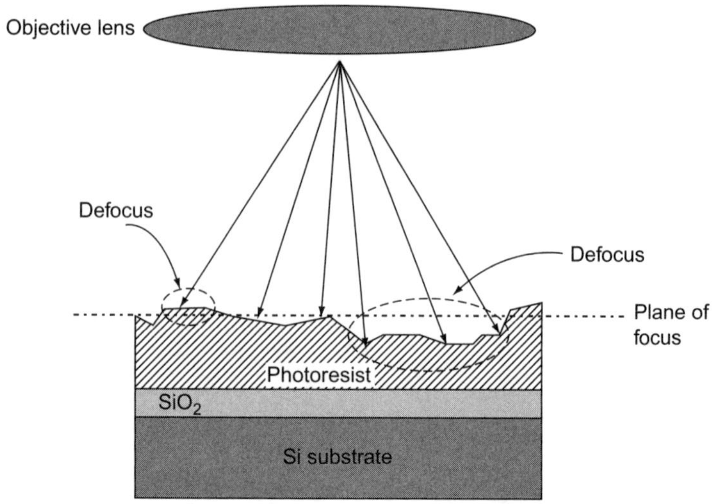

FIGURE 3.11 Defocus caused by resist thickness variation.

Snell's law. Consider the ray diagram of Figure 3.12.[14] For that diagram, Snell's law states that

$$n_i \sin\theta_i = n_t \sin\theta_t \tag{3.3}$$

where n_i and n_t are the refractive indices of (respectively) the incident and transmitted media. A well-known simplification based on first-order theory is that, since θ is very small, $\sin\theta = \theta$. Hence Snell's law becomes $n_i\theta_i = n_t\theta_t$. The angle of incidence and the refractive index of

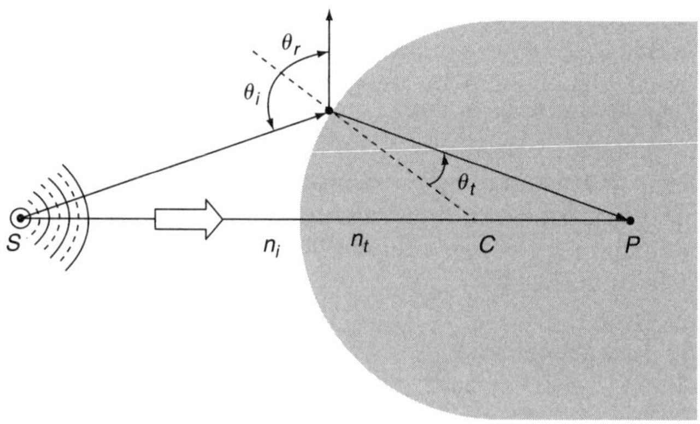

FIGURE 3.12 Understanding Snell's law.

the media under contact describe the light behavior that is used to estimate the front and back focal points of the lens system. This simplification assumes that rays are paraxial. *Paraxial rays* are those that pass close to the axis and thus have very small angle of incidence on the lens.

In practice, rays from the light source that pass through the mask will diffract in different directions and fall on the objective lens. The objective lens now sees rays that are not paraxial, so the assumption behind Snell's law is now invalid. The Taylor series expansion of sin θ can be written as

$$\sin\theta = \theta - \frac{\theta^3}{3!} + \frac{\theta^5}{5!} - \frac{\theta^7}{7!} + \cdots \tag{3.4}$$

In any case, at least the first two terms need to be included when considering rays away from the axis. A incident angles of light rays are higher than the first-order Taylor series approximation of sin θ, Because not all the incident light rays are focused at the same focal point. This brings the third-order theory into the picture leading to the creation of primary aberrations as shown in Figure 3.13. The difference in the focal positions of incident rays causes focus variation.

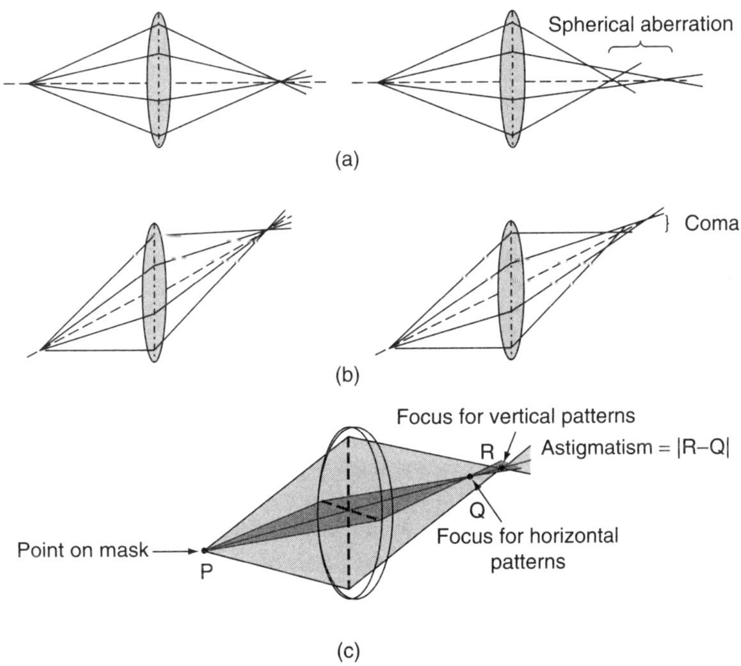

FIGURE 3.13 Defocus due to lens aberration: (a) spherical aberration; (b) coma; (c) astigmatism.

The *optical path difference* (OPD) for each incident beam is defined as the difference in optical path between the current beam and the zero-diffraction beam that passes through the optical axis. This focal variation is termed *lens aberration*, which leads to blurring of the image on a wafer.

There are multiple sources of aberration in a lens, as shown in Figure 3.13. Lens aberrations can be classified into two types: chromatic and monochromatic.[14,15] *Chromatic aberrations* are caused by dispersion of light due to variation in the lens refractive index for constituent wavelengths of light. Longitudinal and transverse aberrations are examples of chromatic aberration, which are not seen when a monochromatic light is used. *Monochromatic aberrations* that can cause defocus include piston, tilt, spherical, coma, and astigmatism aberrations. Piston and tilt aberrations do not model a curvature in the wavefront and hence do not affect the image; they simply cause a small shift in position. The defocus due to piston and tilt aberrations is seldom significant. Spherical aberration causes variation in the position of focal planes of nonparaxial rays. Coma aberrations cause variation in the focal position for rays that are incident at an angle to the lens; these aberrations manifest as asymmetry in the image. Astigmatism is the variation in focus as a function of orientation of the image. This aberration causes shapes in different directions to have relative defocus.

Aberrations can also be classified in terms of whether they are due to (1) manufacturing; (2) the lens type; or (3) the design patterns. Lens manufacturing variations are due to lens usage and can be seen as imperfections on the lens surface, curvature, and/or composition. Improper usage of the lens (e.g., mishandling, incorrect placement tilt) will lead to variations in the patterns formed. Aberrations caused by the design are due to the orientation of patterns during exposure.

A series of lenses is used to perform reduction of the mask image onto the wafer. The step-and-scan approach used in projection printing today scans regions horizontally in one exposure, moves to another region for the next exposure, and so on repeatedly. The region over which the exposure system scans the mask patterns onto the wafer is called a *lens field*. It has been observed that aberration across the lens field may induce variation based on feature position with respect to the center of the lens. Since scanning proceeds horizontally, this type of variation may not be observed on vertical patterns. Because the field is small compared to the wafer, variation within the field is considered to be inconsequential (see Figure 3.14).[16]

Lens aberration causes defocus-induced variation in metal interconnect and gate linewidth. All aberrations due to a lens can be characterized by calculating the optical path difference of beams traveling across the lens. The simplest method, proposed by Zernike,

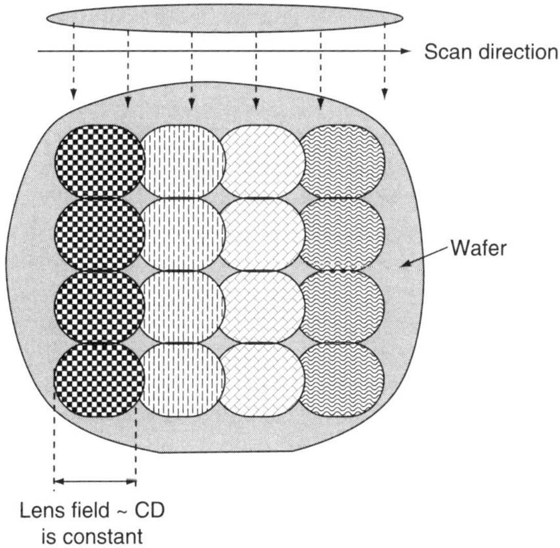

FIGURE 3.14 Lens aberration causing CD variation between different lens fields.

is to represent the OPD surface as a function of all the components of aberration. This function of the orthogonal components can be represented in spherical dimensions as[17]

$$\mathrm{OPD}(\rho, \varphi) = \sum_k s_x Z_x(\rho, \varphi)$$
$$Z_x(\rho, \varphi) = \begin{cases} Z_n^m(\rho, \varphi) = R_n^m(\rho)\cos(m\varphi)\text{:odd} \\ Z_n^{-m}(\rho, \varphi) = R_n^m(\rho)\sin(m\varphi)\text{:even} \end{cases} \tag{3.5}$$

where m and n are nonnegative integers with $n \geq m$, φ is the azimuthal angle in radians, and ρ is the radial distance. The radial polynomials $R(\rho)$ are functions of n, m, and ρ. The orthogonal components (Z_0, ..., Z_k) are the Zernike coefficients, and the coefficients s_x determine the contribution of the orthogonal component to image defocus. The OPD value is positive for patterns in one direction and negative for those in the opposite direction.

Each component models an aberration up to a particular order. The Z_0 polynomial models the piston aberration; Z_1 and Z_2 model image tilt in the x and y directions; and Z_4 and Z_5 model the astigmatism (Z_1, Z_2, and Z_3 do not actually contribute to defocus of the image and have low s_x values). The terms Z_6 and Z_7 represent coma, which causes defocus based on the angle of diffracted rays, and Z_8 models the spherical aberration of the lens. The same set of aberrations at different diffraction orders are repeatedly modeled by higher-order terms Z_9 and above. These polynomials are used by lithography simulators to estimate the impact of lens aberration on the aerial image and resist profiles of the patterns on the mask. Further details can be found in the text by Born and Wolf.[17]

3.2.1.4 Modeling Nonrectangular Gates (NRGs)

Current transistor modeling methodologies assume that transistors on silicon are of the same shape as in layout, with a single channel length and width. But as we move into subwavelength lithography, there is a difference between intent and imprint: poly-metal gates often take a nonrectangular shape on silicon. The channel region is determined by the dopant profile under the gate and by its interaction with the source-drain diffusion regions, so the channel region may not be rectangular even under a rectangular gate.[18] The modeling of such nonrectangular transistors is a complex problem because the dependence of leakage current and threshold voltage on the gate length is nonlinear (see Figure 3.15).[19]

Figure 3.16 shows a drawn gate-mask feature and its printed contour on wafer for 45-nm technology node. Figure 3.17 illustrates the difference between the drawn and printed contours in terms of drive and leakage currents. It is clear from the figure that lithography induces a change in postsilicon transistor characteristics, so discrepancies will arise if one assumes the gate shape to be rectangular for the purpose of presilicon circuit simulation. The first step in modeling nonrectangular gates is to slice the gate into regions of minimum gate length variation. An NRG can be modeled as a set of parallel transistors, where each slice represents a transistor of particular width and length (see Figure 3.18).[20,21] The main drawback of this approach is that the

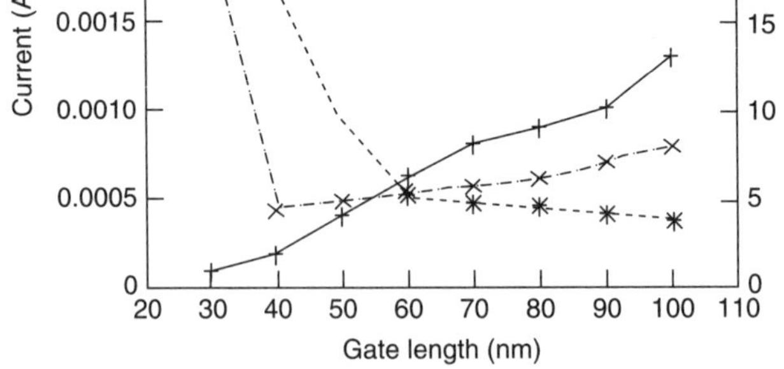

FIGURE 3.15 Variation of ON current, OFF current, and $V_{th} = V_T$ with gate length L_G due to CD variation.

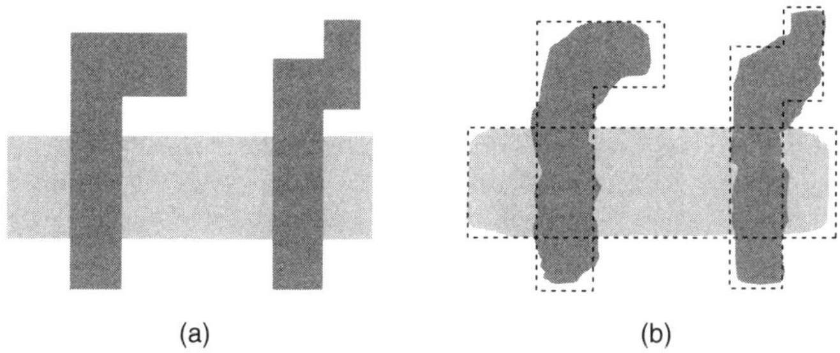

FIGURE 3.16 Gate and diffusion contours: (a) drawn mask; (b) printed contour.

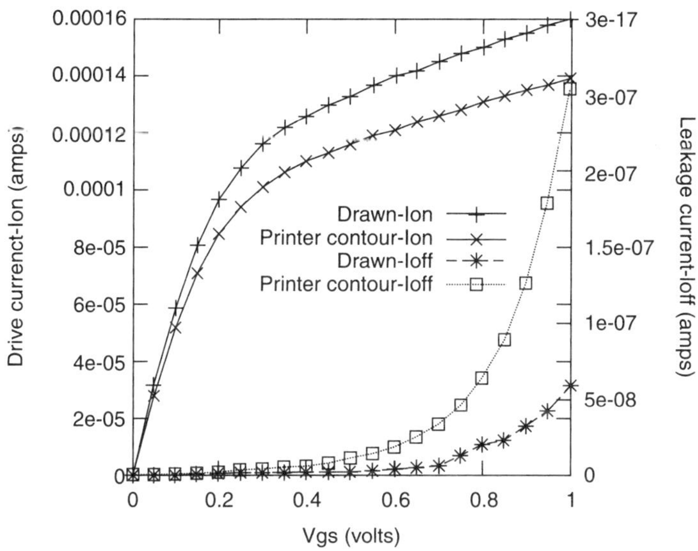

FIGURE 3.17 Variation in drive and leakage currents of drawn and printed contours.

model increases circuit complexity and the transistor count, limiting the size of circuits that can be simulated. Another drawback is that, owing to the effects of narrow width and shallow trench isolation, the standard transistor model may not be applicable to the slices. An alternative approach is to model the gate as a single rectangular transistor that predicts the characteristics of the printed contour for one particular region of operation—for example, cutoff or saturation.[21] This methodology, known as *equivalent gate length* (EGL) modeling, can be used to simulate circuits. The EGL approach offers a prudent compromise in circuit simulation without sacrificing accuracy or simulation performance. Another approach to modeling properties of nonrectangular poly-metal gates is based on three-dimensional device simulation.[22] However; such simulations are impractical for anything larger than small library cells. Many such models were proposed to model the NRG using SPICE-based simulations.[18,19,20,23-28] The NRG

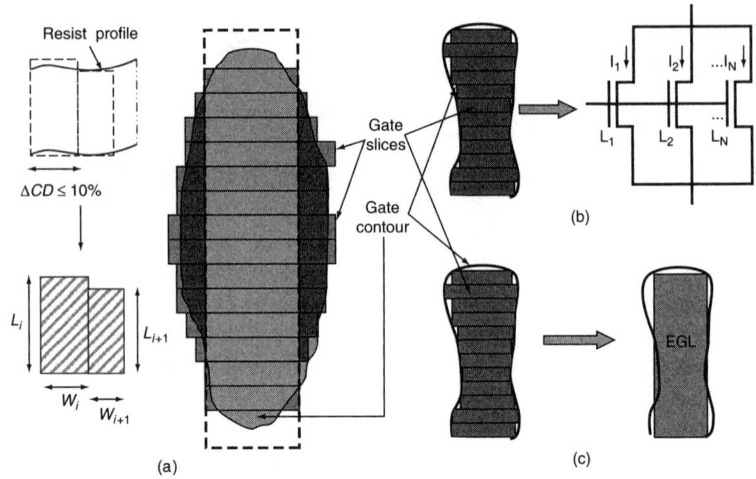

FIGURE 3.18 Modeling the nonrectangular gate (NRG): (a) slicing of gate resist profile; (b) model suggesting a transistor for each slice; (c) single equivalent gate model (EGL) for modeling ON and OFF operations of the device.

model offers a viable CAD methodology for analyzing devices affected by lithography-induced imperfections.

3.2.2 Line Edge Roughness: Theory and Characterization

Line edge roughness (LER) is defined as the variation in resist pattern edges that arises from the complex interaction of exposure and etching process parameters.[29-37] With top-surface imaging, LER for chemically amplified resists with deep ultraviolet imaging in sub-100-nm features has been found to vary between 5 and 10 nm.[38] A variation of up to 10 nm on each edge amounts to equivalent linewidth variation in interconnect and poly-metal gate lines (Figure 3.19). With the scaling of feature sizes down to below 50 nm, LER has a sizable effect on linewidth tolerance. Variation in LER has been under intense scrutiny because of its effect on interconnect resistance, device characteristics, design limitations, and resolution of the imaging system.

Line edge roughness variation is caused by a complex interaction of several factors that occur during the manufacturing process. Many researchers have attempted to identify the source of these variations and quantify the LER effect. Even today, the cause of LER variation cannot be narrowed down to a single source, so we will look at some probable causes that have been identified experimentally. These include:

- Mask roughness
- Aerial image contrast
- Molecular structure of the resist

- Shot noise caused by absorption fluctuations between different locations
- Mixing of resist polymers
- Development processes

Mask roughness is a direct cause of LER on silicon. All errors in the mask are demagnified (by MEEF, the mask error enhancement factor) on the wafer during the exposure process. An error of 10 nm due to mask roughness can lead to a 5-nm edge variation on the wafer in a 4X reduction imaging system with an MEEF factor of 2. So, just as in final wafer feature printing, mask fabrication errors must be controlled by constant monitoring of final linewidth tolerance limits.

The aerial image of the mask pattern falls on the resist to create chemical reaction within the resist. The sharpness of this aerial image affects the final resist profile. Sharpness of an aerial image is quantified by the exposure system's achievable level of contrast. As defined in the previous chapter, *contrast* is the rate of change of resist thickness with exposure dose. Experimentation using exposure systems with varying contrasts have been used to study the effect of aerial image contrast on line edge roughness. Variation in LER is inversely proportional to the contrast of the aerial image. Thus LER variation is low when the contrast is high, and LER variation is high when the contrast is low (i.e., when the intensity slope is not steep). This effect is illustrated schematically in Figure 3.20.[35] In short, maintaining good contrast for features throughout the design leads to a reduction in LER.

The molecular weight dispersion of the resin present in the photoresist polymer has an effect on dissolution properties and on LER. The *molecular dispersion* of polymer is defined as the ratio of the "weight" average molecular weight M_w and the "number" average molecular weight M_n. The polymer becomes inhomogeneous in the presence of high levels of molecular dispersion, resulting in a rough

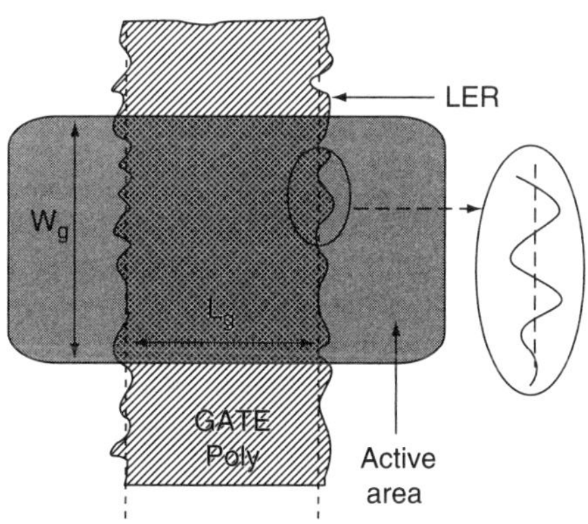

FIGURE 3.19 LER variation in poly-gate patterns.

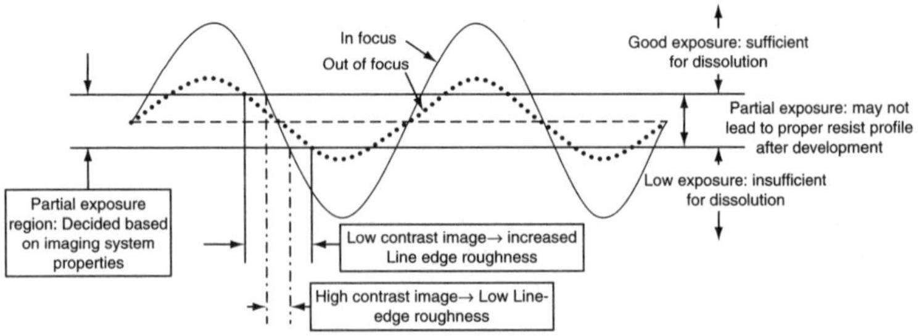

FIGURE 3.20 LER dependence on aerial image contrast.

profile edge.[32] Higher M_w/M_n ratios also lead to less acid diffusion, which increases sidewall roughness in the resist. The resolution of the photoresist polymer depends on the acid diffusion that occurs during the PEB stage.

Shot noise is a type of electrical noise characterized by variation in the number of high energy photons present in an optical device, which leads to statistical fluctuations in light intensity.[31,32] Shot noise increases with average amplitude of light intensity. The use of chemically amplified resists with illumination systems that constitute of a large number of photons increases the level of shot noise present in the resist. Light intensity fluctuations due to shot noise cause irregular exposure of mask features onto the wafer, leading to LER. It has been shown that a variation of 200 ions within a 100-nm^2 area of the wafer can cause LER-induced linewidth variation of 6 nm.[22,35]

Sulfonate and sulfonium salts are typically used as photoacid generators (PAGs) in chemically amplified resists today. Acid chemistry between the salt and resist polymer upon exposure ensures the proper diffusion and photoacid generation within the resin. Inhomogeneous mixing of resist polymer with the salt can lead to improper acid dissolution, which causes LER variation. Sensitivity of the resist to exposure depends on the quantum yield of acid generation from the sulfonium salt.[32-35] Contrast, which directly affects LER variation, relies on the PAG-induced chemical reactions with the salt. The interaction between resist polymer and the developer solution also affects the impact of LER on the feature. Differences in sidewall roughness between dense and isolated lines have been found to be caused by the flow of developer solution into these areas. Because the resist development stage is an isotropic process, the erosion rate of the solvent-free post-PEB resist polymer determines the level of edge roughness.

With the introduction of sub-50-nm gates, the impact of LER on device performance and reliability is a major concern. Line edge roughness variability further reduces the small linewidth tolerance available for today's poly-metal gate features on silicon; that is, an increase in LER erodes linewidth tolerance. The effect of LER is observed in both gate patterns and metal interconnect patterns in the wafer. The linewidth distribution measured at multiple locations of a die is plotted in Figure 3.21, where the 3σ variation was found to be 8 nm.[22] The effects of LER variation on metal interconnect linewidth include reduced current conduction through the metal (due to increased resistance) and higher susceptibility to an electromigration failure. The major concern with gate LER is that gate length variation may lead to improper operation of the transistor. Gate length variation affects the I_{ON} and I_{OFF} of the device, causing leakage and changes in propagation delay.

Various LER modeling techniques have been proposed to aid in measuring the impact of LER variation on device parameters. An LER model is typically formed by first obtaining linewidth and LER information from scanning electron microscopy (SEM) images.[29] When measuring linewidth and LER data from an SEM image, the distance between sample points is specified so as to yield a balance between resolution and throughput. The length of the inspection area and the number of scans of the wafer determine the amount of variation in the data obtained. These data points are used to fit a model for predicting variation; see the example fitted model in Figure 3.22.[35] Although LER is irregular in nature, Fourier analysis reveals periodicity in LER variation. However, the LER period is not constant over any inspection region. A statistical model with varying

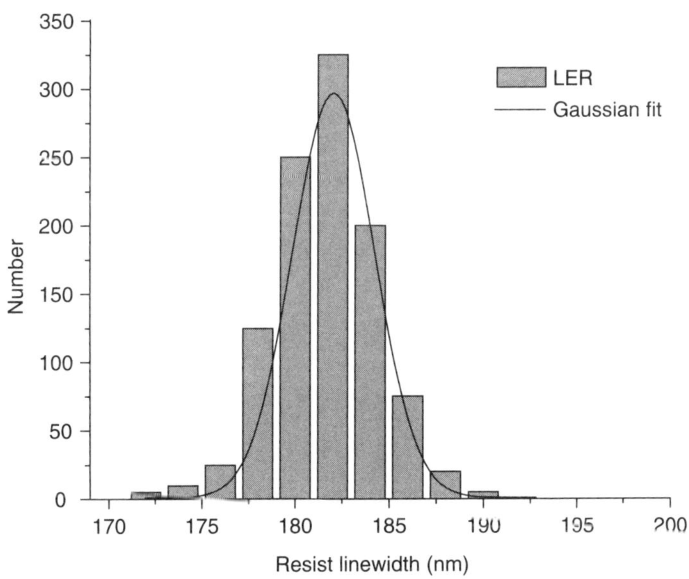

FIGURE 3.21 Linewidth distribution measured at multiple locations of the die; 3σ line edge roughness (LER) variation of 8 nm.

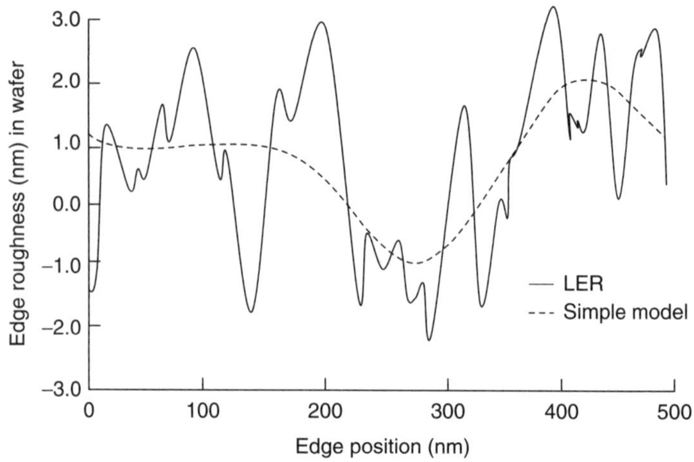

Figure 3.22 Graph reflecting a simple model of LER variation.

probability for LER periodicity is used to predict line edge roughness using the correlation between different sample points. The LER period probability for a particular type of resist is plotted in Figure 3.23.[37]

LER variation will become more prominent as we aggressively scale toward 32-nm and 22-nm devices. Methodologies to prevent LER from occurring—or to mitigate its sources—are needed to reduce variations in gate length and V_T in future technology generations.

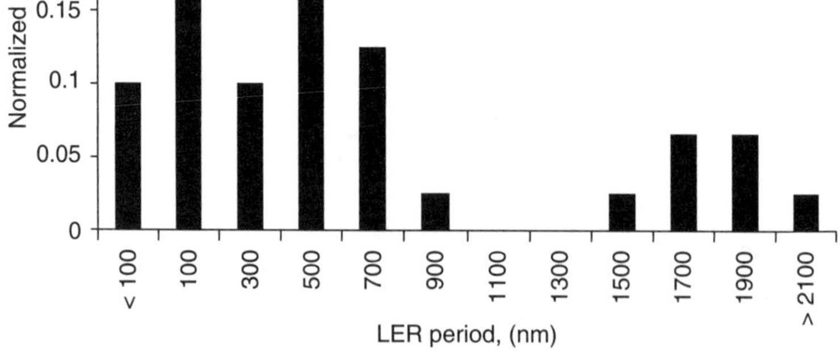

Figure 3.23 Different periods of LER variation and their probability of occurrence.

3.3 Gate Width Variation

The width of a MOSFET is given by the region over which the polysilicon-metal gate overlaps the underlying active area (see Figure 3.24(a)). The active area (aka the diffusion region) lies beneath the gate and is a doped region of silicon created through patterning and ion implantation processes. For the gate width to be rectangular, variations must be minimized in the processes of (1) patterning the gate and (2) diffusion patterning and creation of rectangular oxide spacers through the STI process. Imperfection in either of these processes may cause gate width variation. Because gate width W_G is directly proportional to the drain current of the transistor, W_G variation hinders performance. Gate patterning problems caused by proximity effects (as explained in Sec. 3.2.1.1) can lead to pullback of gate ends; this was shown in Figure 3.8. Pullback causes irregularity and reduction in gate width. Gate width variation due to pullback is included in the nonrectangular models described in previous sections.

It is typically assumed that the diffusion region is rectangular and hence has little effect on transistor operation. But as shown in Figure 3.24(b), diffusion regions do not remain rectangular on silicon. Similar diffraction effects occur when thediffusion region is being patterned, causing the phenomenon known as *diffusion rounding*. The rounding commonly occurs at L- and T-shaped features, depending on how the contacts are placed on the diffusion region.[28] The type of source and drain contacts to the power supply rails depend on whether the transistors are in series or parallel. The shape of the diffusion region that bends toward the source or drain region also

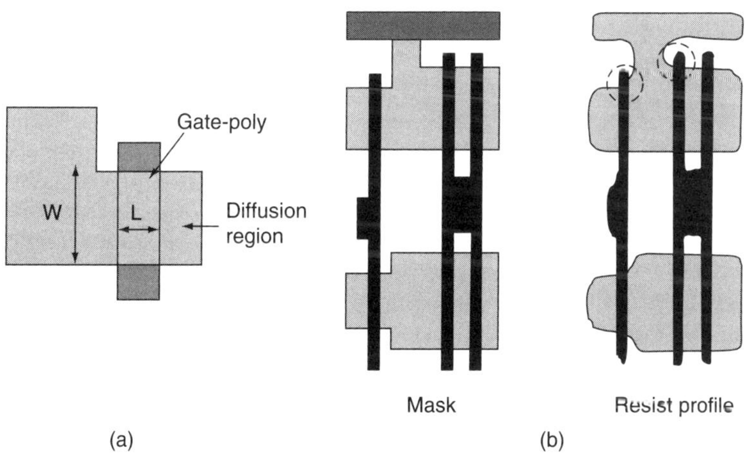

FIGURE 3.24 (a) Poly-gate width and length; (b) diffusion rounding of the gate alters its width and length.

depends on the series or parallel nature of transistor arrangement. The extent of diffusion rounding is a function of the distance between the poly and the irregular diffusion region. Poly regions extending over the diffusion are not a problem because they are used for intracell connections.

Another factor that plays a vital role in determining the shape of the diffusion region is the creation of STI wells between active areas. The aim of shallow trench isolation is to isolate active areas within a die to prevent them from interacting with each other. Isolation is ensured by growing a gate oxide (usually SiO_2) inside a cavity etched in silicon. The stages in STI creation include: wafer oxidation, nitride layer deposition, trench patterning, anisotropic etching, oxidation of liner, chemical vapor deposition of the trench oxide fill, planarization using chemical-mechanical polishing, and finally wet etching to strip the nitride and wafer oxide. The wafer is first oxidized, and this is followed by a CVD-based nitride layer deposition. Trench regions are patterned using lithography and etched using the anisotropic dry etch process. This creates rectangular trench regions ready for the next stage of oxidation and CVD-based trench oxide deposition. The oxide grown or deposited is planarized using the CMP process. The drawback of this type of planarization is that it depends on the oxide removal rate and the underlying pattern density. Regions with differing pattern density lead to the formation of dishes and eroded areas (as explained more fully in Sec. 3.5). Wet etching of the regions not covered by nitride lead to removal of some active areas at the STI boundaries. This removal renders the active area beneath the gate pattern nonrectangular, which causes variation in W_G. Gate width variation due to line end pullback and diffusion rounding will affect the transistor drain current. Gate width variation also changes device geometry, causing parasitic capacitances to vary.

3.4 Atomistic Fluctuations

Ion implantation and diffusion are techniques used to introduce impurities into the silicon wafer. The total number of dopant atoms introduced into the channel region to control the device's threshold voltage V_T was in the hundreds until the advent of 90-nm CMOS technology. With ever-increasing reduction in channel length, it has become a challenge to control the doping profile with fewer than 100 atoms introduced into a nanometer-scale channel. Because the number of atoms is so low, any small change in the channel region can create a significant change to the V_T of the device. Doping today is predominantly done using ion implantation followed by thermal annealing. Dopant profiles created by these techniques are not deterministic, and the distribution of atoms in the channel region is random (see Figure 3.25).

FIGURE 3.25
Dopant distribution in a device.

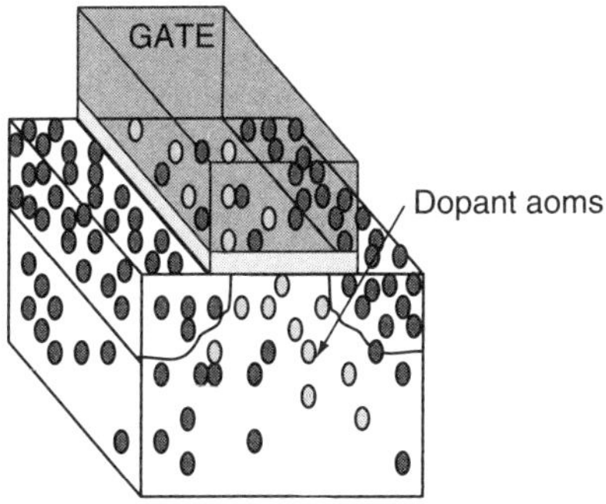

The MOSFET is said to be ON when there is a conducting path formed between the source and the drain. With the introduction of a random set of dopant atoms into the channel, various regions of the channel form nonuniform conducting paths between the source and the drain when the gate voltage V_G equals or exceeds V_T. This conducting path is formed in regions where the relative percentage of impurity atoms is low. If we consider these discrete regions to be cubes in three-dimensional MOSFET channel space, then the presence or absence of impurity atoms in a cube determines the probability of conduction along that path: the higher the number of impurity atoms, the lower the probability of conduction. Now, for a cube that contains a number a_c of atoms at threshold voltage level, the probability of the number of dopant impurity atoms being less than a_c (thus enabling conduction) can be modeled as a Poisson distribution, with average impurity concentration K, as follows:[39]

$$P_{\text{cube-con}} = \sum_{a=0}^{a=a_c} \frac{K^a e^{-K}}{K!} \tag{3.6}$$

The value of $P_{\text{cube-con}}$ must be high to ensure conduction of majority carriers across that cube.

The Poisson distribution, which is used to model the number of atoms around the channel region, gives a good approximation of the conduction behavior. This atomistic variation in number and position of dopant atoms leads to variation in the threshold voltage V_T of the device. This variation is a characteristic of an individual channel region and bears no correlation with other channel regions. To illustrate the effect, Figure 3.26[40] plots the statistical variation in V_T for identical MOSFETs placed as a minimum spaced array. It has been shown that doping near the source-drain regions controls the threshold voltage, so retrograde doping is used to minimize V_T variation in the channel. Transistors with undoped channel regions

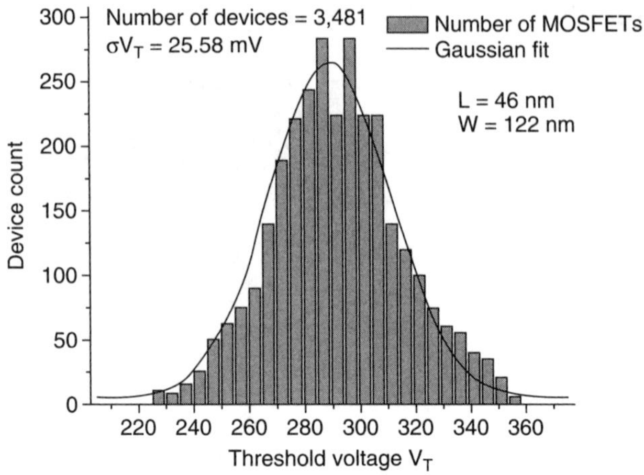

FIGURE 3.26 Statistical variation in V_T as measured from various identical MOSFETs in the die.

are used in silicon-on-insulator (SOI) devices, for which V_T is set by back-gate biasing or gate-metal work functions.[41]

Deviation in threshold voltage is modeled by a Gaussian distribution that depends on the doping (N_A), the channel area ($L_G \times W_G$), and the oxide thickness (t_{ox}). The spread in V_T can be described by the following relation (also see Figure 3.27):[40]

$$\sigma_{V_T} \propto \frac{q t_{ox}}{\varepsilon_{ox}} \sqrt{\frac{N_A W_d}{L_G W_G}} \tag{3.7}$$

The process of gate oxide deposition is a regular and well-controlled procedure. But with the need to increase device performance, gate oxide thickness has been reduced to just a few layers of atoms. The wafer is oxidized and planarized to attain the required thickness, after which lithography is used to print gate patterns and then unprotected areas (other than the gate) are etched to form gate oxides. Improper planarization in the CMP process can cause the number of atoms in the oxide to vary, which can lead to reduced mobility in certain regions of the die. Too thin a gate oxide layer results in oxide tunneling, which leads to device failure. The number of gate oxide atoms varies randomly from device to device and therefore is modeled statistically. As gate length and gate oxide dimensions approach angstrom units, atomistic and quantum variations will have to be accurately modeled to perform proper analysis on presilicon designs.

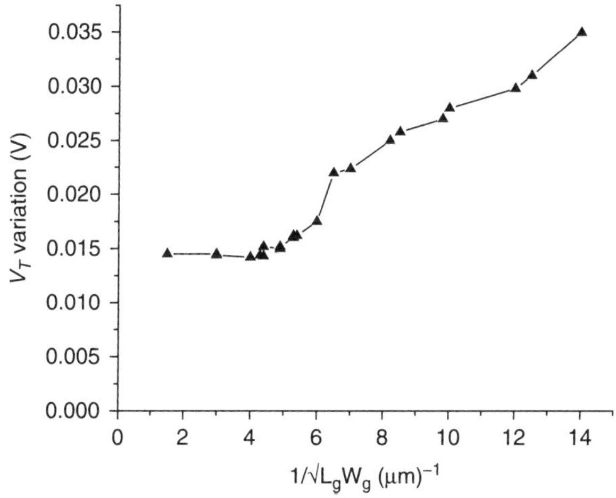

FIGURE 3.27 V_T variation with device area.

3.5 Thickness Variation in Metal and Dielectric

Variation in metal and dielectric thickness is a result of improper planarization. The purpose of planarization is to level the surface after metallization and oxidation processes. With continued aggressive scaling of transistor feature sizes, the importance of chip-level planarity has increased. Without proper planarization, there would be severe focus problems in lithography when creating upper layers. Mask-layout uniformity produces better planarity and thus is highly desired for better manufacturing and parametric yield.

There are several planarization techniques that have been used in semiconductor manufacturing, such as spin-on-glass (SOG), reverse etch back (REB), and chemical-mechanical polishing (CMP). The SOG technique involves the use of a special chemical compounds to coat the surface of the wafer, which are cleaned before the application of SOG materials. Silicate-based compounds are typically used to fill holes during planarization. The SOG material is poured onto a chuck that is holding the wafer and spinning at high speed. The thickness of the SOG layer varies depending on the viscosity of the fluid.

Reverse etch back is a process in which metal is deposited on the wafer completely in the first step. Because the typical deposition step is not uniform, it results in some regions having higher metal deposits. In the second step, these regions with higher metal deposits are removed to form a uniform layer with specified thickness. The REB process requires the use of additional masks. In semiconductor manufacturing today, SOG and REB techniques are not favored because they require complex control of parameters during the planarization process. Furthermore, the required extra mask steps lead to extra cost. For these reasons, CMP is the planarization method most used in industry today.

Chemical-mechanical polishing is a wafer planarization technique that is widely used to satisfy local and global planarity constraints.[42] Unlike the previously used SOG and REB approaches, CMP has been the choice for multilevel metal and oxide planarization for VLSI design processes. Figure 3.28 is a photograph of a metal polishing station, and a simplified schematic is drawn in Figure 3.29.

A wafer is held upside down by a wafer holder using vacuum suction. The holder presses the wafer onto a polishing pad that is spun at a constant speed. At the same time, a chemical compound known as *slurry* is applied continuously to the polishing pad. This slurry is a chemical with suspended abrasive (aluminum and silica) solids that interacts with the wafer to make it softer. The polishing pad itself is also an abrasive surface, which aids in the material removal process. The interaction of mechanical pressure, rotation, and chemical abrasion leads to planarization of the wafer surface. Copper CMP requires extra stages of planarization to remove barrier layers.

In nano-CMOS VLSI circuit manufacturing, the quality of the photolithography, etching, metallization, and other manufacturing

Figure 3.28 CMP planarization station. (*Courtesy of IBM.*)

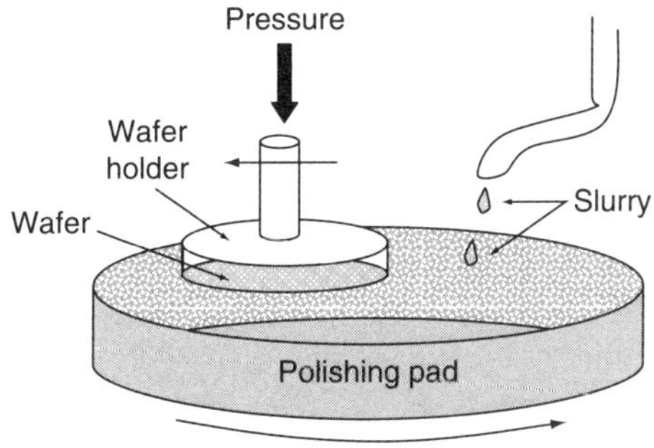

FIGURE 3.29 Schematic diagram of chemical-mechanical polishing (CMP).

steps depends on the target layout. In order to reduce the amount of variations across these manufacturing steps, uniformity in layout pattern density is preferred. Uniformity in patterns also reduces parameter control requirements throughout the process. Experimental data have confirmed that post-CMP wafer topology is highly dependent on mask layout pattern density. The Preston equation, which relates the metal-removal rate to the polishing pad speed and pressure, shows that the planarity of the surface is affected by the pattern density on the wafer.[42] The effective density is estimated by averaging density over a surrounding region. This region is determined by the *planarization length,* which is fixed for each CMP process based on particular properties of that process. The thickness of the oxide and metal after CMP depends not only on the neighboring pattern density but also on the thickness of underlying metal layers and interlayer dielectric (ILD); this is shown in Figure 3.30(a).[43] Variations in thickness of lower metal layers add up to the variation on higher metal layers and ILD thickness. Metal/ILD thickness variation leads to defocus during subsequent patterning, which in turn leads to higher CD variation and other defects. Two additional types of variations that can be caused by improper CMP planarization are material erosion and dishing (see Figure 3.30(b)). *Erosion* is defined as the difference between the post-CMP and pre-CMP thickness of metal or ILD on the wafer. *Dishing* is the reduction in material thickness above spaces and line shapes. Both dishing and erosion are also caused by pattern density variation in the layout.

Material thickness variation due to CMP can be caused by factors that include errors in setting the polishing pad and wafer holder speed, poor polishing pad condition, and incorrect slurry composition—any of which can cause improper abrasion. Because the polishing pad is an abrasive surface that helps in material removal, an unconditioned pad will reduce the rate of material removal. The slurry functions not only to planarize but also (through convective heat transfer) as a cooling agent between the wafer and the polishing pad.[44] Any variation in the slurry composition could impair its cooling

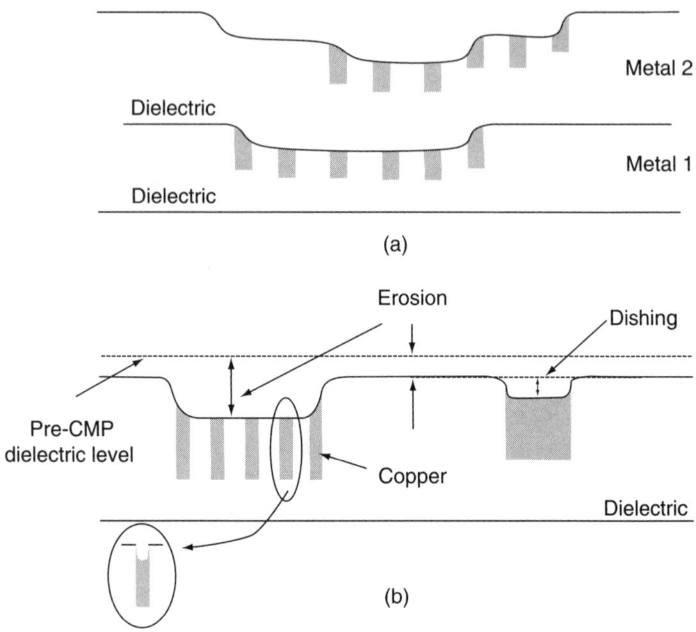

FIGURE 3.30 Compounding effect of CMP, erosion, and dishing.

function, which would cause the pad to heat up. The polishing pad becomes soft when it gets hotter than the specified temperature, and this increases its area of contact with the wafer. Such variations are random in nature and are usually better controlled as the manufacturing process matures. All CMP-induced variations that depend on pattern density lead to change in interconnect capacitance and resistances that directly affect the performance and reliability of the design. The planarization length affects the region over which neighboring features affect the CMP planarity.

Consequently, modeling CMP for oxide planarization or metallization boils down to estimating the pad pressure and pattern density.[45] Several approaches have been suggested for estimating post-CMP oxide thickness. A computationally manageable CMP model was proposed by Stine and colleagues; see Figure 3.31.[42] This model uses the following formula to estimate the interlayer dielectric thickness z at a point on the wafer:

$$z = \begin{cases} z_0 - \left(\dfrac{Kt}{\rho_0(x, y)} \right) & ; \quad \dfrac{t < (\rho_0 z_1)}{K} \\[2ex] z_0 - z_1 - Kt + \rho_0(x, y) z_1 \,; & \dfrac{t > (\rho_0 z_1)}{K} \end{cases} \tag{3.8}$$

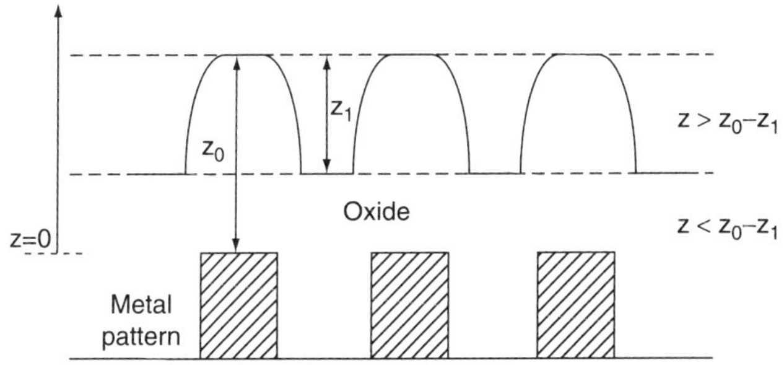

FIGURE 3.31 Modeling copper planarity.

Here K is the blanket polishing rate, z_0 and z_1 are (respectively) the "up" and "down" area thicknesses shown in the figure, t is the polishing time, and $\rho_0(x, y)$ is the initial pattern density. Because t is almost always greater than $\rho_0 z_1/K$, the thickness is equal to the second (lower) expression in Eq. (3.8) Observe that the parameters z_0, z_1, t, and K are constant for a given process, which makes the final oxide thickness dependent on the underlying pattern density.

Ouma and colleagues[46] provided a comprehensive model that includes the effect of pad deformation during planarization. The ILD thickness is no longer just a function of the local pattern density; instead, it is a weighted function based on location within the layout. The weighting function proposed is given as $w(x, y) \approx \exp\{x^2 + y^2\}$, an elliptical function that is obtained by using an elastic material placed normal to the pad surface.[44,47] The planarity of metal and oxide layers underneath has a compounding effect on the current planarization level. Multilevel oxide layer thickness is obtained by considering the pattern density of the underlying layer:

$$\rho(x, y{:}m) = \begin{cases} \left[\rho_0(x, y{:}m) + \dfrac{z_{m-1}}{z_m}\rho(x, y : m-1)\right] w(x, y) & m > 1 \\ \rho_0(x, y{:}m) \cdot w(x, y) & m = 1 \end{cases} \tag{3.9}$$

where z_m and z_{m-1} denote the oxide thickness in the current and the underlying metal layer, respectively. These values are constant for a given process. The term $\rho_0(x, y : m)$ denotes the local pattern density of the current metal layer, and $\rho(x, y : m - 1)$ is the final pattern density of the metal layer underneath. Equation (3.9) captures the effective pattern density by considering the weighted value $w(x, y)$ of pattern densities in the stack below.

The focus and exposure dose of a particular metal patterning stage depend on the thickness of the metal and oxide underneath. As metal or oxide thickness varies, it becomes a contributing factor in changing the focus and dose during photolithography. As discussed previously, change in focus is called defocus and leads to linewidth variations.

There have been several approaches to estimating the pattern density of layout regions across a die. Those available in literature are targeted primarily to finding the maximum or minimum density regions in the layout. A standard practice is to consider windows of fixed dimension when calculating pattern density.[43-49] In this approach, the layout is divided into a grid, and only the windows whose boundaries are on the grid are checked for maximum or minimum density. This technique suffers from inaccuracy in the calculation of density for regions whose boundaries are not on the grid. Multi-window and sliding-window approaches have been suggested to overcome this limitation.[43-49] The details of these algorithms are beyond the scope of this text; for additional details, see the review paper by Kahng and Samadi.[44]

3.6 Stress-Induced Variation

The active areas of devices are separated by STI filled with silicon dioxide. Shallow trench isolation induces strain on silicon and thereby alters carrier mobility. Thus, MOSFET characteristics are, in part, a function of STI-induced stress. Transistors of same gate length and width are typically considered to posses similar characteristics. However, because STI stress varies with the length of diffusion (LOD) and with the distances between active areas and poly-gate lines (see Figure 3.32), the characteristics of otherwise similar transistors tend to diverge. In active areas, the STI-induced compressive stress builds up gradually during the trenching process in response to difference in thermal coefficient of expansion between silicon and oxide.[50] Oxidation of STI sidewalls during various other stages in the process also increases stress in the material.

With increasing demand for faster processing power, foundries have incorporated new process techniques to improve the mobility of majority carriers in silicon. Foundries today have incorporated stress-engineering-based stress memorization techniques on different regions of the device to improve the drive current by altering the transistor's mobility.[51-54] The contact etch stop layer (CESL) and the epitaxial growth of silicon germanium (eSiGe) in source-drain recesses incorporate tensile and compressive stress to improve device mobility. The effect of different types of stress—including STI stress—on contemporary nMOS and pMOS devices is illustrated in Figure 3.33.[55] The stresses also change with the orientation in which they come into effect. Stress on pMOS devices is tensile in both longitudinal and lateral orientations, but this is not the case with n-channel MOSFETS.

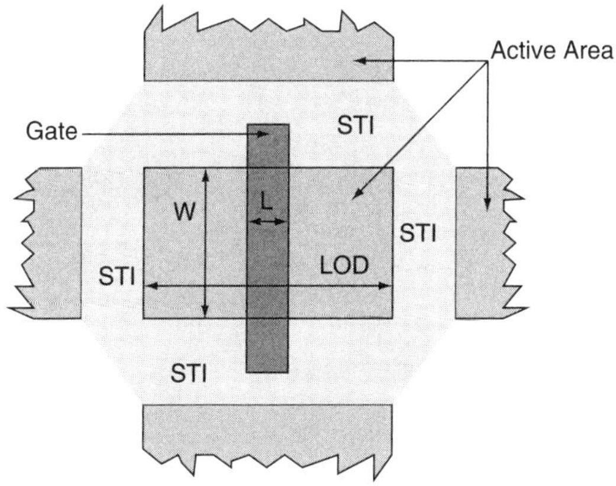

FIGURE 3.32 Layout pattern showing MOSFET active and STI areas.

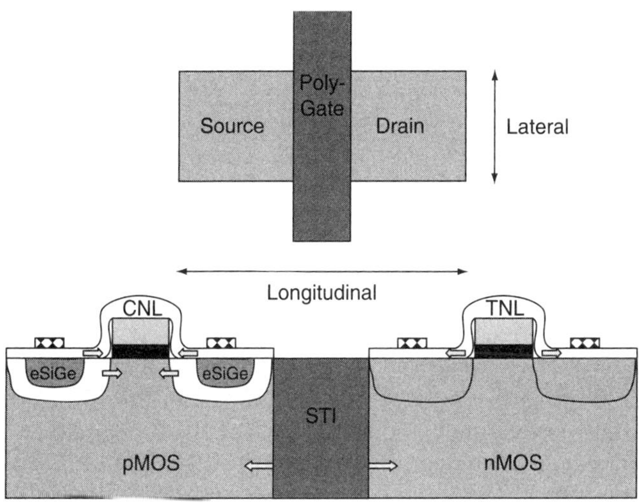

	pMOS	nMOS
Lateral	Tensile	Tensile
Longitudinal	Compressive	Tensile

FIGURE 3.33 Effects of lateral and longitudinal stress on nMOS and pMOS devices.

Figure 3.34 shows where stress is applied and how it propagates within a device. To improve mobility of the transistor channel, an epitaxial growth of silicon germanium (eSiGe) is introduced into cavities made in the source and drain regions. The cavity depth of epitaxial SiGe in source and drain regions influences mobility directly. Gate lines within a standard cell are typically placed at minimum

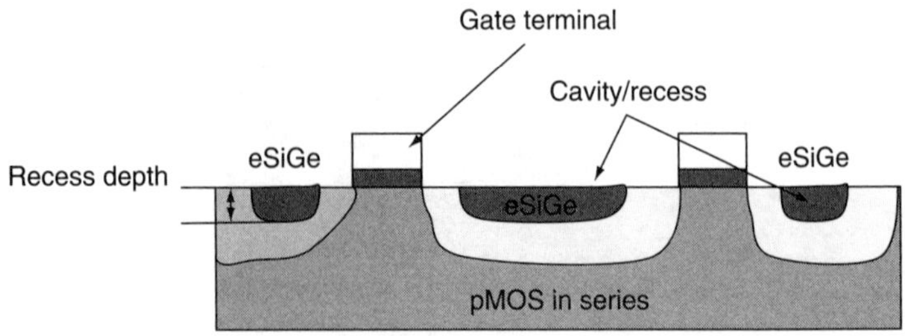

Figure 3.34 Epitaxial growth of SiGe layer in cavities within the source and drain regions of pMOS devices.

spacing in compact layouts. At such spacing, stress-induced mobility due to eSiGe cavities varies rapidly with a small change in spacing between gate polygons. Changes in distance between gate linewidths can be attributed to proximity and other printability issues. In addition to affecting stress-induced mobility, variation in gate linewidths induces other first-order effects, including variation in channel length, channel width, and device threshold voltage.

The contact etch stop layer (CESL) is used to prevent erosion of the gate and its oxide during etching for placement of metal contacts. The CESL film applies a stress that depends on the proximity of the film to the channel region and its volume. One technique for incorporating longitudinal stress in nMOS and pMOS devices is to deposit a nitride liner on top of the gate oxide. Longitudinal stress is tensile in nMOS devices and compressive in pMOS devices. Tensile nitride liner (TNL) in nMOS increases electron mobility, reducing the device's high-to-low transition time. Compressive nitride layer (NL) in pMOS increases hole mobility, reducing the input's low-to-high transition time. Current process technologies use a dual-line approach whereby a highly tensile Si_3N_4 layer is deposited over the wafer and regions are etched over the pMOS; then, a Si_3N_4 compressive liner is deposited and etched over the nMOS region. The amount of stress is controlled by the region of overlap of the gate oxide and the nitride layer. These two parameters are controlled by such aspects of the layout geometry as contact pitch, contact area, and contact-to-gate area. Because the stress develops gradually during the fabrication process, its effect on mobility is not the same for two standard cells of identical size and drive strength.

Mobility variation in all of these stress techniques is partly controlled by layout dimension and spacing between regions of the standard cell: the active area, contacts, and gate polygons. These effects are systematic and can be modeled by simplified design rules to help increase mobility. The development of stress during the fabrication process cannot be tightly controlled and hence is not systematic. Two standard cell layouts with similar dimensions can generate entirely different mobility profiles. Today's designs must deal not only with the partial randomness of mobility but also with stress-induced leakage, which can be devastating to a device. Improper design rules and poor stress engineering can lead to reduced mobility and increased leakage.

3.7 Summary

Variability in the process parameters for current and future CMOS devices is of critical concern. In this chapter we have discussed important sources of variations and their effects. A key observation is that, even though manufacturing processes introduce variability, the variations are a strong function of layout attributes such as pattern size, orientation, density, nesting, and isolation. We also showed that many components of the variation can be modeled in terms of layout attributes. Such components are considered to be systematic, whereas the unmodeled components are considered to be random. Thus, design for manufacturability is an exercise in shaping layouts with the purpose of improving manufacturing and parametric yield while minimizing variations and unpredictability.

References

1. S. Nassif, "Delay Variability: Sources, Impacts and Trends," in *Proceedings of International Solid-State Circuits Conference*, IEEE, San Francisco, 2000, pp. 368–369.
2. M. Chudzik et al., "High-Performance High-k/Metal Gates for 45nm CMOS and Beyond with Gate-First Processing," in *Proceedings of VLSI Technology Conference*, IEEE, Kyoto, 2007, pp. 197–198.
3. S. B. Samaan, "The Impact of Device Parameter Variations on the Frequency and Performance of VLSI Chips," in *Proceedings of International Conference on Computer-Aided Design*, IEEE, San Jose, CA, 2004, pp. 343–346.
4. S. Reda and S. Nassif, "Analyzing the Impact of Process Variations on Parametric Measurements: Novel Models and Applications," in*Proceedings of Design Automation and Test in Europe*, IEEE, San Francisco, 2009, pp. 373–379.
5. International Business Strategies (IBS) Report 2006, http://www.ibs.net/
6. S. Nassif, "Within-Chip Variability Analysis," in *Proceedings of IEEE Electron Devices Meeting*, IEEE, San Francisco, 1998.
7. S. Borkar, T. Karnik, S. Narendra, J. Tschanz, A. Keshavarzi, and V. De, "Parameter Variations and Impact on Circuits and Microarchitecture," in *Proceedings of Design Automation Conference*, IEEE, Anahiem, CA, 2003, pp. 338–342.
8. C. A. Mack, *Fundamental Principles of Optical Lithography*, Wiley, New York, 2008.
9. R. Socha, M. Dusa, L. Capodieci, J. Finders, F. Chen, D. Flagello, and K. Cummings, "Forbidden Pitches for 130nm Lithograph and Below," *Proceedings of SPIE* **4000**: 1140–1155, 2000.
10. S. Kundu, A. Sreedhar, and A. Sanyal, "Forbidden Pitches in Sub-Wavelength Lithography and Their Implications on Design," *Journal of Computer-Aided Materials Design* **14**: 79–89, 2007.
11. D. G. Flagello, H. Laan, J. B. Schoot, I. Bouchoms, and B. Geh, "Understanding Systematic and Random CD Variations Using Predictive Modeling," *Proceedings of SPIE* **3679**: 162–176, 1999.

12. J. W. Bossung, "Projection Printing Characterization," *Proceedings of SPIE* **100**: 80–84, 1977.
13. A. B. Kahng, S. Mamidi, and P. Gupta, "Defocus-Aware Leakage Estimation and Control," in *Proceedings of International Symposium on Low Power Electronics and Design*, IEEE, San Diego, CA, 2005, pp. 263–268.
14. E. Hecht, *Optics*, Addison-Wesley, Reading, MA, 2001.
15. K. Lai, I. Lalovic, B. Fair, A. Kroyan, C. Progler, N. Farrar, D. Ames, and K. Ahmed, "Understanding Chromatic Aberration Impacts on Lithographic Imaging," *Journal of Microlithography, Microfabrication, and Microsystems* **2**: 105–111, 2003.
16. A. B. Kahng, C.-H. Park, P. Sharma, and Q. Wang, "Lens Aberration Aware Placement for Timing Yield," *Proecedings of ACM Transactions on Design Automation of Electronic Systems* **14**(16): 16–26, 2009.
17. M. Born and E. Wolf, *Principles of Optics*, Pergamon Press, Oxford, U.K., 1980.
18. H. Fukutome, T. Aoyama, Y. Momiyama, T. Kubo, Y. Tagawa, and H. Arimoto, "Direct Evaluation of Gate Line Edge Roughness Impact on Extension Profiles in Sub-50nm N-MOSFETs," paper presented at IEEE Electronic Devices Meeting, San Francisco, 2002.
19. A. Sreedhar and S. Kundu, "Modeling and Analysis of Non-Rectangular Transistors Caused by Lithographic Distortions," in *Proceedings of International Conference on Computer Design*, IEEE, Lake Tahoe, NV, 2008, pp. 444–449.
20. Artur Balasinki, "A Methodology to Analyze Circuit Impact of Process Related MOSFET Geometry," *Proceedings of SPIE* 5738: 85–92, 2004.
21. Ke. Cao, Sorin Dobre, and Jiang Hu, "Standard Cell Characterization Considering Lithography Induced Variations," in *Proceedings of Design Automation Conference*, IEEE, San Diego, CA, 2006, pp. 801–804.
22. Seong-Dong Kim, H. Wada, and J. C. S Woo, "TCAD-Based Statistical Analysis and Modeling of Gate Line-Edge Roughness Effect on Nanoscale MOS Transistor Performance and Scaling," *Transactions on Semiconductor Manufacturing* **17**: 192–200, 2004.
23. W. J. Poppe, L. Capodieci, J. Wu, and A. Neureuther, "From Poly Line to Transistor: Building BSIM Models for Non-Rectangular Transistors," *Proceedings of SPIE* 6156: 61560P1–61560P9, 2006.
24. Sean X. Shi, Peng Yu, and David Z. Pan, "A Unified Non-Rectangular Device and Circuit Simulation Model for Timing and Power," in *Proceedings of International Conference on Computer Aided Design*, IEEE, San Jose, CA, 2006.
25. A. Sreedhar and S. Kundu, "On Modeling Impact of Sub-Wavelength Lithography on Transistors," in *Proceedings of International Conference on Computer Design*, IEEE, Lake Tahoe, NV, 2007, pp. 84–90.
26. Ritu Singhal et al., "Modeling and Analysis of Non-Rectangular Gate for Post-Lithography Circuit Simulation," in *Proceedings of Design Automation Conference*, IEEE, Anaheim, CA, 2007, pp. 823–828.
27. Puneet Gupta, Andrew Kahng, Youngmin Kim, Saumil Shah, and Dennis Sylvester, "Modeling of Non-Uniform Device Geometries for Post-Lithography Circuit Analysis," *Proceedings of SPIE* **6156**: 61560U1–61560U10, 2006.
28. Puneet Gupta, Andrew B. Kahng, Youngmin Kim, Saumil Shah, and Dennis Sylvester, "Investigation of Diffusion Rounding for Post-Lithography Analysis," in *Proceedings of Asia-South Pacific Design Automation Conference*, IEEE, Seoul, 2008, pp. 480–485.
29. J. Nakamura et al., "Resist Surface Roughness Calculation Using Theoretical Percolation Model," *Journal of Photopolymer Science and Technology* **1**: 571–576, 1998.
30. J. A. Croon et al., "Line Edge Roughness: Characterization, Modeling and Impact on Device Behavior," paper presented at IEEE Electronic Devices Meeting, San Francisco, 2002.
31. Shiying Xiong, J. Bokor, et al., "Is Gate Line Edge Roughness a First Order Issue in Affecting the Performance of Deep Sub-Micro Bulk MOSFET Devices?" *IEEE Transactions on Semiconductor Manufacturing* **17**(3): 357–361, 2004.
32. M. Yoshizawa and S. Moriya, "Study of the Acid-Diffusion on Line Edge Roughness Using the Edge Roughness Evaluation Method," *Journal of Vacuum Science and Technology B* **20**(4): 1342–1347, 2002.

33. G. W. Reynolds and J. W. Taylor, "Factor Contributing to Sidewall Roughness in a Positive-Tone, Chemically Amplified Resist Exposed by x-Ray Lithography," *Journal of Vacuum Science and Technology B* **17**(2): 334–344, 1999.
34. D. R. McKean, R. D. Allen, P. H. Kasai, U. P. Schaedeli, and S. A. MacDonald, "Acid Generation and Acid Diffusion in Photoresist Films," *Proceedings of SPIE* **1672**: pp. 94–103, 1992.
35. S. C. Palmateer, S. G. Cann, J. E. Curtin, S. P. Doran, L. M. Eriksen, A. R. Forte, R. R. Kunz, et al., "Line Edge Roughness in Sub-0.18-μm Resist Patterns," *Proceedings of SPIE* **3333**: 634–642, 1998.
36. J. P. Cain and C. J. Spanos, "Electrical Linewidth Metrology for Systematic CD Variation Characterization and Causal Analysis," in *Proceedings of SPIE Optical Microlithography*, San Jose, CA, 2003, pp. 35–361.
37. K. Shibata, N. Izumi, and K.Tsujita, "Influence of Line Edge Roughness on MOSFET Devices with Sub-50nm Gates," *Proceedings of SPIE* **5375**: 865–873, 2004.
38. "ITRS 2007," in *International Technology Roadmap for Semiconductors Report*, http://www.itrs.net (2007).
39. R. W. Keyes, "Effect of Randomness in the Distribution of Impurity Ion on FET Thresholds in Integrated Electronics," *Journal of Solid-State Circuits* 10: 245–247, 1975.
40. K. Bernstein, A. E. Gattiker, S. R. Nassif et al., "High-Performance CMOS Variability in the 65-nm Regime and Beyond," *IBM Journal of Research & Development* **50**(4/5): 433–449, 2006.
41. Hamid Mahmoodi, S. Mukhopadhyay, and Kaushik Roy, "Estimation of Delay Variations Due to Random-Dopant Fluctuations in Nanoscale CMOS Circuits," *Journal of Solid-State Circuits* **40**(9): 1787–1796, 2005.
42. B. Stine et al., "A Closed-Form Analytic Model for ILD Thickness Variation in CMP Processes," in *Proceedings of Chemical Mechanical Polishing for VLSI/ULSI Multilevel Iinterconnection Conference*, IMIC, Santa Clara, CA, 1997, pp. 266–273.
43. T. Tugbawa, "Chip-Scale Modeling of Ptern Dependencies in Copper Chemical Mechanical Polishing Processes," Ph.D. dissertation, MIT, Cambridge, MA, 2002.
44. A. B. Kahng and K. Samadi, "CMP Fill Synthesis: A Survey of Recent Studies," *IEEE Transactions of Computer-Aided Design of Integrated Circuits and Systems* **27**(1): 3–19, 2008.
45. R. B. Lin, "Comments on Filling Algorithms and Analyses for Layout Density Control," *IEEE Transactions on Computer-Aided Design of Integrated Circuits Sytems* **21**(10): 1209–1211, 2002.
46. D. Ouma, D. Boning, J. Chung, G. Shinn, L. Olsen, and J. Clark, "An Integrated Characterization and Modeling Methodology for CMP Dielectric Planarization," in *Proceedings of IEEE-IITC*, IEEE, San Francisco, 1998, pp. 67–69.
47. A. B. Kahng, G. Robins, A. Singh, H. Wang, and A. Zelikovsky, "Filling and Slotting: Analysis and Algorithms," in *Proceedings of ACM/IEEE International. Symposium on Physical Design*, Monterey, CA, 1998, pp. 95–102.
48. A. B. Kahng, G. Robins, A. Singh, and A. Zelikovsky, "New Multilevel and Hierarchical Algorithms for Layout Density Control," in *Proceedings Asia South Pacific Design Automation Conference*, IEEE, Wanchai, 1999, pp. 221–224.
49. H. Xiang, K.Y. Chao, R. Puri, and M. D. F. Wong, "Is Your Lay-Out Density Verification Exact?—A Fast Exact Algorithm for Density Calculation," in *Proceedings of ACM/IEEE International Symposium on Physical Design*, Austin, TX, 2007, pp. 19–26.
50. S. M. Hu, "Stress from Isolation Trenches in Silicon Substrates," *Journal of Applied Physics* **67**(2): 1092–1101, 1990.
51. C. Ortolland et al., "Stress Memorization Technique (SMT) Optimization for 45nm," in *Proceedings of Symposium on VLSI Technology*, IEEE, Honolulu, HI, pp. 78–79, 2006.
52. L. Smith et al., "Exploring the Limits of Stress Enhanced Hole Mobility," *IEEE Electron Devices Letters* **26**(9): 652–654, 2005.
53. R. Arghavani et al., "Strain Engineering in Non-Volatile Memories," http://www.semiconductor.net/article/208246-Strain_Engineering_in_Non_Volatile_Memories.php (2006).

54. V. Moroz and I. Martin-Bragado, "Physical Modeling of Defects, Dopant Activation and Diffusion in Aggressively Scaled Si, SiGe and SOI Devices: Atomistic and Continuum Approaches," *Proceedings of Material Research Society Symposium* **912**: 179–190, 2006.
55. V. Chan et al., "Strain for CMOS Performance Improvement," in *Proceedings of IEEE Custom Integrated Circuits Conference*, IEEE, San Jose, CA, 2005, pp. 667–674.

CHAPTER 4

Manufacturing-Aware Physical Design Closure

4.1 Introduction

The quality of patterns printed on wafer may be attributed to factors such as process window control, pattern fidelity, overlay performance, and metrology. Each of these factors plays an important role in making the process more effective by ensuring that certain design- and process-specific parameters are kept within acceptable variation. Quality of image transfer from mask to silicon is a function not only of manufacturing process parameters but also of design quality. This is where design for manufacturability (DFM) plays an active role, involving improvements to the quality of physical design. This is done using rules, guidelines, and simulations. Design rules and DFM guidelines themselves are obtained through simulation and experimentation using control structures, as shown in Figure 4.1.

The foundry communicates a set of physical design rules known as *restricted design rules* (RDRs) to the designers. As an alternative, the foundry may publish a set of design guidelines; these are not checked during the design rules check process but are considered to be good design practices. Restricted design rules are obtained either through simulation or through actual silicon observation. Simulation is useful for establishing forbidden pitches, the control of interconnect and gate linewidth, and the placement of via and contacts. The experimentation process is expensive, but is more comprehensive in terms of assessing the impact of etch, line edge roughness, overlay structures, and so forth.

Manufacturing an IC under a new process technology involves the use of control structures to measure the effectiveness of design rules and variabilities of the process. Many critical measurements are performed through experimental observation. The line edge roughness

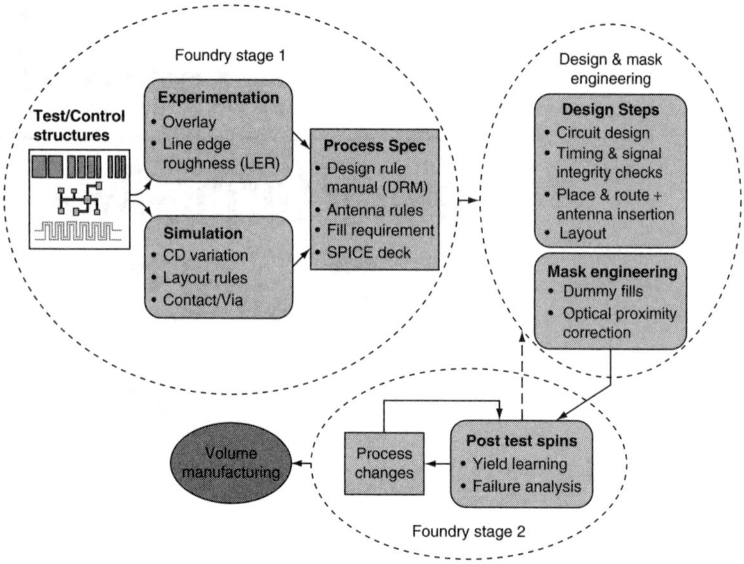

FIGURE 4.1 Transfer of process specification and design database between the foundry and the design house.

of both resist and overlay are measured through careful experimentation. Control structures help create a database of measurements that are used to formulate the process specification to be handed to the design house. The process specifications as shown in Figure 4.1 include (1) a design rules manual (DRM) containing all the layout rules to be used; (2) fill information that indicates the minimum and maximum allowable regions to be filled during routing; (3) antenna diode insertion rules; and (4) circuit simulation (SPICE) models, parameters, and process corner information. In the design phase, these four specifications are used to produce a design that satisfies performance objectives, area and power constraints, and all layout design rules. The DRM contains the geometric design rules that must be satisfied during the layout generation process. Layout generation may be based on a number of design methodologies, including "sea of gate," standard cell-based, semicustom, and fully custom design. The usage of automation and tools varies accordingly, but the layout must be RDR compliant regardless of which design process is used.

In a completed CMOS circuit, the source drain of a transistor drives the gate terminal. However, during an intermediate manufacturing step, a metal line connecting the gate terminal may not yet be connected to a driver. Because the etch process involves application of an electric field, this metal line may act as an antenna, developing an electrical potential that exceeds the maximum allowable voltage on a gate terminal, causing the gate oxide to break down. Foundries publish so-called antenna rules to prevent occurrence of such effects. The usual solution falls into one of three categories: (1) inserting a Zener protective diode to limit the line voltage; (2) inserting jumpers; or (3) breaking up a line into multiple layers or segments.

Traditional design flow ends with the production of a placed, routed, and antenna-inserted design. In the traditional design process, mask engineering typically takes care of dummy fills and optical proximity correction (OPC). This procedure is predicated on the assumption that the dummy fill and OPC may be completed in one step. However, with the advent of deep subwavelength lithography and the increased range of optical interaction between adjacent features, the OPC process may fail if it includes both dummy fills and interconnect lines. The only available solution may then be to change the original layout, which makes the layout-fill-OPC convergence an iterative process. In an iterative design, a layout designer is not shielded from the details of fill and OPC process. Hence, this traditional DFM step must now be performed by the designer using design tools (see Figure 4.1).

Yield learning and failure analysis is performed on the manufactured dies. During this process, repeated patterns of process abnormalities such as defects and patterning issues are observed. In many cases, these issues are resolved by changes to manufacturing process parameters. Successive process tuning is done to improve yield and reduce failures. Issues that are not resolved by process tuning are often associated with a mask defect. In such cases, changing a mask or two may solve the problem. However, this procedure requires accurately diagnosing the defects and correlating repeated defects to masks. Sometimes, neither process tuning nor mask changes can fix a problem. Then the physical design must be reassessed and layout modifications implemented. Such a step may also be associated with the introduction of new design rules, as indicated by the dashed connecting line in Figure 4.1. This causes an unwarranted increase in the chip's time to market and reflects poorly on a foundry.

Aside from physical design rules, a foundry is also responsible for publishing circuit simulation models and parameters. Many foundries have embraced public transistor models such as BSIM, BSIM-SOI, et cetera. In that case, the foundry need only supply the equation parameters. When equation parameters are supplied, the foundry must also specify parameters for various process corners; this will enable circuit simulation at multiple process corners and thus encapsulate the range of manufacturing process variation. In the bipolar design days, the range of such process variation spanned ±3 standard deviations of the process. However, with today's higher overall level of process variation, such a wide range can lead to large differences in gate delays under slow and fast process corners. A wider range is particularly problematic for CMOS transistor sizing, which occurs during the circuit optimization step. It invariably leads to upsizing of transistors and increasing the number of iteration steps during circuit performance optimization. Allowing parameters to vary by a full $\pm 3\sigma$ places time constraints on the optimization steps in terms of converging to a solution. For this reason, process specifications today are abstracted to provide only a $\pm 1.5\sigma$ range in circuit simulation, so that designers can produce an effective design within the prescribed tape-out time. With the advent of DFM-based flows (see Figure 4.2), complete information of process variation is made available to layout design engineers who perform OPC and other layout modifications. However, this procedure can complicate the abstraction of circuit simulation parameters. If, say, circuit simulation corners come from a single source, then consistency may be maintained. But if parameters such as transistor length come from multiple sources (e.g., supplied by the foundry or extracted from the layout), then inconsistencies may arise across the design process.

We mentioned at the start of this chapter that the quality of image transfer from mask to silicon is a function of the manufacturing process *and* the design quality. An important change that has occurred in the design process is the implementation of model-based design

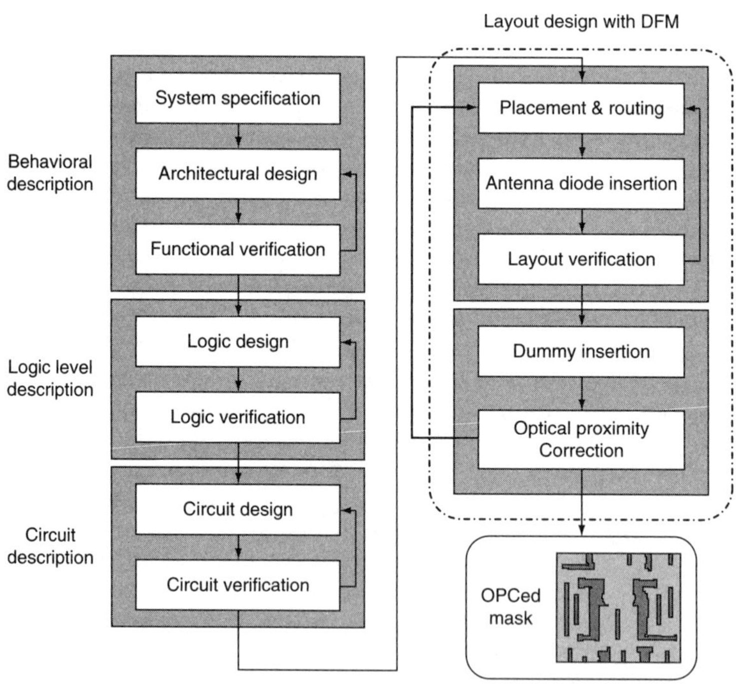

FIGURE 4.2 Typical design for manufacturability (DFM) flow.

rules check (MB-DRC). In this methodology, an approximate but fast model is used to predict the quality of the mask image on silicon, then design decisions are made accordingly. Over time, these models have become more comprehensive and may even include a statistical range of variations that correspond to the manufacturing process. This information is quite helpful in producing highly manufacturable layouts. For layouts that are less manufacturable, postprocessing of layout may allow fast yield estimation based on statistical predictions of the line edge range.

Errors in the fabrication may be caused by improper wafer handling as well as by nonuniformities in the photolithography process stemming from variations in mask placement and alignment, scanner vibrations, thickness of the resist coat, and postexposure baking (PEB) temperature. These nonuniformities can be systematic or random, and there can be lot-to-lot, wafer-to-wafer, and within-die variations. For effective control of linewidth and process latitude (i.e., variation), all the errors described so far must be tightly controlled by the characterization and modeling of each cause-and-effect relationship in the process. When considering the possibility of a correlation between errors in the process and their complex interaction, linewidth control is of utmost importance for overall process and design improvement. All the parameters mentioned previously can be tied to two fundamental components of a photolithography system: focus and exposure dose. In order to simplify the control of process latitude, a window of focus and dose variation is defined for each process. This so-called focus-exposure matrix, which is more simply referred to as the *process window*, is used as a metric to control process-specific line errors. More details are provided in Sec. 4.2.

Information transfer that provides intricate details about process variability is required to enhance a design and adapt it to manufacturing constraints. For early technology generations (i.e., above 250-nm technology nodes), where variability was not as intense, layouts drawn using geometric design rules were considered to be "golden." The layouts underwent postsilicon analysis, the results of which were used to tune the design for improved timing and power and for reduced noise.

The multitude of variations observed in current nanometer-scale designs, together with the level of pattern density in layouts, has lead to design rule explosion in today's DRMs. Lithographic printability issues have created a divergence between design pattern intent on the mask and the imprint on the wafer. Because all analysis and optimizations performed on the design depend on patterns printed on the wafer, simple DRM-based rules will not suffice to quantify all the variations in the process. Furthermore, simply providing information about the nominal shape on wafer is not sufficient. For tool engineers using computer-aided design (CAD), it is preferable to work with contours that represent features on the wafer and their intrinsic ranges of variations. Thus variations in the process are replicated by CAD-based models to help designers analyze their designs after lithography. Some tools require more lithographic knowledge (to infer simulation results) than a typical designer wishes to learn. In order to insulate layout designers from intricacies of the lithography process, many DRC tools today incorporate changes to the design without any input from the circuit designer. Resolution enhancement techniques (RETs) are an example of such changes, which are described more fully in Sec. 4.3. These tools do not require designer knowledge about the lithographic process and its variability yet still provide in-depth information about electrical parameter variation for timing and power optimization. Newer analysis and layout modification tools have been incorporated into the design flow to create layouts of high manufacturability.

The methodology of using information obtained from the foundry to create effective, manufacturable designs is known as design for manufacturability. Various DFM techniques used today help create printable polygons with reduced variability in design timing, power, leakage, and other electrical parameters, resulting in overall yield improvement. As demonstrated in Chapter 1, DFM is associated with monetary benefits because yield is highly dependent on manufacturability. Companies follow changes to DFM methodology closely, since each additional step or modified technique can affect the number of mask steps and the overall yield. As mentioned previously, CAD tools have incorporated DFM-based analysis and layout modification techniques to existing design rules check tools.

Although lithography is not tied to the variability of all electrical parameters in a design, many parameters can be controlled by effective lithography techniques. This chapter provides a brief description of process window analysis, resolution enhancement techniques, and traditional and modern rule checks; it also offers some insights into the advanced processing techniques used to help mitigate lithography-induced variability.

4.2 Control of the Lithographic Process Window

Variability is associated with nearly all manufacturing steps. The design community is more concerned with the overall effect than with decorrelating individual sources of variation, although these may well be of interest to process engineers.

Among the many sources of lithographic process errors, the most important are those due to focus changes and exposure dose changes. Dose and focus errors determine the process variability. The process window characterizes variation in focus and dose in terms of depth of focus and exposure latitude. *Depth of focus* (DOF) can be defined as the range of focus errors that the process can tolerate while still producing acceptable patterns on the mask. Chapter 3 described how focus errors in lithography can be caused by (among other things) wafer misalignment, surface modulations induced by chemical-mechanical polishing, and vertical or horizontal displacement due to lens aberration. Focus errors typically induce changes in the resist profile as well as other second-order effects. The resist profile is estimated by modeling the resist as a trapezoid that depends on three parameters: base width, sidewall angle, and resist thickness. (See Figure 3.5 for an illustration of the resist profile used to measure the overall linewidth of a feature.) It is possible to ascertain the variation in these parameters with focus, information that is used in process window analysis. Also considered are other second-order effects due to focus variation (e.g., resist development variation and etching issues).

Exposure latitude is defined as the range of exposure dose values that the resist can withstand while still producing a good representation of the mask pattern on the wafer. Exposure latitude also contributes to an effective etch process, since proper development of the resist depends on the incident light intensity. The amount of light energy that falls on the wafer changes with each scanner step and also differs from wafer to wafer. Other variations that cause the same type of changes as caused by dose error are photoresist development time, PEB temperature, and resist dissolution. Such variations are modeled as contributions to dose errors in order to simplify the characterization of process variability (and also to reduce the number of independent sources of variability). In short, good estimates of resist profile behavior and imaging process tolerance can be found by using just two parameters: dose and focus.

Because exposure dose is a second-order effect of defocus, variation in focus and dose are typically considered together when modeling the process. The response of the process (i.e., the resist profile) is obtained by simultaneously varying both exposure dose and focus to obtain what is known as the *focus-exposure matrix* (FEM).[1] A *Bossung plot*, which displays the linewidth variation for different focus and dose values, is shown in Figure 4.3.[2] A similar plot can be obtained for various exposure dose values at discrete focus points. (A Bossung plot also gives linewidth variation at different pitches, as shown in Figure 3.10.) Feature pitch variation induces changes in linewidth for dense and isolated features, a phenomenon known as *isodense bias*. For larger defocus values, densely spaced features tend to increase in linewidth, forming a "smile" in the Bossing plot; whereas sparsely spaced features tend to decrease in linewidth, forming a "frown" (see Figure 4.3).

The Bossung plot can also be drawn as contours depicting focus and dose variation for fixed linewidth values, as shown in Figure 4.4.[2] All the plots displayed so far can be similarly obtained for the other two parameters of the resist profile, sidewall angle and resist thickness. A process window plot records the variation in dose and

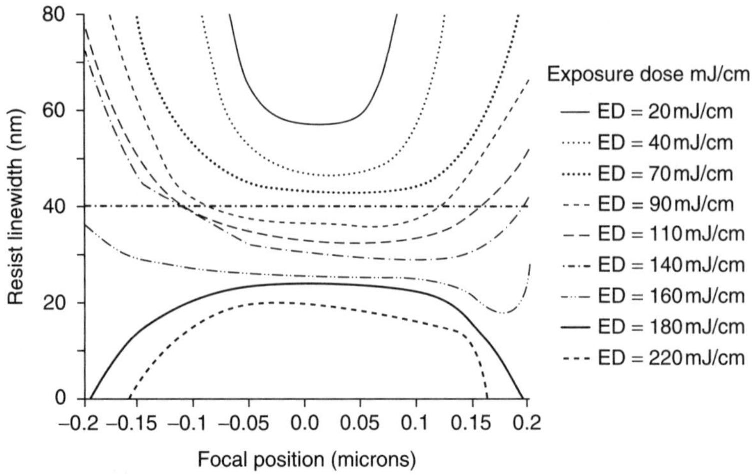

FIGURE 4.3 Bossing plot: simulating the effect of focus and exposure on the resist linewidth.

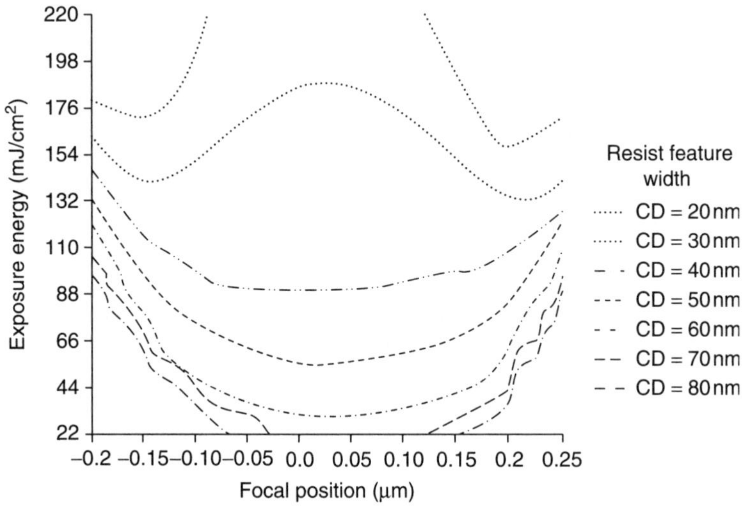

FIGURE 4.4 Resist feature width represented as contours of variation in focus and exposure energy.

focus for a given value of each of the three parameters that define the resist profile. So, for a given linewidth tolerance of ±10 percent, the focus and dose values that keep the linewidth within specification are plotted first. These two curves (one for +10% and one for −10%) define the critical dimension (CD) bound. Next, the resist thickness and sidewall angle tolerance are stipulated, and the corresponding curves for dose and focus variation within specification are plotted. Overlapping these contour curves in a single graph reveals the focus-exposure window for the current process, as shown in Figure 4.5.[1] The exposure latitude and depth of focus is obtained from the process window plot of the common (overlapping) region. This region is marked by a rectangle or an ellipse that encloses the intersection area based on type of focus and dose variation.

When variation in the independent process parameter is systematic, a rectangle is drawn within the overlapping region to obtain the process latitude. The height of the rectangle gives the exposure latitude for all focus values, and the width of the rectangle defines the depth of focus for different discrete exposures (see Figure 4.6). In this case, every point in the rectangle can be used as a process corner to obtain resist profiles within specification. If the variation of focus and dose is random, then this variation occurs with a given probability; as a result, the values fall within an ellipse fit to the overlapping region. This area defines the process latitude wherein a large number of process corners can be used without observing any extremity in the resulting resist profiles. Figure 4.7[1] compares the range of process corners that can be used during manufacturing when the process window is fit with a rectangle versus an ellipse. Because of the isodense bias described previously, process windows for dense and isolated line have very little overlap.[1] This is a cause for concern and calls for techniques to increase this overlap.

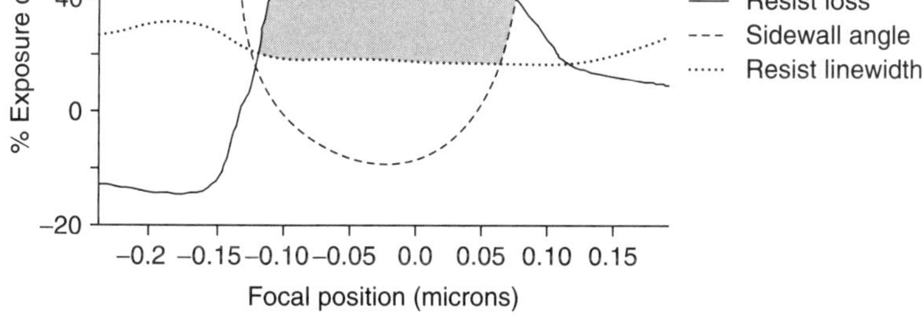

FIGURE 4.5 The focus-exposure process window constructed from contours of linewidth, sidewall angle, and resist loss within specification; the overlapping shaded area constitutes the overall process window.

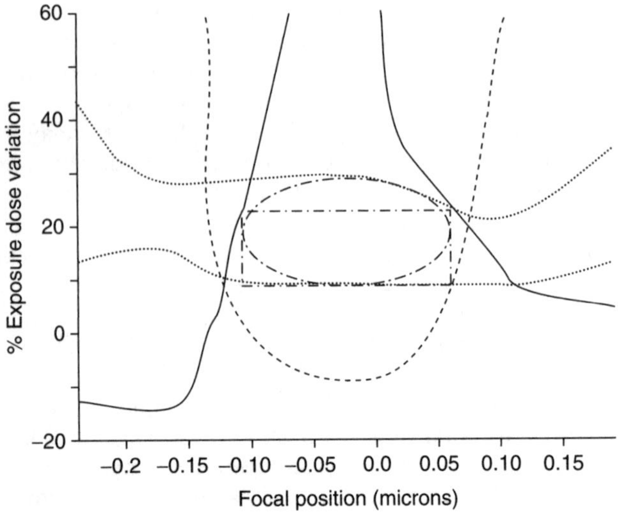

FIGURE 4.6 Process window fitted with maximum rectangle and maximum ellipse.

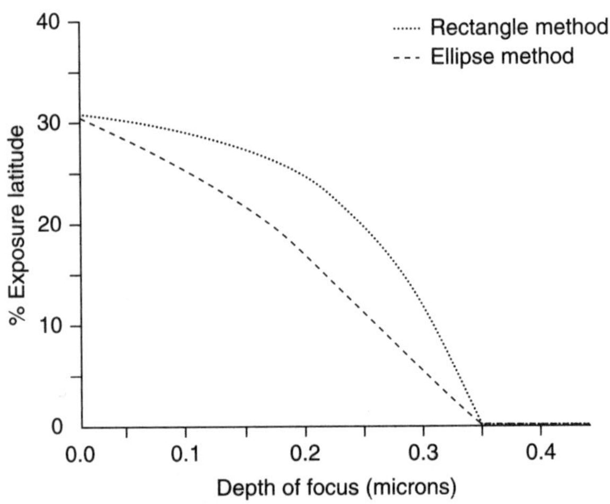

FIGURE 4.7 Exposure latitude versus depth of focus (DOF) for rectangle and ellipse method of establishing process window (cf. Figure 4.6).

Process window analysis has established that controlling the focus and dose of a process is a critical aspect of obtaining distortion-free resist profiles on the wafer. The best focus and dose "sweet spot" is obtained by interpreting the plot of DOF versus exposure latitude. Centering the process on the derived value is vital for maximizing tolerance under process variability. This centering is ensured by metrology techniques, which are described in Chapter 5.

4.3 Resolution Enhancement Techniques

A simple illumination scheme was shown in Figure 2.7. An illumination source of wavelength λ is incident on a set of patterns etched on a chrome-on-glass mask. The light waves are diffracted and are projected onto a resist-coated wafer. The illumination system most often used today is an excimer laser of wavelength 193 nm. The projection system consists of a series of lenses that reduce the image on the mask by a factor of 4 or 5 while projecting it onto the wafer (see Figure 2.21). At the current technology node, the minimum feature width (i.e., 45 nm)—which is also referred to as the system resolution—is far less than the wavelength of the light source used. As explained in Sec. 3.2.1, limits due to optical diffraction cause printability problems for features whose width is less than half that of the source wavelength (in this case, less than 90 nm). Because no flare-free light source of shorter wavelength has been found, problems remain in the transfer of mask patterns to wafer. Resolution enhancement techniques were proposed to improve the fidelity of features projected onto the wafer.

There are four key characteristics of an electromagnetic wave: wavelength (λ), which is constant for an imaging system; amplitude; direction of propagation (**k**) and phase (ω). Resolution enhancement techniques target the latter three properties of the diffracted electromagnetic wave to improve the overall resolution of the system. Improving resolution leads to printability of increasingly smaller features.

There are four principal aspects of resolution enhancement:

1. Optical proximity correction (OPC)
2. Subresolution assist features (SRAFs)
3. Phase shift masking (PSM)
4. Off-axis illumination (OAI)

Optical proximity correction modifies the amplitude of the electromagnetic wave by making changes to the features present on the mask. By increasing or decreasing certain mask features, OPC controls the amplitude of the diffracted waves. Because OPC-induced changes are made to the layout pattern and not to the mask itself, this is categorized as "soft" RET. *Subresolution assist features* expand the process window by adding extra features that improve the diffraction pattern. *Phase shift masking*, as the name suggests, changes the phase of the diffracted wave in order to enhance resolution and contrast. Neighboring patterns are assigned alternate phases, which improves the resolution of each pattern. This process is categorized as "mask"

RET because it requires changes to the mask properties. Finally, *off-axis illumination* is used to print features that are oriented in a particular direction on the mask. Because OAI prescribes what lenses to use for a given feature orientation, this technique is categorized as "lens" RET. The use of resolution enhancement techniques have been instrumental in enabling continued feature scaling with each technology generation.

4.3.1 Optical Proximity Correction

Proximity effects between adjacent features affect the profile of the resist on wafer. As discussed in Sec. 3.2.1.1, proximity effects cause unwanted interference of diffraction patterns, which can lead to changes in feature width. Such effects are especially severe for feature widths of less than half the source wavelength. Optical proximity correction involves changing mask features to improve printability in the presence of proximity effects. In essence, OPC divides each polygon into segments and then removes features from (or adds features to) each segment in the mask pattern to minimize nonuniformity and ensure that the patterns printed on wafer closely match those of the mask. Figure 4.8 shows a comparison—with and without OPC—of resist images on silicon for layout. Clearly, the mask pattern is reproduced more accurately after OPC. One obvious approach to obtaining ideal predistortion of the mask is to use an inverse transform on the target image. Using inverse lithography to create masks is a complex process because it requires accurate modeling of wave diffraction and three-dimensional resist dissolution. The main purposes of OPC are to enhance the patterns imaged on the wafer and to reduce the variability in postsilicon parameters.

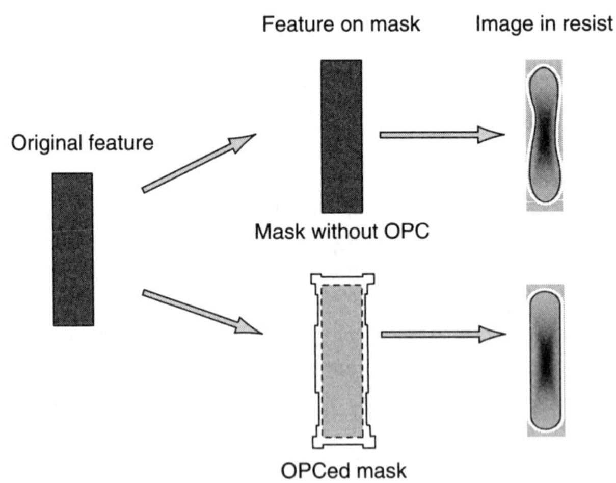

FIGURE 4.8 Resist images with and without optical proximity correction (OPC).

Optical proximity correction employs two types of approaches, rule-based and model-based, to changing patterns on the mask. In the *rule-based* approach, geometric rules are applied to the layout in order to identify regions that are vulnerable to proximity effects. Rule-based OPC is similar to design rule checks; however, the OPC algorithm not only flags a bad region but also modifies the patterns in that region until it complies with specification. Simulations are performed on various mask patterns to derive the geometric rules used to perform OPC. Pointers from actual experimental data are also used to augment the rule set.[3] Rule-based OPC techniques have been used since the advent of 250-nm technology. The geometric OPC rules are typically based on interactions between features adjacent to each other. A pattern's region of influence on other patterns is called its *optical diameter*. It is no longer sufficient to consider only the nearest neighbor, because layouts are becoming denser as feature sizes shrink. A feature's optical diameter subsumes much more than the nearest neighbor, so designers must consider the effect of all neighboring patterns within this range of optical influence on the diffraction pattern of the main feature. As a result, rule-based OPC techniques cannot accurately predict required corrections to the mask patterns. This fact led to the development of model-based OPC.

The *model-based* approach to OPC is a complex algorithmic technique that involves simulation of electromagnetic wave propagation and of various process steps in order to find and correct suspect features in a mask. The simulation is typically performed by computing a weighted sum of process and optical parameters based on previously computed look-up tables. (See Chapter 2 for more details on lithography simulation.) Model-based OPC divides polygons into small segments whose individual resist profiles are obtained from look-up tables. The basic mathematical principle behind these OPC techniques is the convolution of mask patterns with precomputed diffraction kernels. The construction of diffraction kernels is based on models of lithographic imaging system and process stages. These kernels take into account the following parameters: type of illumination source (refer to Figure 2.11); lens and pupil functions of the imaging system; resist contrast; and the photoresist polymer diffusion rate. Since these parameters remain constant for a given manufacturing process, the kernels are precomputed and stored in tables to be used during the OPC process.

Figure 4.9 shows the model-based OPC flow, in which the given layout is modified recursively; this involves adding or removing polygons based on quick simulation of the features and then comparing how closely the simulation matches the required image. The procedure continues until a preset level of closeness to the ideal image is achieved in all regions of the layout. Thus, model-based OPC uses repeated simulations to guide the modification of small

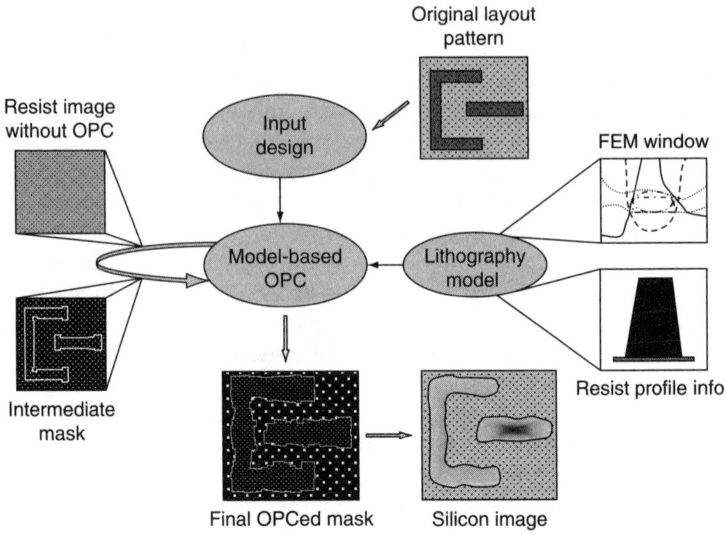

FIGURE 4.9 Model-based OPC flow.

regions of mask features, choosing the best solution based on a cost metric. Because this technique is so intensive computationally, it is not performed on entire layout masks. Model-based OPC is highly parallelizable when it is based on layout selection and also when parallel, uncorrelated modifications can be made to regions of the layout. Optical proximity correction is performed individually for each metal layer because there are no interlayer optical interactions. The linewidths of geometrical one- and two-dimensional shapes in the mask are increased or decreased by OPC to counteract the effect of pitch-dependent linewidth variation, also known as through-pitch variation (see Figure 3.10). Simulation of lithography can anticipate problems in two-dimensional features that are due to proximity effects, including line end shortening and corner rounding. These problems lead to reduced yield and cause defects in the metal layer, necessitating changes through OPC. Line end shortening is corrected by adding *hammerhead* structures, which increase the printability of line ends. Corner rounding occurs at both line ends and at places where a line changes direction; *serif* structures are added to these region to compensate for proximity effects. As shown in Figure 4.10, serifs form a protrusion (polygon added) for rounding at line ends but form an intrusion (polygon removed) to compensate for rounding at other locations.

Optical proximity correction is typically performed on the layout just before the mask is handed off to the mask shop. This means that designers rely on OPC to help produce nearly ideal images on silicon.

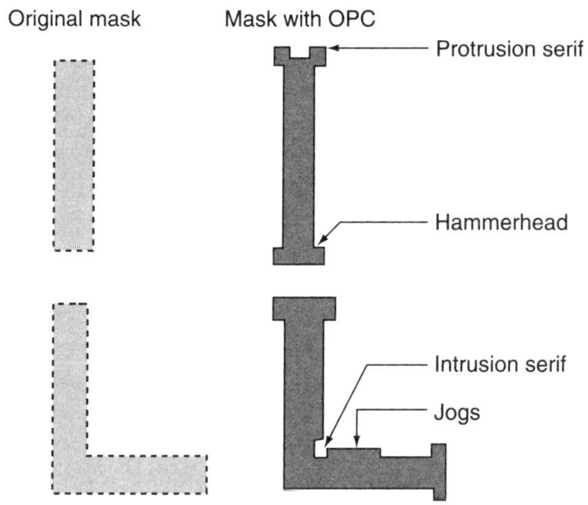

Figure 4.10 Serifs, hammerheads, and jogs added to the original mask by the OPC process.

The final mask after OPC is transferred to mask manufacturers for further processing. Masks are manufactured by using an E-beam setup to write the layout pattern onto a glass mask. The OPC-induced changes to mask features will, of course, increase the mask writing time. Unlike the rule-based approach, model-based OPC incorporates a multitude of small changes to the features; therefore, the time and consequent expense of mask writing increases the cost of manufacturing the mask. Any change in a mask feature is referred to as a *shot*. Shot count (see Figure 4.11) is directly related to the cost of the

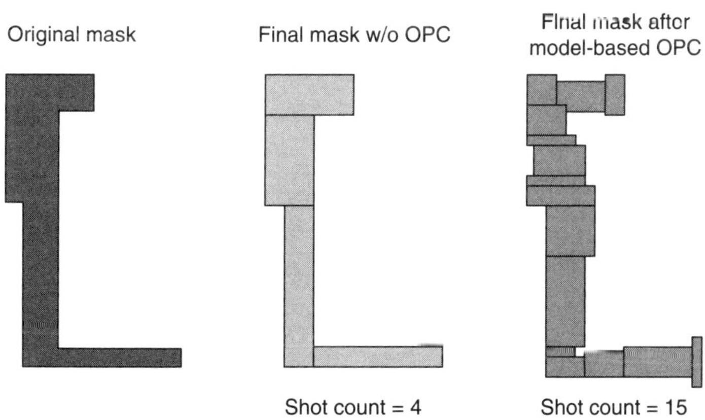

Figure 4.11 Shot counts at the mask-writing stage for layout with and without model-based OPC.

mask-writing stage in the manufacturing process. Because companies designing application-specific integrated circuits (ASICs) typically need two or three mask tape-out cycles for a design, mask costs have skyrocketed with the increased shot count due to model-based OPC. A balance between rule-based and model-based OPC is needed to keep mask manufacturing costs under control.

4.3.2 Subresolution Assist Features

Optical proximity correction modifies the amplitude of the diffraction pattern of mask features that cause linewidth reduction, line end shortening, and corner rounding. Special features such as jogs, hammerheads, and serifs are added to reduce change in the linewidth of such features. During this process, OPC increases the overlap of process window between isolated and dense features. Tight control of CD variation and increased process latitude for isolated features remain problems that cannot be solved with OPC techniques. Subresolution assist features constitute a new resolution enhancement technique that aims to increase the process window of isolated features by adding extra features that improve the diffraction pattern of the main features. The SRAFs are assist features or scatter bars that are drawn adjacent to mask polygons in order to enhance the diffraction pattern. These assist features are of lower resolution (see Figure 4.12) and are not printed on the wafer, but they aid in modifying the diffraction pattern of the main feature. The SRAFs cause destructive interference due to phase difference, which improves the contrast of the image being formed on the wafer (see Figure 4.13). The phase depends on the pitch between the main feature and the SRAF.

Increasing the number of SRAFs can further improve the pattern, but the relation is not monotonic because the effects saturate owing to interactions of the SRAFs. Process latitude improvement relies on the original layout's provision of unobstructed regions for optimal SRAF placement. Neither the size nor the number of SRAFs placed adjacent

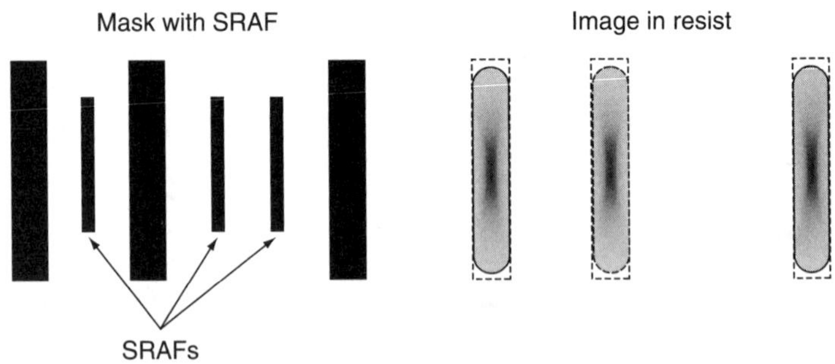

FIGURE 4.12 Subresolution assist features (SRAFs)–features of lower resolution, not printed on the wafer, that enhance the diffraction pattern.

to mask patterns is continuous, so the process latitude variation has a discontinuity when there is a quantized jump in the size of SRAF used to improve the central feature (see Figure 4.14).[4] Apart from space constraints set by the original layouts, SRAF placement in modern fabrication is controlled by design rules for SRAFs, SRAF shapes optimized for two-dimensional (2-D) features, and better

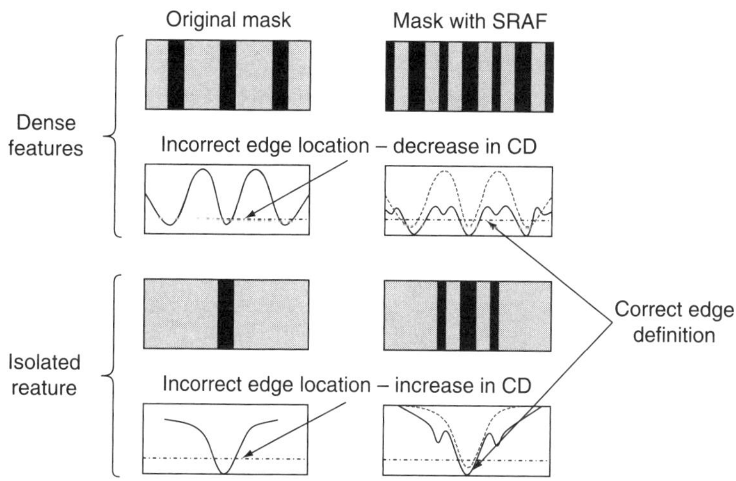

FIGURE 4.13 Destructive interference between diffraction patterns of original mask polygons and SRAF features.

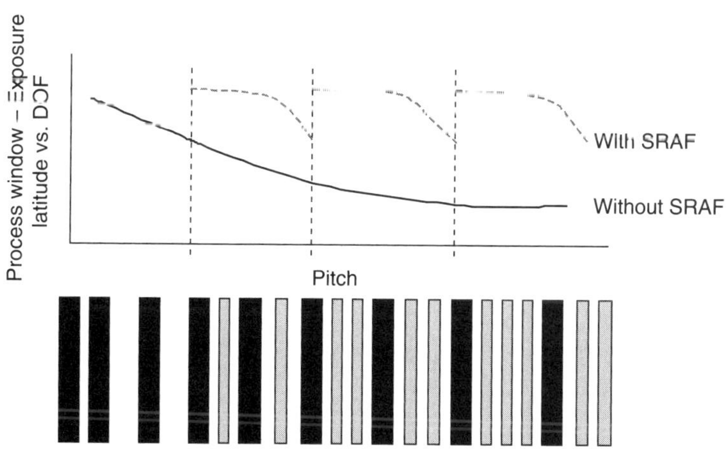

FIGURE 4.14 Variation in chip process latitude with and without SRAF insertion; the ideal number of SRAFs cannot be added here, resulting in lower process latitude.

placement engines.[4] Unlike the tried and tested SRAFs shown in Figure 4.13, real-world layouts involve 2-D features for which the SRAF structures and placement are more complicated. Two-dimensional SRAFs (see Figure 4.15) for enhancing printability of corners and line ends must be defined and optimized based on yield results. Designers can also use a ranking of SRAF placement options based on parameter optimization for chip timing, leakage control, and density enhancement.

4.3.3 Phase Shift Masking

The generic mask structure used in photolithography is chrome-on-glass (COG). Chrome-on-glass masks consist primarily of chrome patterns etched on a base material of glass. The most basic type of COG mask is the binary image mask (BIM). This type of mask has patterns and spaces that enable only two types of transmission through the mask. Chrome-filled regions form the mask patterns that have zero transmittance, while the nonchrome regions have high transmittance. The BIM mask creates slits that cause diffraction patterns to be formed on the image plane. Masks that use glass for the background material are called *light-field* masks. Conversely, *dark-field* masks are filled with chrome throughout and the patterns are formed by removing the chrome.

Now observe the formation of diffraction patterns in the light transmitted through a BIM mask, as shown in Figure 4.13. Depending on the type of photoresist being used, certain regions react to light and certain regions do not. Soluble regions of the resist are removed by development and etch processes. The sharpness of the image depends on the resist properties and the shape of the diffraction pattern. It can be observed that, for patterns whose width is smaller than the light source wavelength, the linewidth is severely limited. Moreover, according to the Rayleigh criterion (i.e., $R = k_1(\lambda/\mathrm{NA})$), the

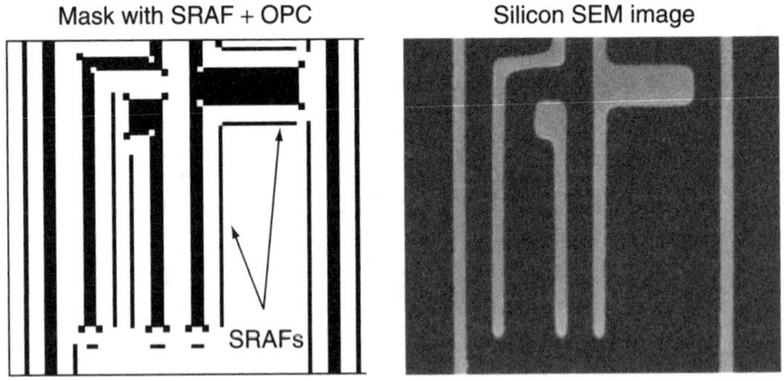

FIGURE 4.15 SRAFs for two-dimensional (2-D) features.

minimum resolvable resolution (with $k_1 = 0.5$) is a function of two fixed parameters of the process: source wavelength and numerical aperture. With no foreseeable change in the illumination department, using an argon fluoride 193-nm light source for imaging features smaller than 45 nm is bound to cause pattern formation problems in the future. Patterns placed at minimum pitch can cause interference, leading to the improper intensity profiles shown in Figure 4.16(a).

One method of overcoming such a fundamental limitation is to use RETs that can manipulate the phase of the incident wave to cause favorable diffraction patterns on the wafer. Phase shift masking is an RET used to enhance diffraction patterns for features in the subwavelength regime. This resolution enhancement technique modulates the phase of the incident wave as it transmits through the patterns in the mask. Special phase shifters are used to modulate the wave, creating a phase difference between neighboring mask features. Destructive interference occurs for features with opposite phases, which improves image intensity and contrast on the wafer. Figure 4.16(b) illustrates how adding phase shifters to certain regions of the mask causes a 180° phase difference with respect to adjacent features.

The two types of PSM techniques used in mask manufacturing are *alternating* PSM (AltPSM) and *attenuating* PSM (AttPSM). These PSM techniques are described for light-field pattern masks in the text that follows, but a similar technique can also be implemented for dark-field patterns. Alternating PSM involves creating a phase

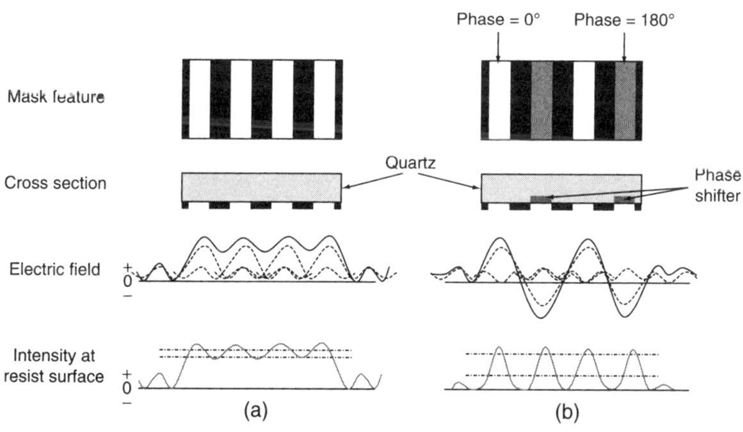

FIGURE 4.16 Light waves passing though openings of different phase cause destructive interference of the diffraction patterns, leading to better pattern transition: (a) no destructive interference between same-phase patterns, which leads to poor pattern contrast; (b) the mask produced using phase shift masking (PSM) yields improved pattern contrast.

difference on both sides of a dark region in order to increase pattern contrast. Alternate light features are assigned 0° and 180°. Because light passing through adjacent high-transmittance regions are out of phase, the resulting destructive interference forms zero-intensity curves at these locations. The alternating phase regions are selectively etched on the mask, creating an optical path difference of $\lambda/4$ to produce a proper 180° phase change for the incident light. Each dark feature is flanked by light features of opposite phases. The PSM technique enables processes to create features of smaller resolution with increased process latitude and contrast (see Figure 4.17).[5]

An important issue in PSM is the problem of phase assignment in layouts. Phase assignment requires that no light feature have more than one phase assigned to it, and a layout is considered *phase assignable* if it satisfies this condition. Otherwise, the layout is not phase assignable and hence changes to the existing layout are necessary. Two common patterns that lead to phase conflict are shown in Figure 4.18. The required changes typically involve transforming the feature into a noncritical one by increasing either the feature width or the spacing between patterns. The results is an overall reduction in design pattern density.

The attenuating PSM technique applies the concept of alternating PSM to create a phase difference that causes destructive interference to enhance image contrast. The difference with AttPSM is the change in transmittance of dark patterns and the utilization of same-phase light-field patterns. Dark-field patterns in a binary mask have zero

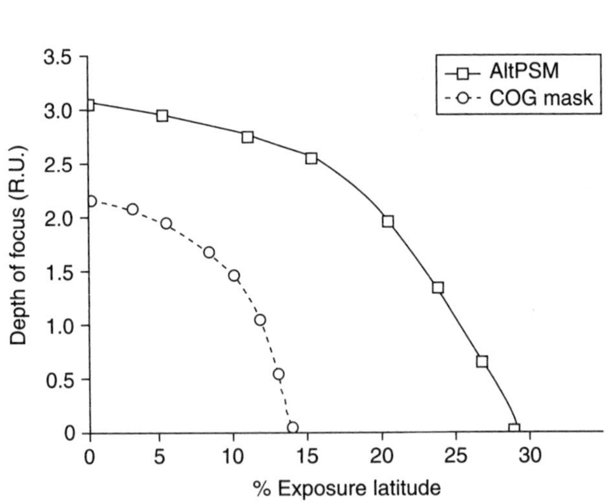

FIGURE 4.17 Alternating PSM yields greater process latitude than that of a binary, COG mask.

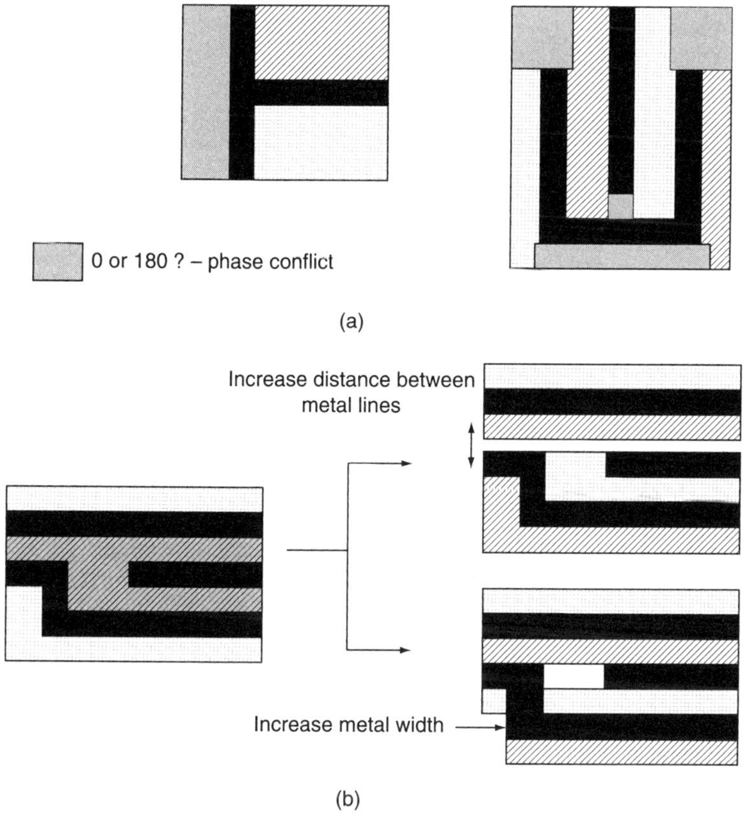

FIGURE 4.18 (a) Two common layout patterns with phase assignment conflicts; (b) solutions to the phase assignment problem.

transmittance, which prevents any light from passing through them. With the AttPSM technique, these patterns transmit 7 to 15 percent of the light energy incident on them. Resist exposure is prevented because the light energy that passes through the transmitting dark patterns is very small. However, the increased transmittance of dark patterns causes a 180° phase change for the light that does travel through it. Since all light-field patterns have the same 0° phase, the phase change causes destructive interference with the fully transmitted light, which leads to improved contrast at pattern edges.[5] The result is an increase in image contrast, as shown in Figure 4.19.[5] The transmittance factor is a key parameter for enhancing high-density patterns, and its benefits include reduced mask costs and simpler design rules. Attenuating PSM is most beneficial in imaging isolated contacts and trenches, which require a low k_1 as well as a high NA.[6]

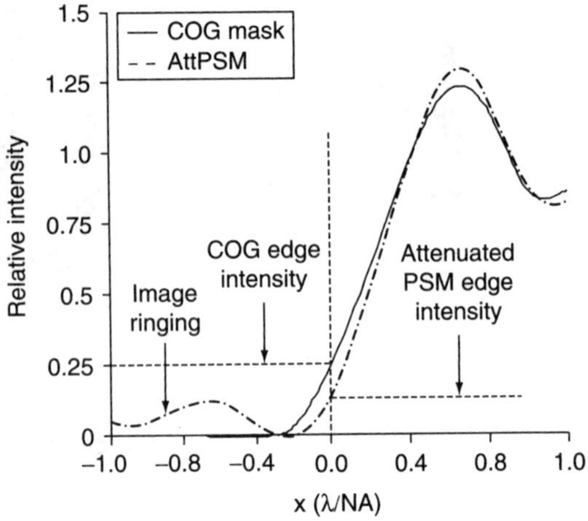

FIGURE 4.19 Effect of attenuating PSM on intensity profile.

Because higher performance and better control of variations are directly tied to the need for features with lower resolution and enhanced contrast, PSM has become an indispensable tool for designers in the mask manufacturing process. Therefore, CAD tools that create layout features must be capable of producing designs that are phase assignable. The disadvantage of a layout with phase conflicts is a severe relaxation of pattern density, which leads to increased area. Thus, phase-shifting RETs unavoidably require designers to trade off improved printability against design density.

4.3.4 Off-Axis Illumination

We have remarked that the RETs discussed so far attack the amplitude (OPC, SRAFs) and phase (PSM) of the wave incident on the mask. These techniques have led to the enhancement of such mask feature attributes as linewidth, process latitude, and image contrast. However, off-axis illumination manipulates the angle and direction of the light wave incident on the mask plane. As the name implies, this resolution enhancement technique is based on reducing (or eliminating) the effects of the on-axis component of illumination. Off-axis illumination increases the imaging system's depth of focus by using alternative pupils (see Figure 4.20).[7]

By tilting the illumination system, light rays incident at an angle causes diffraction patterns to be spatially shifted from the optical axis; see Figure 4.21.[2] This spatial shift allows higher-order diffraction patterns to be formed on the projection lens system, which improves the depth of focus of the mask features. A tilt in opposite direction causes similar effects. Pupils can be used to produce the required angle and direction of illumination. Figure 4. 22 shows several types

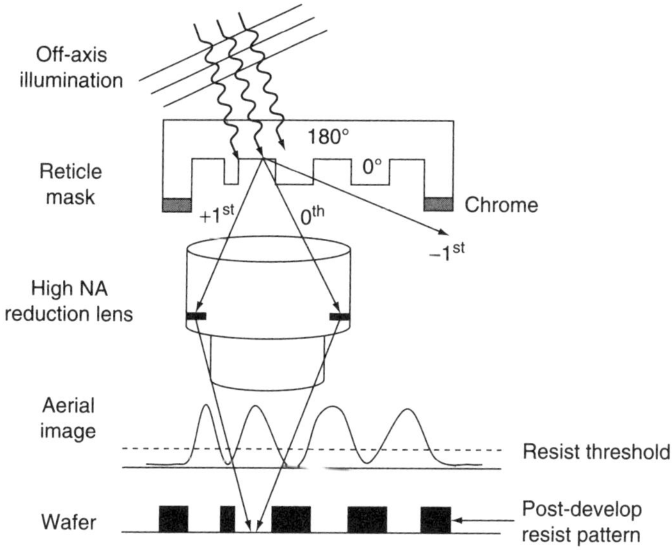

FIGURE 4.20 Off-axis illumination (OAI).

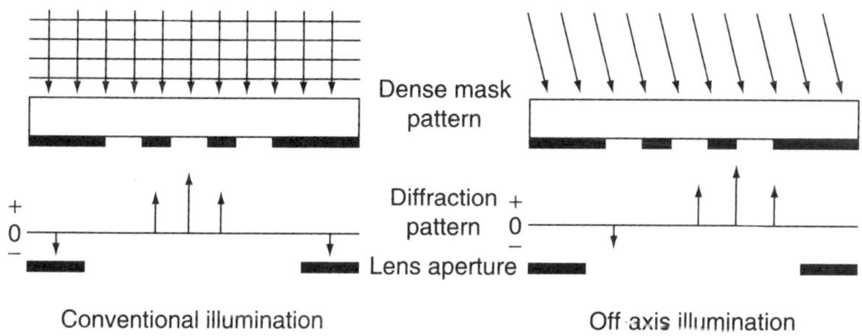

FIGURE 4.21 Off-axis illumination causes a spatial shift in the diffraction pattern; patterns of higher order pass through the aperture.

of pupil filters. Monopole pupils can capture individual effects, and dipole pupils can be used to produce a combined effect. Pupil orientations can be changed to increase the depth of field of a particular orientation. As shown in Figure 4.23,[8] dipoles offset by 90° capture variations in x direction and y direction separately and thus are used to print (respectively) horizontal and vertical lines with increased DOF.[9] Other pupils, such as the quadrupole and quasar, are typically used to enhance OAI-based depth of field for technologies below 65 nm.[10] Pitch-based DOF optimization can be obtained by choosing the best OAI pupil for the pattern set being printed.

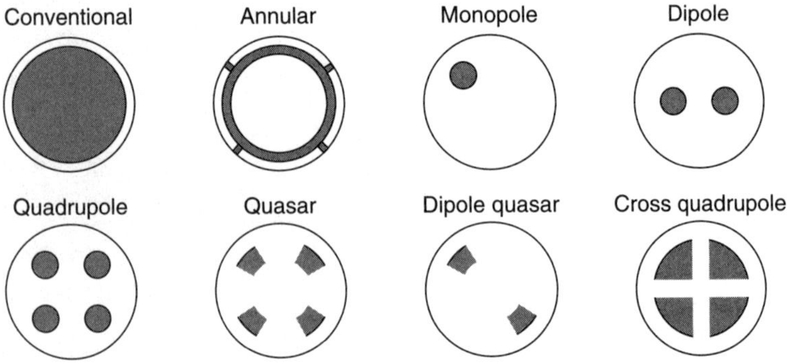

FIGURE 4.22 Pupil filters.

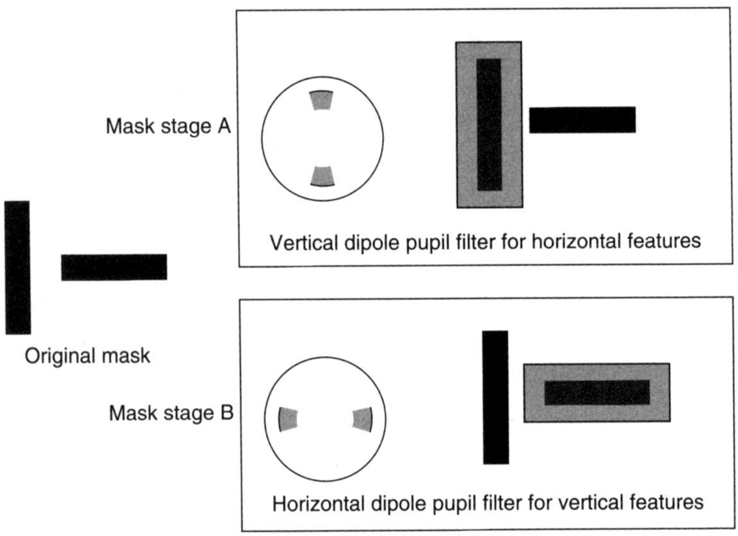

FIGURE 4.23 Alternating dipole pupil filters to print vertical and horizontal patterns using off-axis illumination.

4.4 Physical Design for DFM

Physical design solutions have been in use for drawing layouts, standard cell placement, interconnect routing, parasitic extraction, and various other requirements. The physical design layout forms the interface between the high-level logic graphs that represent circuits and the abstract geometries that represent metal lines on silicon.[11] For technology beyond 65 nm, parameter variability control has become the most important challenge for physical design.

Analyzing and modeling variability are also part of DFM methodologies that take advantage of information from the foundry. Because the economics of the semiconductor industry is tied to effective mask manufacturing and high-yield processes, physical design methodologies incorporating DFM have become indispensable. In this section we see how physical design tools have incorporated techniques to help mitigate problems arising in today's design for manufacturability.

4.4.1 Geometric Design Rules

The classical application of physical design tools is performing design rules check to ensure the physical and electrical manufacturability of a design. Design rules specify geometric distances, between 1-D and 2-D layout features, that must be preserved to ensure design imaging within specifications. All design rules are written into the design rules manual, which is constantly updated through feedback from manufacturing failure analysis. The DRM consists of two generic subsets of rules, one for polygons of the same layer and another for those between different layers. Rules such as spacing, pitch, linewidth, and minimum bounding box for shapes are listed for comparison among polygons in the same mask layer. Overlay rules that dictate the distance between contact and diffusion edge, metal thickness around contacts, and gate to contact distances are stipulated in order to avoid misalignments between different layers during photolithography. These rules also convey electrical specifications, since the contact thickness and distances from gate layers define effective current paths and stress levels in designs.

All geometric rules consider only the effect of the nearest neighbor to the current polygon. But as technology scales, these rules become grossly insufficient for estimating the actual interaction between polygons. The effect of neighbors beyond the nearest one is no longer local and must be considered within a region of influence of the diffraction pattern. The number of possible shapes within the radius of influence around each polygon is high because of the increased layout density seen in current technology nodes. Geometric design rules cannot accurately characterize all the interactions of the contours as a simple function of the distances between them. Although DRM suggests various rules for such contours, these tools are not up to the task of quantifying and preventing variation in the design. Hence, in the subwavelength domain, compliance to geometric DRC rules does not guarantee manufacturability. Layouts that are DRC compliant can still face printability problems, leading to variation in electrical parameters or even to the formation of defects.

4.4.2 Restrictive Design Rules

Geometric design rules are not binary; that is, a part will not necessarily fail just because it violates the rule. Nonetheless, the yield function for tools that perform geometric DRC on designs is simplified to a step function as a conservative design practice. But for layouts in the subwavelength regime, such simplified yield functions do a poor job of representing actual postlithography circuit behavior, as shown in Figure 4.24.[12] Yield is a far more complex function, and even a DRC-compliant layout will produce suboptimal yield as a result of subsequent errors in lithographic imaging and processing.

Many manufacturing and design specifications are written into the DRM. Most of these requirements are converted to abstract design geometric rules. Continued technology scaling has led to denser layouts being created, which increases the number of interfeature and interlayer interactions and so requires more design rules to verify the layout for specification conformity. This translates into an increase in complexity of CAD tools that rely on the DRM manual, and the result is an ambiguous and insufficient set of rules to be checked for verifying a design. To alleviate the sheer volume of DRM specifications, a new set of rules—*restricted design rules* (RDRs)—were proposed. The RDRs add specific criteria, such as standard orientation for a single cell or line, a limited number of allowed pitches for lines, uniform layout regularity, and limited narrow width lines on critical

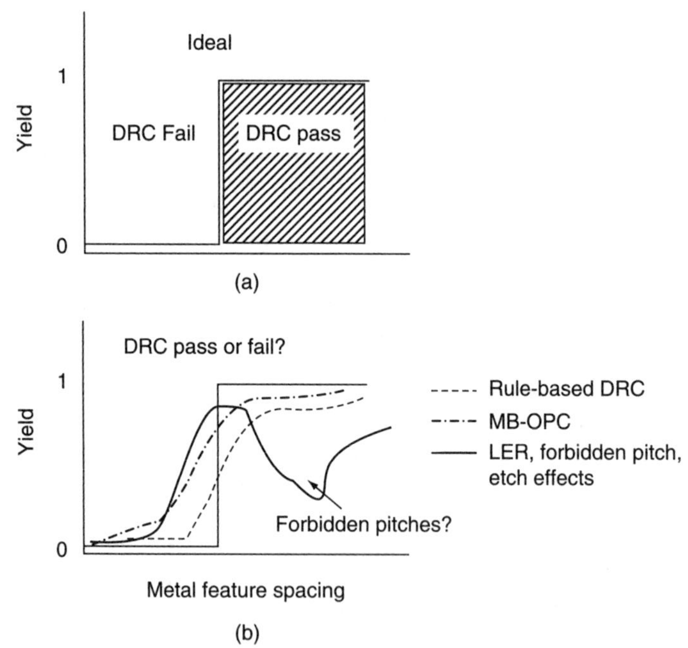

FIGURE 4.24 (a) Yield as a step function of geometric design rule (GDR) dimensions; (b) in subwavelength regimes, yield is no longer a step function of GDR dimensions.

features. These new specifications reduce the influence of DRM on layout and have been found to reduce the 3σ interconnect linewidth variation.[13] The regularity obtained by the use of RDRs reduces linewidth variation across the chip. The main disadvantage of RDRs is their inability to make predictions for 2-D device features, since all calibrations are performed on 1-D features.

4.4.3 Model-Based Rules Check and Printability Verification

The rules check procedures embodied in DRMs and RDRs are classified as rule-based DRC techniques because they use rule tables to flag errors that can cause manufacturing problems. As the number of interpolygon and interlayer contour interactions increased, rule-based DRC was supplanted by model-based DRC tools that use complex predictive models to identify lithography "hotspots" in densely packed sub-90-nm layouts.

The most prevalent such technique used by designers today is the use of complex lithography modeling tools to forecast design hotspots. The supplied design is run through a "black box" stage that mimics the process and estimates hotspot locations and variability at sensitive regions of the circuit. The model typically performs lithography simulation after application of OPC and other resolution enhancement techniques. The process accurately predicts hotspots based possible process instabilities, but it is computationally intensive. Another technique (suggested by Gennari and Neureuther)[14] detects hotspots by pattern matching of two-dimensional structures. This technique uses bitmap images to identify 2-D features while rapidly performing image-based pattern matching. The bitmap images include not only regular polygons, such as L-shapes and T-shapes, but also nonrectangular polygons derived by performing edge extraction on scanning electron microscopy images. The table of bitmap images and the layout are provided as inputs to the hotspot detection engine. The output consists of a ranked list of manufacturing vulnerabilities for exact and nearly exact matches, a listing that enables quick identification of 2-D configurations likely to produce low or borderline yield. Other types of hotspot detection mechanisms include techniques that incorporate foundry-issued information on process irregularities.[15,16] These tools target specific regions of the layout that have been identified during the failure analysis stage and mark them as process hotspots. A good example of this is the double via insertion checkers. It had been observed that via resistance increased with technology scaling, and this led to a requirement for using double vias in critical locations. Hotspots caused by these double vias can be detected by using model-based via resistance prediction.

With lithography hotspot detection, hotspot removal becomes a necessary addition. Various techniques have been proposed to detect and remove hotspots based on simple modifications of polygons and spacing. One such method, the *lithography compliance checker* (LCC), aims to verify layouts by using lithography simulation.[17] In general, LCC tools begin verification from primitive standard cell-sized regions and work up to higher regions in the layout. Hotspots are identified at the full-chip level after considering all neighborhood polygon interactions. The three main components of an LCC engine are OPC, verification of marginal conditions through lithography simulation, and hotspot judgment. Optical proximity correction is the first stage in the LCC process. Next is lithography simulation to verify the layout and find marginal conditions; a subset of these are ranked and marked as litho hotspots. The final stage is to modify the layout and to reanalyze, based on the designer's input, whether these regions are actual physical or electrical hotspots. The layout modification and reanalysis is performed by simple CAD tools. Any further analysis is based on user input. Hotspot detection is performed at standard cell level in ASIC designs to allow layout flexibility during modifications when hotspots arise at higher levels of layout hierarchy. If a layout is LCC clean, it means the designer is satisfied with the electrical characteristics of the design parameters. It also guarantees that the design complies with all lithographic imaging and other process tolerances. Hence care must be taken during hotspot identification and reanalysis so that no unclean layouts can pass LCC. Such hotspot checking at various design houses have yielded impressive results: hotspot mitigation rates in excess of 80 percent with acceptable run times.[18,19]

Layout printability verification (LPV) is implemented as a full-chip process simulation performed on post-OPC layouts. This engine typically groups all residual OPC errors into printability classes that are ranked in terms of their severity. The ranking is based on indexes such as mask CD distribution, process window, and yield. Full-chip simulations are rare before the tape-out stage, since they involve intense calculations that require high computing and storage levels. The LPV engine is based on OPC tools that segment the design into points at which simulations must be performed to check printability. Typical printability errors include necking and bridging errors that passed OPC. Because simulation at a nominal process corner is not enough for accurate verification, simulations with process parameters at multiple process corners are performed. The simulation engine incorporates lithography simulation while considering such process parameters (and their variability) as dose, focus, overlay, resist thickness, and critical dimension.

Figure 4.25 illustrates the printability checker engine using process parameter distributions to provide hotspot analysis. Figure 4.26 shows hotspots at different process corners. It can be seen that necking and bridging do not occur under normal conditions (process corner C) but do occur at more extreme corners. One of the greatest benefits of printability verification is obtaining yield values on a full-chip level for various process corners. This provides valuable

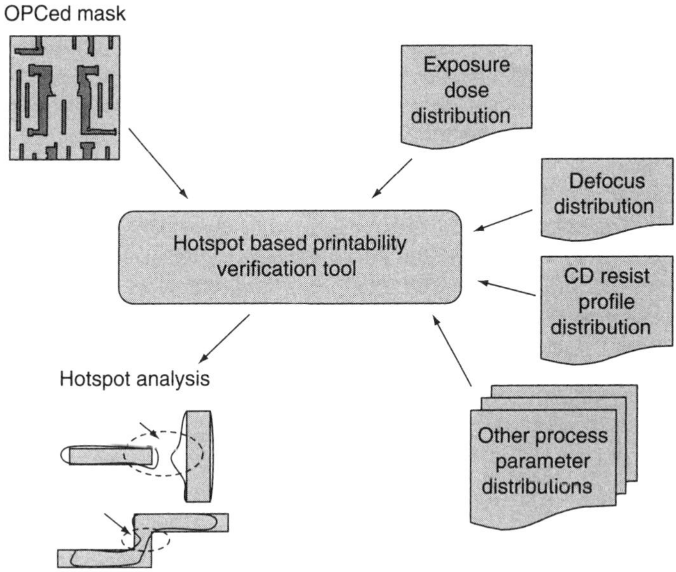

FIGURE 4.25 Verifying printability by checking process parameter variabilities.

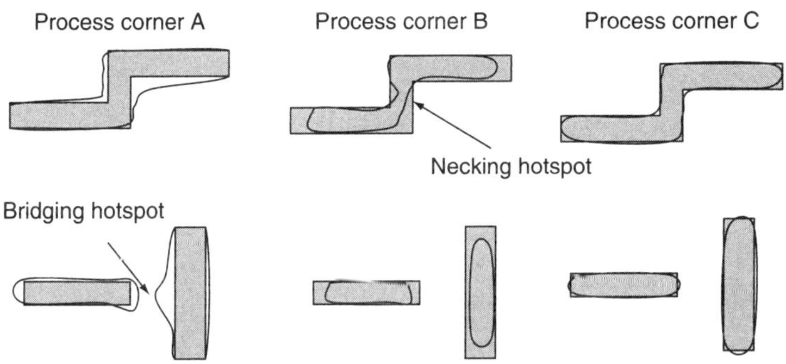

FIGURE 4.26 Necking and bridging defects that can arise at different process corners; simulations based on variation of input parameters.

information for the designer and the process engineer regarding which process corners will lead to the best yield and performance.

4.4.4 Manufacturability-Aware Standard Cell Design

Resolution enhancement techniques such as OPC, PSM, and OAI have been applied to chip-level analysis and layout modification for increased printability. Conventional OPC involves running lithography simulation on flattened layouts. This has proven to be the most accurate and manufacturability-aware technique for modifying layouts, but it is extremely time-consuming and—because dense OPC is not performed throughout the layout—may not produce the best results within standard cells. Manufacturability effects due to scaling are most pronounced in standard cell layouts. Common effects (e.g., change in linewidth, improper contact connections, poly-gate length variations, diffusion rounding) are quantified as either gate delay or leakage that can affect overall circuit performance. It is widely recognized that standard cells are central to the IC design process. The impact of manufacturing variations on cell performance is critical to design, so layouts must be analyzed and compensated appropriately. One approach is to perform cell-level RET-aware characterization to ease the OPC load at the full-chip level.[8] A physical design flow that includes RET-aware techniques for DFM is shown in Figure 4.27.[8]

Standard cells are drawn based on the DRM, which dictates the properties of metal lines, polysilicon region, active areas, contacts, and vias. Then the model-based rule checkers described previously, including LCC and LPV, are used to analyze the standard cell layout for hotspots under the given manufacturing specifications. Standard

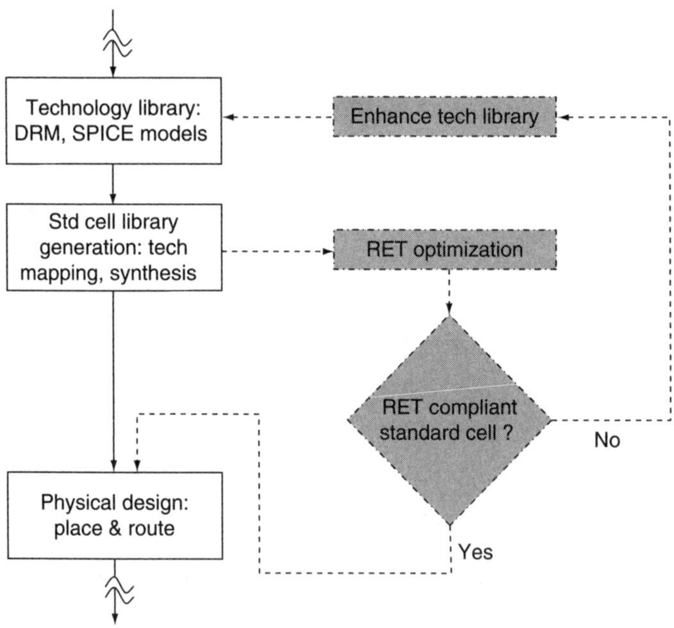

FIGURE 4.27 RET-aware standard cell library characterization; dashed lines indicate new flow.

cells are also evaluated at multiple process corners and conditions in order to create a complete library characterization. This procedure is similar to characterizing process corners based on V_T and gate sizing for overall circuit timing and power analysis. Hotspots are fixed after performing resolution enhancement techniques such as OPC and PSM. The modified layout is now stored as a characterized standard cell for use in circuit layout design and other stages in the design flow.

In Figure 4.28,[20] a yield-loss mechanism is shown where a poly line extends over the diffusion region. By increasing the extension of the poly line, yield can be improved. Statistics on misalignment and edge placement error (EPE) reveal that metal overlap on contacts can cause yield errors due to the reduced width; see Figure 4.29.[20] Therefore, an increase in metal overlap leads to improved contact and reduced resistance. The use of off-axis illumination enhances the resolution of patterns placed at a certain pitch range (these are "contacted" pitches) but not for other pitches (the "forbidden"

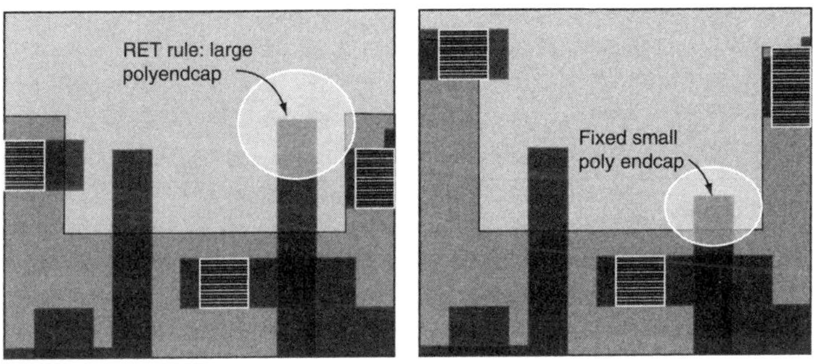

FIGURE 4.28 Poly extension rule: increasing gate extension near diffusion.

FIGURE 4.29 Increased metal over contact overlap.

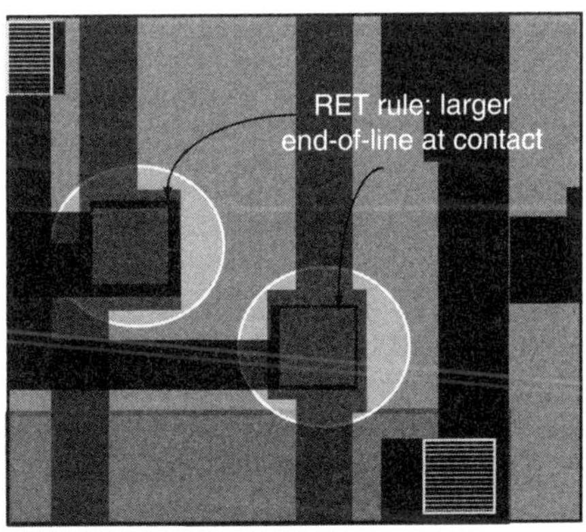

pitches). Within the diffusion region, poly lines can be modified so they will be in the contacted pitch range. For poly lines that are not in the contacted pitch range but are adjacent to active areas, dummy features are inserted to counteract the effect of diffraction; see Figure 4.30.[21]

Standard cell performance and yield are a function of three critical metrics: poly-gate length, gate width, and contact coverage.[20] Litho hotspots can trigger variation in these metrics, so next we describe some conditions that can lead to such hotspots.

The differences between cell-level and full-chip OPC tend to be large when standard cells are assumed to be independent of the environment. To reduce this disparity, dummy features (aka SRAFs; see Sec. 4.3.2) are added at cell boundaries. These dummy features are also added at the top and bottom and near poly line end contacts, as shown in Figure 4.31.[22] Intracell PSM is also performed to enhance the resolution of features. Phase shifting with 0° and 180° regions requires the neighboring geometries to be opposite in phase. Since the standard cells of the current technology generation are highly compact, it is difficult to assign alternating phases. This phase assignment conflict is resolved by moving features or by increasing

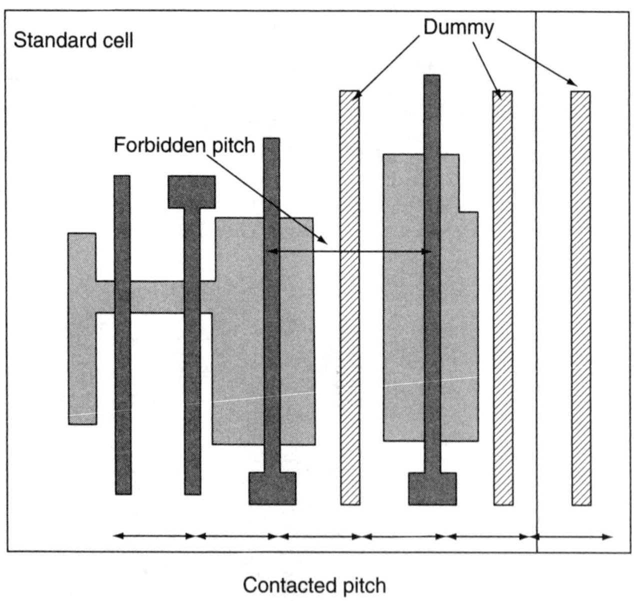

FIGURE 4.30 Poly features placed at contacted and forbidden pitches within a standard cell; dummy features are added between active areas and around cell boundary.

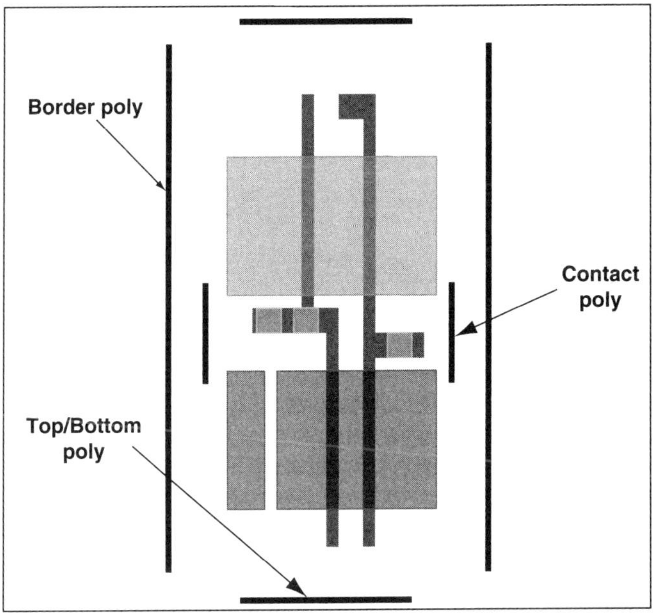

FIGURE 4.31 Three types of dummy poly features.

the width of gate regions outside the diffusion region and metal lines (cf. Figure 4.18). Diffusion rounding at regions where contacts are placed can cause yield problems. For a sample standard cell layout, the gate linewidth variation with and without manufacturability-aware changes are graphed in Figure 4.32.[21] These plots demonstrate the importance of incorporating an RET-aware methodology for standard cell characterization.

Typical DFM flows incorporate modifications to standard cell designs based on information from the manufacturing side. These standard cell layouts are used to ease the full-chip OPC process. In addition, when properly used for analysis, they can enhance the capacity of simulations to predict postsilicon circuit performance and reliability failures.

4.4.5 Mitigating the Antenna Effect

The antenna effect that results from layout structures is a phenomenon that can cause yield and reliability problems to arise during intermediate steps of the CMOS manufacturing process. Antenna effect is also known as plasma-induced gate oxide damage. As the name suggests, such problems occur when charge accumulates on metal lines; the result is a transistor with increased voltage at gate, which causes gate oxide breakdown.

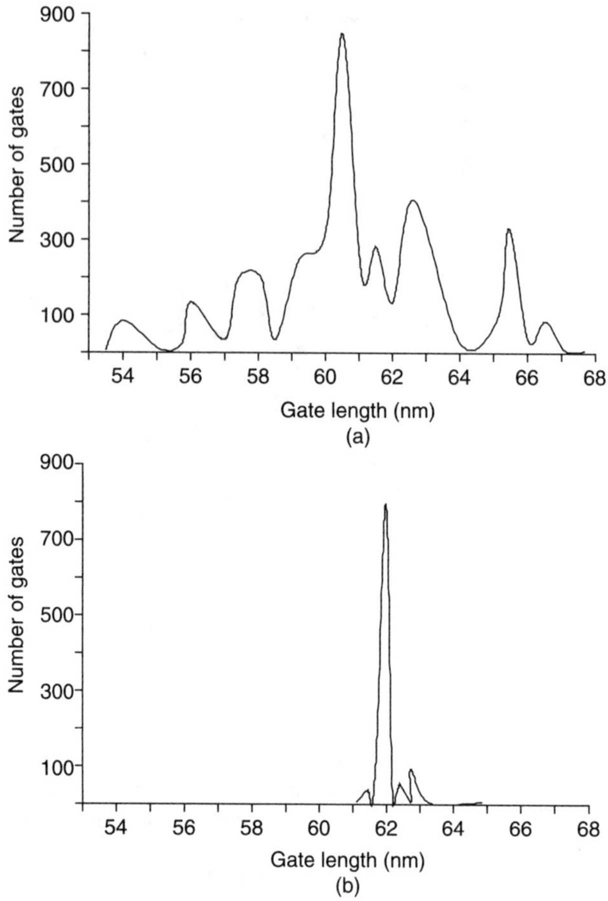

FIGURE 4.32 Gate length variation for a standard cell over the entire chip area: (a) without dummy feature; (b) with dummy feature inserted.

An example is shown in Figure 4.33(a).[23] In this case, the gate terminal is connected to the net at the lowest metal layer, which then travels through other layers to connect to a diffusion region. Figure 4.33(b) shows the state of the wafer after patterning of the first metal layer. The connection to the gate from the diffusion is open: the gate is connected to a dangling line that accumulates charges, causing breakdown of the gate oxide. Such accumulations may occur during the reactive ion etching process. This effect is associated with intermediate steps in manufacturing of the device. When the chip is fully fabricated, the gate terminal is connected to a diffusion region that acts as protection diode(s), limiting the voltage level. Reducing the length of a dangling wire connected to a gate reduces the possibility of damage to the gate; this is one of the antenna rules.

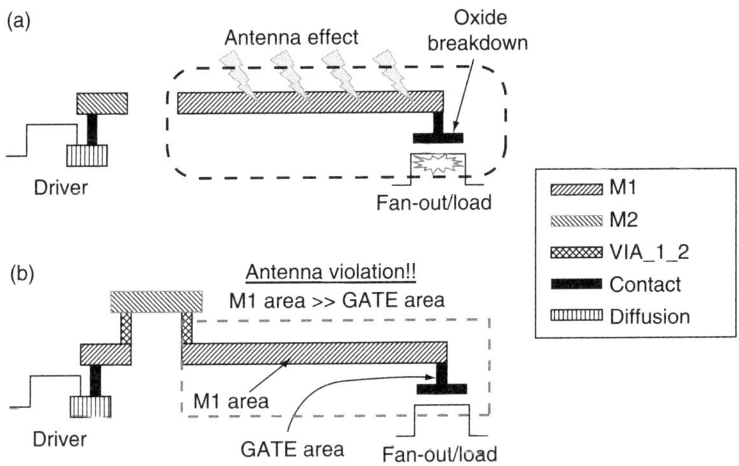

FIGURE 4.33 Antenna effects: (a) gate breakdown during construction; (b) violation of antenna rule.

Antenna rules are provided in the DRM document to avoid antenna effects. The effect on device reliability depends on the antenna dimensions. Hence, the rules typically provide an allowable ratio of metal area to gate area for each interconnect layer. Antenna rules are used by routers that assign metal layers to the nets. Techniques used by routers to perform antenna effect mitigation include change in routing order,[24] antenna diode insertion,[23] and jumper insertion.[25] Because antenna effects are primarily due to the presence of lower metal layers connecting to the gate terminal, routing techniques can prevent this by allowing only higher metal layers to be connected to the gate terminal, as shown in Figure 4.34(a).[26,27]

The technique of antenna diode insertion creates a diffusion region close to the gate terminal that can form a diode, thus limiting gate oxide breakdown (see Figure 4.34(b)). This diode insertion technique can be implemented as standard cell methodology to completely remove all antenna violations throughout the layout, thus relieving the router from having to consider antenna effects at all. The main drawback of this approach is that the extra capacitance of the inserted diode increases cell delay. For this reason, antenna diodes are inserted only on critical nets that are vulnerable to antenna effects.

Jumper insertion involves the minimization of lower layer metal area near the gate terminal. This is managed, as shown in Figure 4.34(c), by moving affected portions of the node to upper interconnect layers. Doing so decreases the ratio of lower layer metal interconnect area to the gate area.

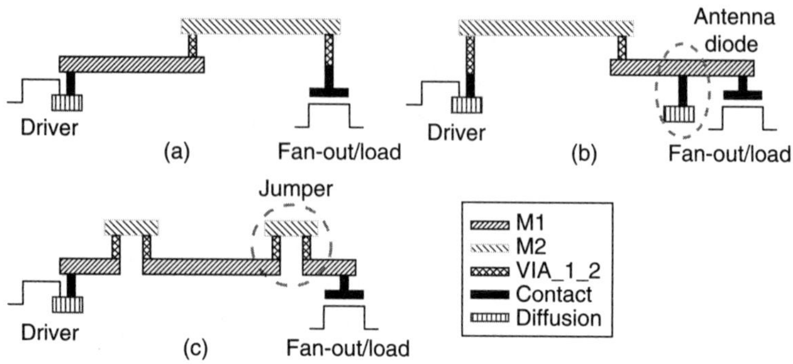

FIGURE 4.34 Mitigation of antenna effect: (a) change in routing order; (b) antenna diode insertion; (c) jumper insertion.

4.4.6 Placement and Routing for DFM

Other than DRC, placement and routing form the major workhorse of physical design tools used today. Lithography-aware placement and routing have become an integral part of the ASIC design flow because they optimize design constraints under a specific lithographic variability.

Placement tools model the impact of a process parameter on the position and orientation of the standard cell within the layout. Given data on process parameter variability, placement algorithms implement new cost functions to be minimized while obtaining an optimum placement. As seen in Sec. 2.3.3 (see Figure 2.20), modern lithography systems use the step-and-scan approach to expose regions of the wafer individually.[22] These small regions or fields are exposed from one side to the other. Lens aberration parameters—as quantified by Zernike's coefficients (see Sec. 3.2.1.3), which capture divergence of nominal on-axis light behavior—change during scanning. This induces CD error for lines that are oriented in the same direction as the scanning (horizontal, in this case). The variations in average CD for different types of standard cells are plotted in Figure 4.35.[28] Linewidth change in standard cells results in an input-to-output delay; this change in delay is plotted in Figure 4.36.[28] Information on the delay change due to aberration-induced linewidth variation can be used to optimize the placement of standard cells, thus maximizing design timing yield. The process flow for this technique is illustrated in Figure 4.37.[28]

Routers have increasingly adopted novel flows that incorporate design rules based on process knowledge. These flows use improved process windows that create better rules for optimizing design constraints. These new routing strategies include restricted pitches for each metal layer—to eliminate forbidden pitches; metal-width-

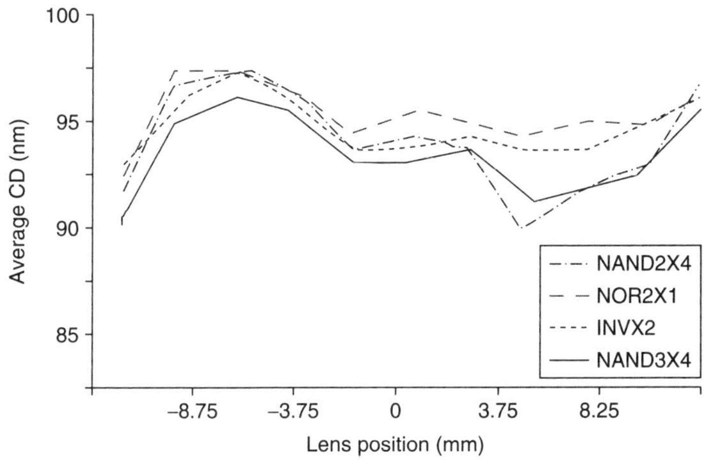

Figure 4.35 Average CD variation across lens field.

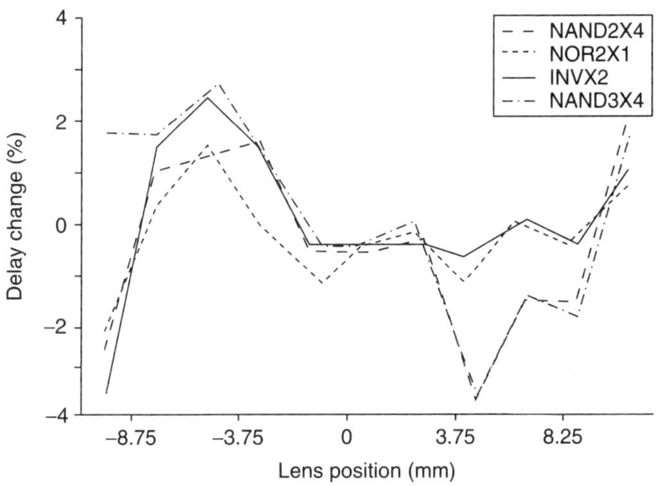

Figure 4.36 Change in average delay with position of the lens center.

dependent spacing—to reduce increased capacitive effects; restricted metal line orientations—to regularize layouts; and increased spacing at vulnerable necking and bridging areas—to prevent catastrophic defects (see Figure 4.38). One approach to litho-friendly routing uses edge placement error to identify hotspots.[29] Edge placement error is defined as the difference in the position of the edge in the mask and on the aerial image, and it is found by using lithographic simulation

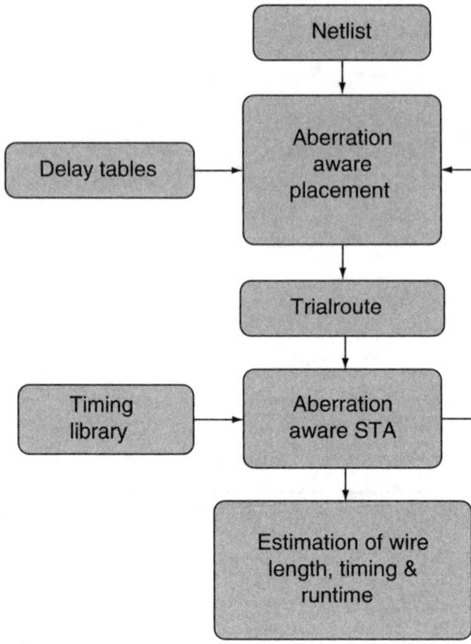

FIGURE 4.37 Lens-aberration-aware placement for timing yield.

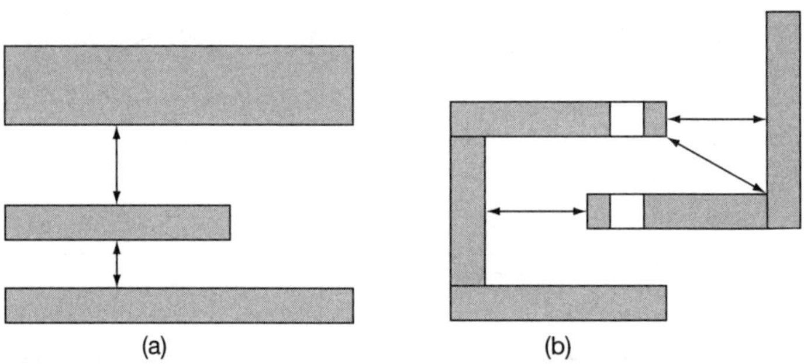

FIGURE 4.38 New rules for effective manufacturability-aware routing: (a) linewidth-dependent spacing rules; (b) increased end-of-line and corner spacing.

based on edge look-up tables (see Sec. 2.4.1.4). This approach, called RET-aware detailed routing or RADAR, is a noniterative process that rips and reroutes after generating blockage data for hotspot regions. The EPE of the new route is again estimated to determine whether the new route or the old one should be kept. A step-by-step flowchart for this approach is shown in Figure 4.39.[29]

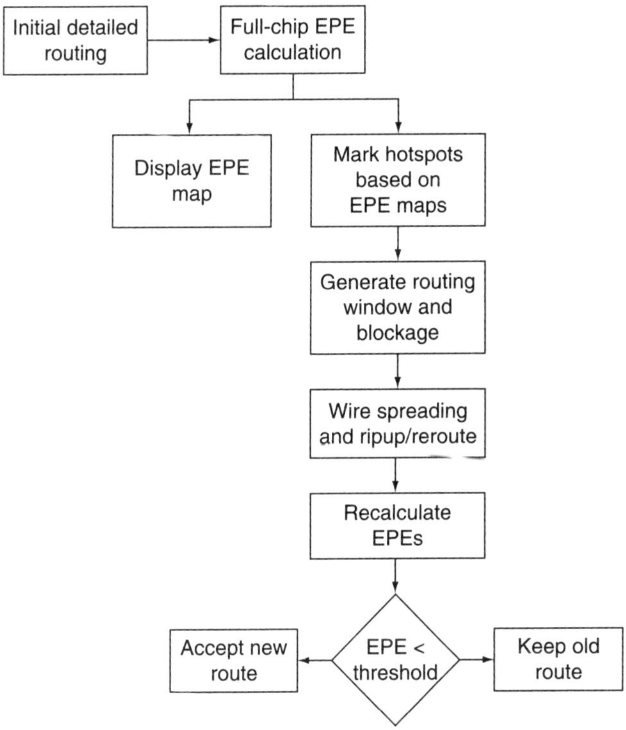

FIGURE 4.39 Flowchart for RET-aware detailed routing (RADAR).

4.5 Advanced Lithographic Techniques

The techniques described in previous sections have been incorporated into the CAD flows employed by various design houses. Yet even with printability enhancement techniques modeled on process information, manufacturability problems persist in today's designs. These problems involve performance degradation, yield, and other economic factors. Mask patterns that are tuned by OPC and other resolution enhancement techniques almost always satisfy the overall constraint of optimizing printability. Today, however, designers require approaches to printability that are based on local constraints pertaining to circuit performance. In particular, the use of nonrectangular devices and interconnect features can lead to performance degradation and increased reliability concerns. There is always the need for improvements in the CAD tools used to predict shapes on wafer so that better-quality designs can be produced. One advanced lithography technique, double patterning, is described next.

4.5.1 Double Patterning

Since the advent of 45-nm technology, a 193-nm light source has been used to transfer images from mask to wafer. As design density grows, the need arises for producing patterns that push the resolution limit. This increased resolution requirement is observed in memory manufacturing. In Rayleigh's equation, $R = k_1(\lambda/\mathrm{NA})$, the k_1 factor controls the resolution of images being printed on wafer. There are only two options for improving resolution with the same set of exposure steps. The first, *immersion* lithography, uses high-index fluids between the projection optics and the wafer with the same illumination source. The second option, *extreme ultraviolet* (EUV) lithography, uses an illumination source of 13.5-nm wavelength.[30] Both of these process techniques feature k_1 values equal to or less than 0.5; however, the practical limit of the k_1 factor is 0.25, which cannot be attained by these methods. Also, because of technical hurdles in manufacturing a flare-free EUV source, the only alternative is to decompose the mask into multiple layers. Such *dual-pattern lithography* (DPL) is now being used in commercial productions. In this approach, a mask is split into two separate masks that are exposed in two separate steps. Hence the pitch size doubles, which enables resolution below 30 nm without any technical barriers (see Figure 4.40).[31] Unlike all other methods, double patterning makes it possible to go below the "limit" value $k_1 = 0.25$ because the minimum pitch constraint is relaxed.[31] Half-pitch resolution as low as 18 nm has been obtained with the combined use of double patterning and immersion lithography. With these techniques come new challenges for mask design and processing flows as well as increased tightening of mask error budgets (owing to overlay requirements).

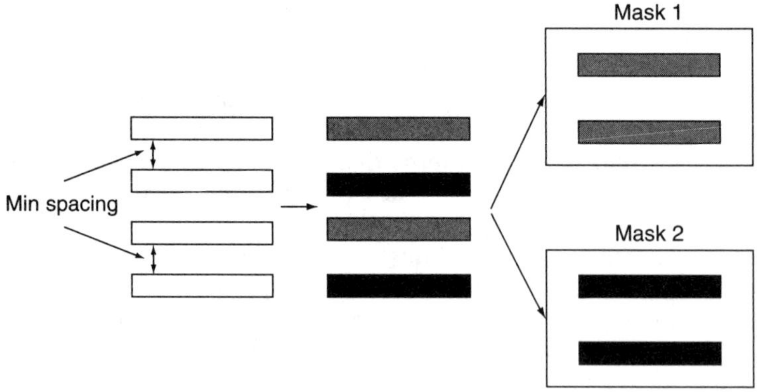

FIGURE 4.40 Dual-pattern lithography (DPL): a single mask is decomposed into two masks, where minimally spaced features are placed on different masks; DPL increases pitch size for each of the resulting masks.

Layout decomposition for dual-pattern lithography is approached differently for positive and negative processes. Positive processes use light-field masks, where dark regions indicate the patterns that need to be printed on the wafer. Decomposition in this case is similar to the phase assignment problem in the alternating PSM technique. Polygons separated by a minimum pitch are assigned different colors, which indicates the need to move one of them to another mask layer[32] (see Figure 4.40). For a negative process, spaces between metal lines are colored alternatively. Then the mask is decomposed to print spaces in two stages. The decomposition technique resembles that used for the positive process, but assigning alternating colors will not produce a solution because spaces do not have a regular boundary. Furthermore, when lines have spaces on either side that are being printed using different masks, the consequent overlay problems can create bottlenecks due to excessive linewidth variation. Figure 4.41 illustrates the difference in decomposition between positive and negative tone processes.

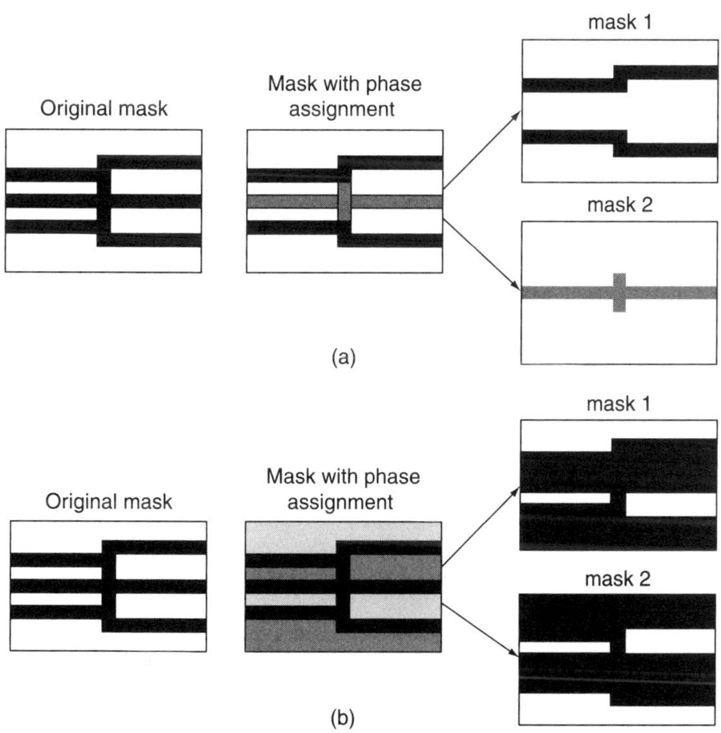

FIGURE 4.41 Phase assignment and mask decomposition for (a) positive tone process and (b) negative tone process.

For the positive process, polygon coloring may not lead to a complete dual-colored solution, as with the example shown in Figure 4.42.[33] This problem arises because of the high pattern density and nonconformity of patterns to a particular orientation in today's designs. Coloring conflicts of this sort are solved by splitting the polygon in two, as in Figure 4.43,[33] to generate *stitches*. However, splitting the polygon moves half of it to another layer, which can create overlay and yield problems. Extra metal or jogs must be added in order to maintain connectivity at stitched locations. After splitting, a few unresolved regions may exist that can be corrected only by a complete redesign of the layout. It is therefore important to solve the problem by simultaneously minimizing the number of conflicts and reducing the number of stitches required. These factors have been used as cost functions in many of the techniques described in the literature.

Several different process techniques are classified as double patterning: double exposure, double exposure–double etch, and self-aligned spacer. *Double exposure* consists of two separate exposure steps with two different photo masks but on a single photoresist layer. This technique involves two masks, two exposure steps, one photoresist coat, and one each develop, etching, and cleaning stage. Double exposure is used for patterns on the same layer that are of different pitches and/or irregular density. Each stage in this technique patterns features that are perpendicular to each other for improved resolution. As shown in Figure 4.44, only one photoresist coating is used for both the exposure stages. The wafer undergoes resist coat

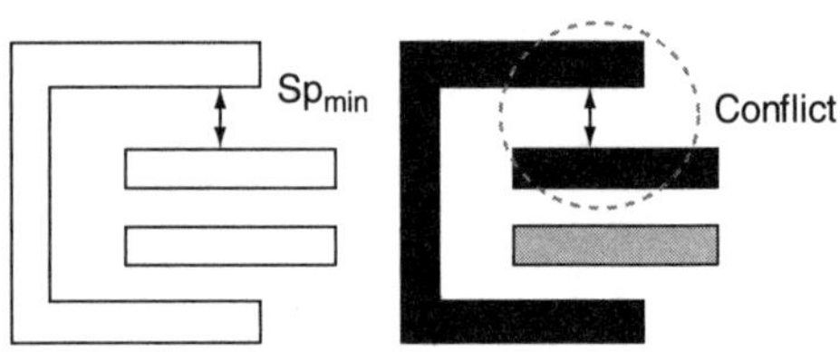

FIGURE 4.42 DPL decomposition: example of a phase assignment conflict.

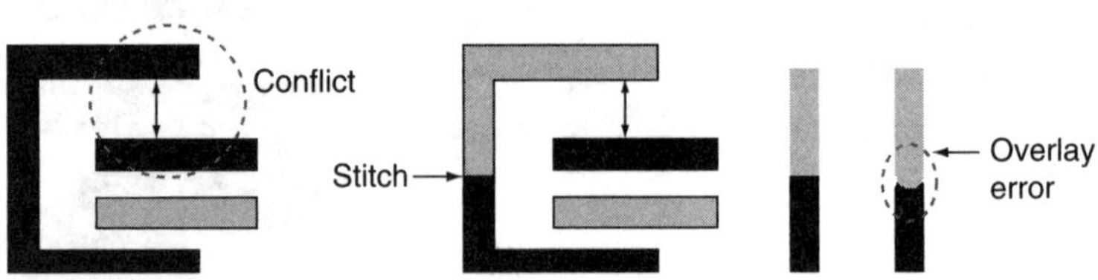

FIGURE 4.43 A feature being split to resolve an assignment conflict; the requisite stitch has led to an overlay error.

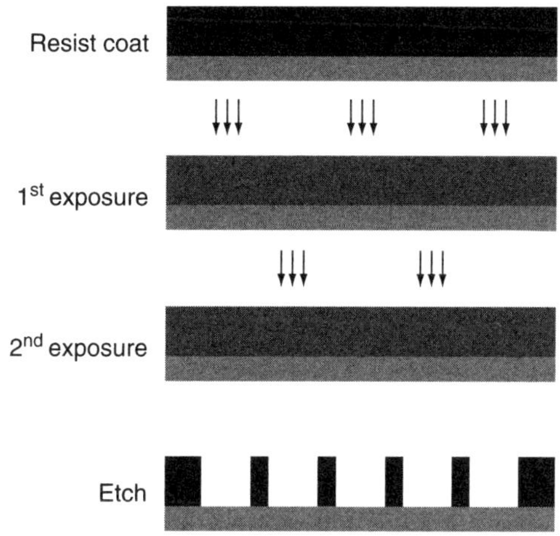

FIGURE 4.44 The double exposure process.

followed by two exposure stages without being moved away form the exposure station. All settings for illumination and the projection system are modified for each wafer. The develop, etching, and resist strip stages finally generate the patterns. The image intensity is the sum of first and second exposure stages. The fundamental resolution cannot go below $k_1 = 0.25$ because the photoresist response is the sum of the individual stage intensities.

The *double exposure–double etch* (or litho-etch-litho-etch) technique is a process that involves two resist coats, two exposures, two development cycles, and two etch and resist strip steps (see Figure 4.45). Unlike with the double exposure technique, here the combination of the two exposures with the other stages in between make the photoresist response a nonlinear function of the illumination intensity. Therefore, a resolution limit below $k_1 = 0.25$ can be achieved. Figure 4.45 shows the processing steps for a positive resist, where lines are patterned; Figure 4.46 shows the steps for a negative resist, where spaces are patterned. For both types of imaging, alternate patterns/spaces of the same color are exposed in two stages. In this technique, two hard mask coats are applied before the exposure stages. After a resist coat, the first exposure transfers one set of patterns to the first hard mask layer. Next, another layer of resist is coated over the patterns so formed, followed by a second exposure that transfers the next set of patterns (in between the patterns already present) onto the second hard mask. Each coat is removed using an appropriate etchant after exposure to create the required pattern. Because there is a delay between the first hard mask coat and the second photoresist coat, variation in pattern formation is a concern. To prevent the first hard mask from etching, the resist surface is hardened to control linewidth variation.

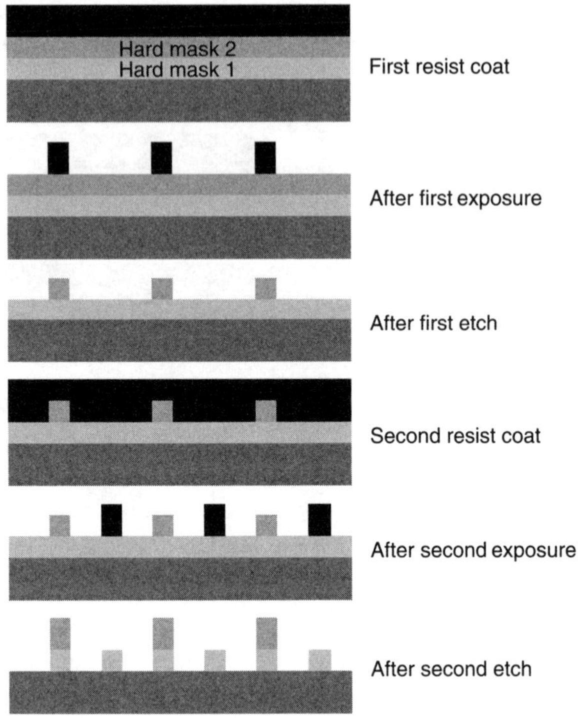

FIGURE 4.45 Double exposure–double etch DPL process: positive tone for line features.

In the *self-aligned spacer DPL* (SADP) process, special film layers (the spacers) formed on the sidewall of patterned features are used to double the density of features printed on the wafer. After the first pattern is printed on the wafer, a layer of spacer material is formed by deposition or by reaction with the prepatterned layer. This stage is followed by an etching step that removes material from all horizontal surfaces but leaves material on the sidewalls, as shown in Figure 4.47. The original patterned layer is removed, leaving two spacers for each line patterned in the first step. This technique doubles the pattern density, and its primary application is to gate patterning at smaller technology nodes. The self-aligned spacer technique avoids overlay-induced linewidth errors, so FinFETs and tri-gate transistors that require narrow gate lengths are ideal candidates for its application.[34] Spacer formation, spacer pattern collapse, and etching results are areas of concern with this technique.

Significant challenges in double patterning are linewidth variability and the formation of defects due to overlay errors. Regular patterns, such as those characteristic of memory cells, can easily be

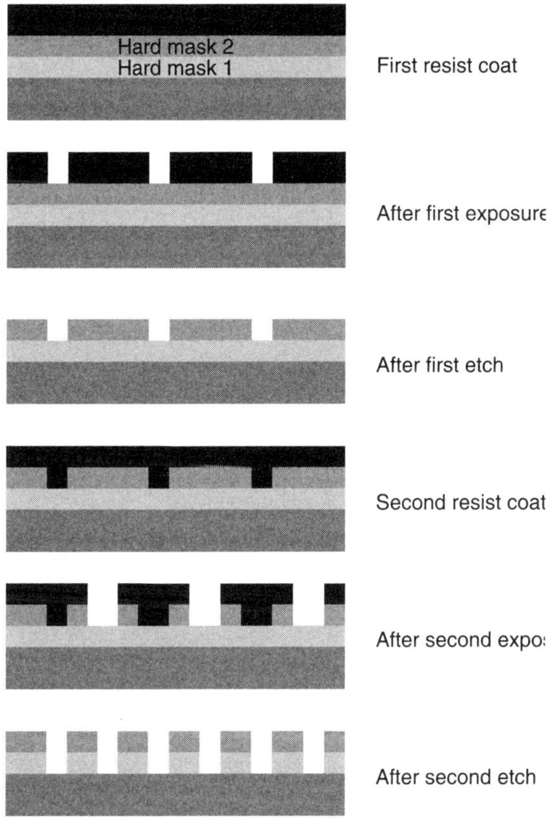

Figure 4.46 Double exposure–double etch DPL process: negative tone for space features.

decomposed into two masks. However, the task is far more complex for logic circuits, which do not exhibit regularity in distance and orientation. Because straightforward two-color solutions cannot be obtained for industrial logic designs, the only other option is to modify the layout by increasing distances between patterns that are colored differently. This leads to an increase in chip area, which adds to the chip's cost. Increased cost due to extra mask and process stages is another obvious concern. As we have seen, process errors may be found at each stage of the lithographic process. With the increased number of stages necessitated by double patterning, the probability of process errors increase and consequently yield is reduced. Another area of concern is foundry throughput, since double exposure may entail a reduction in the number of wafers fabricated per hour. Yet despite all of these limitations, double patterning is viewed by many as a "savior" technique for increasing achievable resolution (by

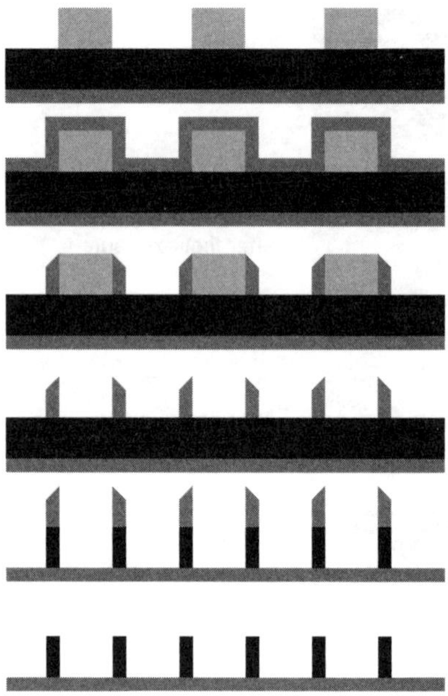

FIGURE 4.47 Self-aligned spacer DPL scheme: process flow.

reducing the minimum pitch constraint). Researchers are experimenting with triple and quadruple patterning technique in attempts to further push the boundaries of resolution.[35]

4.5.2 Inverse Lithography

Model-based OPC with SRAF insertion is today's preferred resolution enhancement technique for increasing printability and the pattern fidelity of mask features. Because a 193-nm laser light source is still being used for designs at the 32-nm technology node, newer techniques that further improve the image transfer process have been developed. *Inverse lithography,* as the name implies, seeks to obtain the inverse of the required wafer image in order to reduce variation between intent and imprint. This concept resembles the image retrieval-and-reconstruction strategy of using the blurred image (seen through the photographic sensor) to reconstruct the original image.[36,37]

Unlike OPC, whose modification algorithm targets the mask pattern, inverse lithography attempts to reconstruct the mask by using information about the required image on wafer and process information. Consider the following variables of the lithography process: α, the mask pattern; ξ, the target pattern on wafer; f, the combined imaging and resist functions; and ω, the final wafer image. In terms of the normal lithographic process, the final wafer image can be described as follows:

$$\omega = f(\alpha) \tag{4.1}$$

Now, by ideal inverse lithography (see Figure 4.48 for the flowchart), the required mask pattern can be written as the inverse of the target pattern on the wafer:[38]

$$\alpha^* = f^{-1}(\xi) \tag{4.2}$$

Here α^* is the optimal pattern required on the mask to create ξ on the wafer. Equation (4.2) does not, of course, accurately describe the real-world scenario of a complex lithography process. Since the resist dissolution process is nonlinear and since the resist contrast is high, several different mask patterns could lead to the same image on the wafer; in other words, there is no direct inverse of the function f. Also, the relation $\xi = f(\alpha)$ does not actually hold in view of the rectangular geometry required for mask manufacturing. So, in order to model the inverse lithography technique more realistically, an iterative optimization program has been devised that uses information on the exposure system and other processing parameters.[39]

In Chapter 2 we learned that the key elements of the lithography system are its exposure process and resist development. The exposure process involves the transfer of the mask pattern onto the wafer through an optical system. The image on the wafer causes resist development based on the intensity of the image profile and the resist contrast. If the mask function (in the spatial domain) is given by $M(x, y)$ and if the image transfer function is given by $Tf(x, y)$, then the aerial image $\mathrm{AI}(x, y)$ of the patterns on the mask is given as the square of the magnitude of the convolution between $M(x,y)$ and $Tf(x, y)$: $\mathrm{AI}(x,y) = |Tf * M(x,y)|^2$. Light incident on the wafer causes photoresist

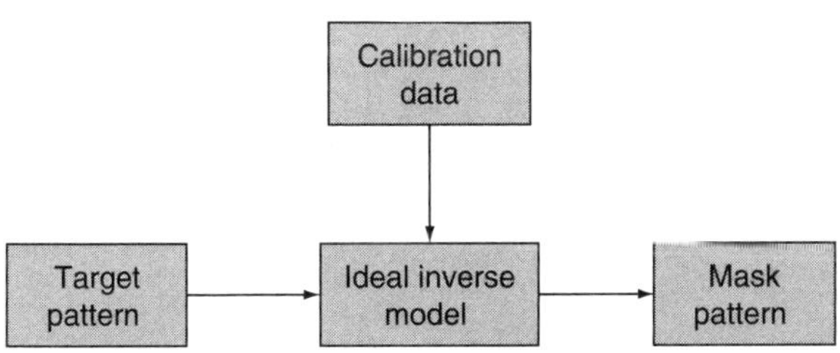

Figure 4.48 Inverse lithography: technology flow.

to react as a function of the light's intensity. This process depends on the dissolution threshold of the resist, so it can be assumed to a sigmoid function $sig(z)=1/\{1+\exp[-a(z-\tau)]\}$ producing the final resist image I(x,y). Now, if the target image is represented as $\hat{I}(x, y)$, then the optimization function can be written as follows:[40]

$$I(x,y)=\left\{1+\exp\left[-a\left(\left|Tf*M(x,y)\right|^2-\tau\right)\right]\right\}^{-1} \tag{4.3}$$

The goal here is to minimize the error between the current image on the wafer and the required image on the mask. The term η (eta)

$$\begin{aligned} &\text{Minimize} \quad \eta=\sum_{x,y}\left(I(x,y)-\hat{I}(x,y)\right)^2 \\ &\text{such that} \quad M(x,y)\in\begin{cases}(0,1), & for\,\text{BIM}\\ (-1,0,1), & for\,\text{PSM}\end{cases}\end{aligned} \tag{4.4}$$

denotes the mean square error value. Because the required mask function could be either binary or phase shifting, the corresponding constraints are also listed. The practical way to solve such an inverse problem incorporates an iterative perturbation algorithm that starts with a suitable guess for the final image. For each perturbation of the initial image, an aerial image will be calculated and compared to the target pattern, and the differences will be noted. The overall goal of this approach is to minimize the differences between the two aerial images. Other optimization criteria can be added to this methodology in order to form a global cost function that can be used in the iterative process. A simplified flowchart of this procedure is shown in Figure 4.49.[39] Process calibration data provides information on the

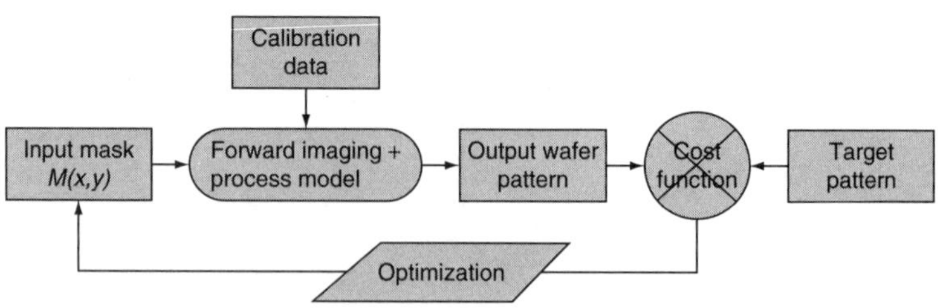

FIGURE 4.49 Practical optimization flow for solving the inverse lithography problem.

imaging system parameters, projection optics, and resist functions, which aids in the creation of a good forward imaging model.

A number of different solutions for inverse lithography technology (ILT) have been proposed in the literature. All these techniques attempt to find the ideal required mask pattern based on the iterative optimization method just described. One such technique pixilates the mask into equal-sized regions that are well below the system's resolution limit. Each discrete pixel is randomly assigned a particular phase to generate the required mask pattern[41,42] (see Figure 4.50).[42] Gradient-based efficiency has been incorporated into the random pixel-flip technique to improve the solutions obtained.[38] Genetic algorithm and simulated annealing techniques have also been suggested as possible solutions.[43]

Another class of techniques divides the layout into different regions that have multiple transmission properties. A technique that closely resembles OPC has also been suggested to solve the inverse lithography problem.[44] Instead of running a script to perform pattern segmentation, this new technique partitions the pattern into different regions according to topography: pattern edge, pattern corner, or pattern end; see Figure 4.51.[44] Iterative movement is directed within a

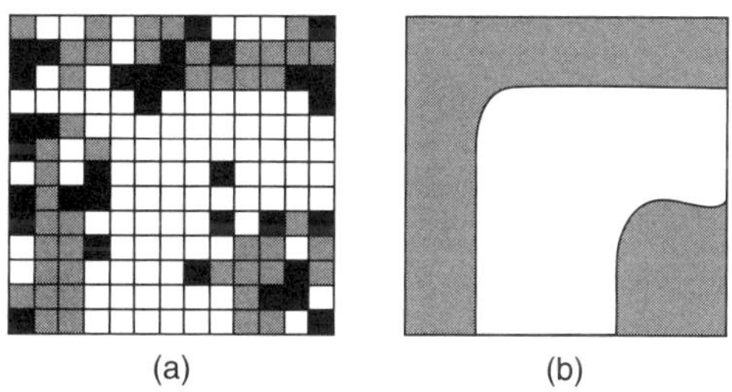

FIGURE 4.50 (a) Pixel pattern using phases −1, 0, 1 (i.e., opaque and two out-of-phase unit transmissions); (b) calculated wafer contour.

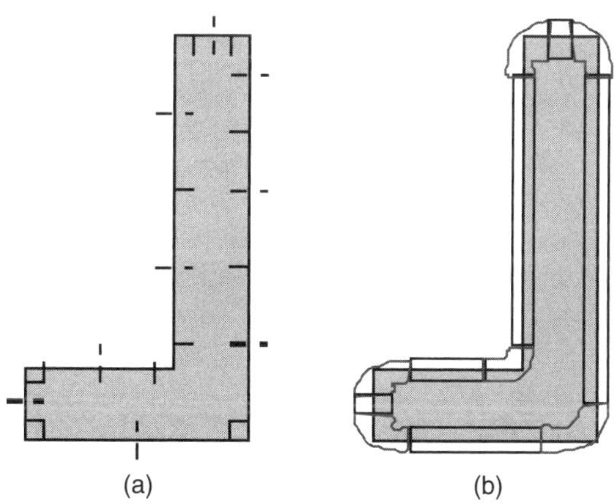

FIGURE 4.51 (a) OPC segmentation and sampling; (b) ILT topography.

particular portion of the image under certain constraints. Because the mask pattern newly created by the image reconstruction technique has been optimized to the target pattern, it offers an alternative to model-based OPC. Another difference between this technique and OPC is the existence of nonrectangular sections in the pattern, as shown in Figure 4.52.[44] Such contours are reprocessed to create edges at 90° and 45° angles.

The main drawback of ILT is that it induces minute feature changes that complicate the mask-writing process.[44] As discussed previously (see Sec. 4.3.1), increasing the number of "shots" renders extensive model-based OPC a nightmare for mask manufacturing, and likewise for masks obtained using inverse lithography. This is a grave problem with pixelized masks. Recreating images with rounded edges and corners on the mask exponentially increases the mask manufacturing cost and time. One way to minimize this problem is to modify the final ILT mask so that its elements include only horizontal, vertical, and other rectilinear features (see Figure 4.52(d)). Inverse lithography technology solutions are being implemented by design houses as an addendum to OPC in order to reduce defects and manufacturing costs and to increase lithographic control.

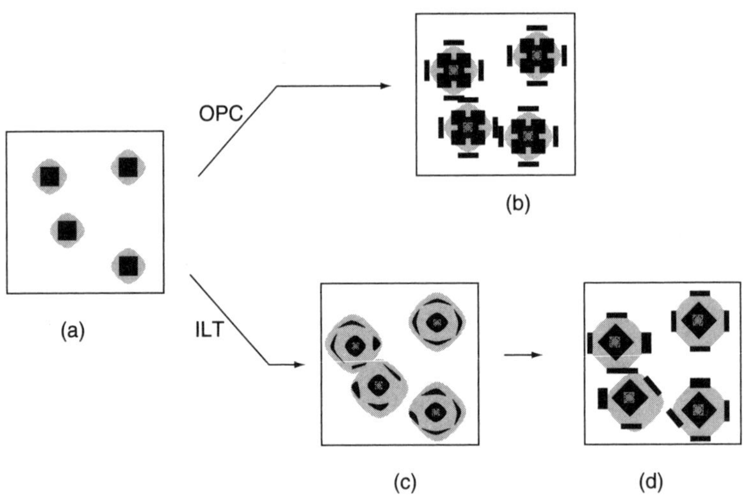

FIGURE 4. 52 OPC and ILT compared: (a) uncorrected mask tile consisting of four semi-isolated vias; (b) mask corrected by conventional segmentation-based OPC and rule-based SRAFs; (c) mask corrected by single iteration of ILT-based on pixel inversion; (d) final simplified mask aligned with 45° and 90° line segments.

4.5.3 Other Advanced Techniques

Investigations have been performed on other techniques to optimize existing lithography simulation setups in order to improve printability while keeping mask costs low. One such example includes the use of free form source shapes based on Source-Mask Optimization (SMO) technique. Traditional well known illumination shapes include conventional, annular, quasar and others as shown in Figure 4.22. It is also known that any irregularity in source shape can lead to increase CD variation. Researchers have been working on another new technique called *source mask optimization* (SMO). Source shapes that do not confine to traditional illumination shapes have been proposed to improve lithographic printability. Advanced illumination systems are used to create such free form shapes that are tied to a particular process[45]. Researchers have shown that by optimizing the shape of the illumination source based on the mask for a particular process node, effective CD control is observed. Various advancements in SMO are being displayed by lithographic system manufacturers and CAD tool development companies alike. Detailed explanation in each such technique is not within the scope of this book.

4.6 Summary

In this chapter we described the design rules manual and how it is created. We then explored the topic of design rules check using rule-based and model-based techniques. It was observed that restricted design rules are typically checked with rule-based methods whereas remaining optical printability issues are checked with model-based methods. We discussed resolution enhancement techniques in detail as well as the extent of their use in today's designs. Design for manufacturability has become a prevalent process in layout generation, so several methodologies that employ DFM were also discussed. Finally, we described dual-pattern lithography and explained how it improves resolution. The general capabilities of available DFM tools and their usage in current designs and methodologies were also explored.

References

1. Chris. A. Mack, *Field Guide to Optical Lithography*, SPIE Press, Bellingham, WA, 2006.
2. Chris. A. Mack, *Fundamental Principles of Optical Lithography*, Wiley, New York, 2007.
3. N. B. Cobb, "Fast Optical and Process Proximity Correction Algorithms for Integrated Circuit Manufacturing," Ph.D. thesis, University of California, Berkeley, 1998.
4. L. W. Leibmann, S. M. Mansfield, A. K. Wong, M. A. Lavin, W. C. Leipold, and T. G. Dunham, "TCAD Development for Lithography Resolution Enhancement," *IBM Journal of Research and Development* **45**(5): 651–666, 2001.
5. A. K. Wong, *Resolution Enhancement Techniques in Optical Lithography*, SPIE Press, Bellingham, WA, 2001.

6. V. Wiaux, P. K. Montgomery, G. Vandenberghe, P.Monnoyer, K. G. Ronse, W. Conley, L. C. Litt, et al., "ArF Solution for Low-k_1 Back-End Imaging," *Proceedings of SPIE Optical Microlithography* **5040**: 270–281, 2003.
7. ASML, "CPL Technology," http://www.asml.com (2010).
8. J. A. Torres and D. Chow, "RET Compliant Cell Generation for Sub-130nm Processes," *Proceedings of SPIE* **4692**: 529–539, 2002.
9. C. Mack, "The Lithography Expert: Off-Axis Illumination," in *Micro-Lithography World,* PennWell, Farmington Hills, MI, 2003.
10. T. Matsuo, A. Misaka, and M. Sasago, "Novel Strong Resolution Enhancement Technology with Phase-Shifting Mask for Logic Gate Pattern Fabrication," *Proceedings of SPIE Optical Microlithography* **5040**: 383–391, 2003.
11. Jie Yang, Luigi Capodieci, and Dennis Sylvester, "Layout Verification and Optimization Based on Flexible Design Rules," *Proceedings of SPIE* **6156**: A1–A9, 2006.
12. K. Lucas, S. Baron, J. Belledent, R. Boone, A. E. Borjon, C. Couderc, K. Patterson, et al., "Investigation of Model-Based Physical Design Restrictions," *Proceedings of SPIE* **5756**: 85–96, 2005.
13. L. Liebmann, A. Barish, Z. Baum, H. Bonges, S. Bukofsky, C. Fonseca, S. Hale, et al., "High-Performance Circuit Design for the RET-Enable 65-nm Technology Node," *Proceedings of SPIE* **5379**: 20–29, 2004.
14. Frank E. Gennari and Andrew R. Neureuther, "A Pattern Matching System for Linking TCAD and EDA," in *Proceedings of ISQED*, IEEE, San Jose, CA, 2004, pp. 165–170.
15. D. Perry, M. Nakamoto, N. Verghese, P. Hurat, and R. Rouse, "Model-Based Approach for Design Verification and Co-Optimization of Catastrophic and Parametric-Related Defects due to Systematic Manufacturing Variations," *Proceedings of SPIE* **6521**: E1–E10, 2007.
16. J. Andres Torres, "Layout Verification in the Era of Process Uncertainty: Target Process Variability Bands vs Actual Process Variability Bands," *Proceedings of SPIE* **6925**: 692508.1–692509.8, 2008.
17. M. Miyairi, S. Nojima, S. Maeda, K. Kodera, R. Ogawa, and S. Tanaka, "Lithography Compliance Check Considering Neighboring Cell Structures for Robust Cell Design," *Proceedings of SPIE* **7379**: 737911.1–737911.9, 2009.
18. J. A. Bruce, E. W. Conrad, G. J. Dick, D. J. Nickel, and J. G. Smolinski, "Model-Based Verification for First Time Right Manufacturing," *Proceedings of SPIE* **5756**: 198–207, 2005.
19. Luigi Capodieci, "From Optical Proximity Correction to Lithography-Driven Physical Design (1996–2006): 10 Years of Resolution Enhancement Technology and the Roadmap Enablers for the Next Decade," *Proceedings of SPIE* **6154**: 615401.1–615401.12, 2006.
20. Paul de Dood, "Impact of DFM and RET on Standard-Cell Design Methodology," in *Proceedings of Electronic Design Processes Workshop*, IEEE, Monterey, CA, 2003, pp. 62–69.
21. H. Muta and H. Onodera, "Manufacturability-Aware Design of Standard Cells," *Transactions of IEICE* **E90-A**(12): 2682–2690, 2007.
22. P. Gupta, F. L. Heng, and M. Lavin, "Merits of Cellwise Model-Based OPC," *Proceedings of SPIE* **5379**: 182–189, 2004.
23. P. H. Chen, S. Malkani, C.-M. Peng, and J. Lin, "Fixing Antenna Problem by Dynamic Diode Dropping and Jumper Insertion," in *Proceedings of ISQED*, IEEE, San Jose, CA, 2000, pp. 275–282.
24. Z. Chen and I. Koren, "Layer Reassignment for Antenna Effect Ninimization for 3-Layer Channel Assignment," in *Proceedings of International Symposium on Defect and Fault Tolerance in VLSI Systems*, IEEE, Boston, 2000, pp. 77–85.
25. L.-D. Huang, X. Tang, H. Xiang, D. F. Wong, and I-Min Liu, "A Polynomial Time-Optimal Diode Insertion/Routing Algorithm for Fixing Antenna Problem [IC Layout]," *Transactions on Computer-Aided Design of Integrated Circuits and Systems* **23**: 141–147, 2004.
26. T.-Y. Ho, Y.-W. Chang, and S.-J. Chen, "Multilevel Routing with Antenna Avoidance," in *Proceedings of International Symposium on Physical Design*, ACM, Phoenix, AZ, 2004, pp. 34–40.

27. Di Wu, Jiang Hu, and Rabi Mahapatra, "Coupling Aware Timing Optimization and Antenna Avoidance in Layer Assignment," *Proceedings of ISPD*, ACM, San Francisco, 2005, pp. 20–27.
28. A. B. Kahng, C. H. Park, P. Sharma, and Q. Wang, "Lens Aberration Aware Placement for Timing Yield," *Proceedings of ACM Transactions on Design Automation of Electronic Systems* **14**(1): 1–26, 2009.
29. Joydeep Mitra, Peng Yu, and David Z. Pan, "RADAR: RET-Aware Detailed Routing Using Fast Lithography Simulations," in *Proceedings of Design Automation Conference*, ACM, New York, 2005, pp. 369–372.
30. ASML, Martin van den Brink, "Shrink, an Expanding (Litho) Market," presentation at Industry Strategy Symposium, Halfmoon Bay, CA, 2007.
31. Mircea Dusa, Jo Finders, and Stephen Hsu, "Double Patterning Lithography: The Bridge between Low k_1 ArF and EUV," in *Microlithography World*, PennWell, Farmington Hills, MI, 2008.
32. Mircea Dusa, John Quaedackers, Olaf F. A. Larsen, Jeroen Meessen, Eddy van der Heijden, Gerald Dicker, Onno Wismans, et al., "Pitch Doubling through Dual Patterning Lithography: Challenges in Integration and Litho Budgets," in *Proceedings of SPIE* **6520**: 65200G.1–65200G.10, 2007.
33. K. Yuan, K.-S. Yang, and D. Pan, "Double Patterning Layout Decomposition for Simultaneous Conflict and Stitch Minimization," in *Proceedings of International Symposium on Physical Design*, ACM, New York, 2009, pp. 107–114.
34. Yang-Kyu Choi, Ji Zhu, Jeff Grunes, Jeffrey Bokor, and Gabor. A. Somorjai, "Fabrication of Sub-100nm Silicon Nanowire Array by Size Reduction Lithography," *Journal of Physical Chemistry B* **107**: 3340–3343, 2003.
35. Christopher Cork, Jean-Christophe Madre, and Levi Barnes, "Comparison of Triple-Patterning Decomposition Algorithms Using Aperiodic Tiling Patterns," *Proceedings of SPIE* **7028**: 702839.1–702839.7, 2008.
36. H. M. Sheih, C. L. Byrne, and M. A. Fiddy, "Image Reconstruction: A Unifying Model for Resolution Enhancement and Data Extrapolation—Tutorial," *Journal of Optical Society of America A* **23**(2): 258–266, 2006.
37. K. M. Nashold and B. E. A. Saleh, "Image Construction through Diffraction-Limited High-Contrast Imaging Systems: An Iterative Approach," *Journal of Optical Society of America A* **2**(5): 635–643, 1985.
38. A. Poonawala and P. Milanfar, "OPC and PSM Design Using Inverse Lithography: A Nonlinear Optimization Approach," *Proceedings of SPIE* **6154**: **3**: 61543H.1–61543H.14, 2006.
39. S. H. Chan, A. K. Wong, and E. Y. Lam, "Inverse Synthesis of Phase-Shifting Mask for Optical Lithography," in *Proceedings of OSA Topic Meeting on Signal Recovery and Synthesis*, Optical Society of America, Washington, DC, 2007, pp. 1–3.
40. L. Pang, Y. Liu, and D. Abrams, "Inverse Lithography Technology (ILT): What Is the Impact to Photomask Industry?" *Proceedings of SPIE* **6283**: 62830X.1–62830X.11, 2006.
41. J. Zhang, W. Xiong, Y. Wang, Z. Yu, and M. Tsai, "A Highly Efficient Optimization Algorithm for Pixel Manipulation in Inverse Lithography Technique," in *Proceedings of ICCAD*, IEEE, New York, 2008, pp. 480–487.
42. Yuri Granik, "Fast Pixel-Based Mask Optimization for Inverse Lithography," *Journal of Microlithography, Microfabrication, and Microsystems* **5**(4): 61543H.1–61543H.14, 2006.
43. A. Erdmann, R. Farkas, T. Fuhner, B. Tollkuhn, and G. Kokai, "Towards Automatic Mask and Source Optimization for Optical Lithography," *Proceedings of SPIE* **5377**: 646–657, 2004.
44. Jue-Chin Yu, Peichen Yu, and Hsueh-Yung Chao, "Model-Based Sub-Resolution Assist Features Using an Inverse Lithography Method," *Proceedings of SPIE* **7140**: 714014.1–714014.11, 2008.
45. David O. S. Melville *et.al.*, "Demonstrating the benefits of source-mask optimization and enabling technologies through experimentation and simulation," in Proceedings of SPIE, Vol. **7640**, 764006-1- 764006-18, 2010.

CHAPTER 5

Metrology, Manufacturing Defects, and Defect Extraction

5.1 Introduction

Semiconductor manufacturing is a complex process that involves concepts from various science and engineering disciplines. Since its start during the late 1940s, semiconductor manufacturing has evolved into an industry whose reach has spread into every facet of life today. From space technology to handheld devices, the number of applications that use semiconductor-based components is constantly on the rise. Simply because the transistor shrinks in size every two years, its processing power enables computing and signal processing applications that were unrealizable previously. Semiconductor manufacturing is the process of fabricating semiconductor-based devices to be used in systems. It involves three basic stages:

1. Wafer production
2. Wafer processing, or the transfer of design to wafer
3. Wafer analysis, testing and packaging

Wafer production is the process of producing thin wafer slices in the form of disks from large silicon ingots. With time, *wafer size* has increased, mainly to increase the number of dies that are produced from each wafer; a *die* is a copy of the design on wafer. After processing, a wafer may contain hundreds of dies. As the number of dies per wafer goes up, the cost per die decreases. *Wafer thickness* has also increased with time to improve mechanical handling. Standard wafer diameter and the corresponding thicknesses are shown in Table 5.1.

Diameter	Thickness
150 mm	675 μm
200 mm	725 μm
300 mm	775 μm
450 mm	925 μm (target)

TABLE 5.1 Standard Wafer Thickness and Diameter

There are reasons to hold down wafer thickness: namely, material cost and heat dissipation. Increased device density has led to problem with thermal density, which is addressed by decreasing thermal resistance. This is accomplished by reducing wafer thickness,which also reduces material cost. However, the large thin wafers complicate the wafer handling and alignment process, thereby increasing susceptibility to manufacturing defects. The wafer thickness standard shown in Table 5.1 results from a compromise between these conflicting goals.

Semiconductor manufacturing involves a number of wafer processing steps. The various steps in wafer processing (e.g., oxidation, lithography, and metal deposition) were described in Chapter 2. After the wafer processing steps are completed, a wafer sort test is used to screen bad dies. This procedure may also be used as a feedback path for manufacturing process tuning. In the next step, a wafer is cut into dies and the good dies are packaged. Depending on the die size, thermal properties, and cost, there are several packaging options available today, including plastic ball grid array (PBGA) and ceramic ball grid array (CBGA). Packaging a die involves connecting or bonding the die pads to external package pins. In addition to hermetic and optical sealing, reducing thermal resistance of the package to facilitate heat dissipation is an important packaging concern. Today's desktop microprocessors can dissipate as much as 100 watts of power. The package must be able to dissipate this heat without raising the die temperature above acceptable levels, usually below 100°C.

During the 1950s, semiconductor feature sizes were of the order of millimeters. The requirement that manufacturing be clean and free of suspended particles could be met without great difficulty. Suspended particles and contaminants contribute to semiconductor manufacturing defects. As designs started to use devices of smaller size, cleanliness became an increasing concern to ensure high yields. Today, with more than a billion transistors of sizes approaching 22 nm, clean rooms for manufacturing has become the highest priority. Clean-room requirements scale with transistor and interconnect dimensions (see Sec. 5.2 for some statistics related to clean-room requirements).

Defects in fabrication stages can be caused by handling, particles and contaminants, equipment irregularities, improper chemical reactions, and patterning issues. Defects caused by the patterning process are classified as *feature-* or *design-dependent* defects, and the rest are classified as *process-induced* defects. Suspended particles and particulates from the chemical-mechanical polishing process are the most common source of defects in semiconductor manufacturing. Although contaminants do not necessarily cause errors in device operation, the probability (for a given particle size) of device error increases dramatically with device and interconnect feature size scaling. Particulates that affect wafers during the fabrication process could lead to opens and shorts, potentially causing design failure. Voids and blobs in interconnects, vias, or gate structures cause defects. Such particulate-induced defects may be intralayer or interlayer. In interlayer defects, particulates may cause shorts between consecutive layers or create voids in multiple layers.

The lithography process involves the steps of photoresist coat, exposure, baking, and development. Each stage requires equipment precision and hence proper control over process parameters. Defects may be formed by an irregular resist coat, improper baking procedure, mask and wafer misalignment, and/or irregular resist development.

The imaging system that controls the exposure process can cause errors due to printability issues. Such issues may involve focus, dose, lens aberration, resist thickness variation, flare, and other problems related to the projection system. Printability errors are often rooted in patterns on the mask. The etching stage may result in irregular surfaces due to the properties of the etchant and the formation of protective layer. Etching can also lead to necking and bulging of patterns in certain regions in the design. Typical photoresist is an inhomogeneous material. Etching such materials tends to produce roughness along etch lines, or line edge roughness (LER). In conjunction with other optical effects, too much LER may contribute to defects. As the number of interconnect metal layers increases, so does the number of lithography steps associated with interconnect formation. Interconnect is deposited by a sputtering process for aluminum or, for copper, by a combination of sputtering to deposit a seed liner followed by an electroplating process. The process for copper is inherently susceptible to opens because of voids that arise during the sputtering of seed liner and the diffusion of copper through oxide.

The manufacturing process calibration method is called *metrology*. Metrology is defined as a battery of measurements of the die that are taken in-situ, in-line, and off-line.[1] *In-situ* metrology is the measurement and process control that is performed using sensors placed inside the analysis chamber. *In-line* metrology is the same performed inside a clean room, and *off-line* metrology is performed outside the clean room. The failure analysis of defective parts forms a major component of off-line metrology, and the material characterization used to fine-tune the process is often done off-line. Metrology is crucial to the manufacturing process because it involves periodic tool calibration based on data analysis. Accuracy, precision, resolution, sensitivity, and stability of the measurements taken are of utmost importance during metrology and calibration. Metrology of every stage of the process can aid in better control. In this chapter we examine various metrology techniques that are used today for process control.

Metrology analyzes the overall process and measures process parameters periodically. Another critical component of manufacturing is analyzing the cause of defects in the wafer itself. Analyzing the causes and manifestations of defects is known as *failure analysis* (FA). Failure analysis chiefly targets device failures due to nonconformity of the process with the required physical, chemical, and electrical specifications. There are two types of such failures: functional and parametric. *Functional* failures render the device unable to perform its intended function. In contrast, *parametric* failures cause device parameters to vary beyond the designed specifications (e.g., an increase in overall circuit timing), even though they continue to behave properly under most conditions. The objective of FA is to discover the mode of circuit operation (i.e., the circuit operating conditions) at the onset of failure, the failure mechanism, and the root cause of the defect. Process control and defect mitigation techniques undergo continuous improvement as a result of failure analysis. The techniques used for FA are discussed in Sec. 5.4.

Failure analysis helps to find the root cause of a failure. The root cause analysis may also point to a mask defect or the need for layout changes. Identifying this cause helps to improve the process as a whole. The yield of a process depends on the control and conformance of various process parameters and equipment to predefined specifications. A change in any of these conditions can have a significant effect on the process yield. Process yields, like process failures, are categorized as being either functional or parametric. The yield is typically measured as the ratio of good dies to all the dies produced in a lot. Thus functional yield is the ratio of functional dies to total dies produced; similarly, parametric yield is defined as the proportion of all dies fabricated that are functional but whose parameters may fall outside of specifications under some conditions. Process yield is directly related to product cost and indicates the effectiveness of the current manufacturing process control. A low functional yield requires extensive failure analysis and changes to the process steps, whereas a fall in parametric yield may not warrant such a decision. Yield models are devised based on FA information to predict the effectiveness of the design under high process variation and, like FA techniques, vary with the type of defect that causes a reduction in functional or parametric yield. Defect formation mechanisms are effectively used to model the yield of a design accurately. Section 5.5 summarizes the literature on yield models driven by particle defects as well as some recent work in the area of patterning-induced yield models.

The aim of this chapter is to introduce the reader to material on the importance of process control. This is accomplished through a detailed review of defect formation theory, metrology, failure analysis, and yield-modeling techniques.

5.2 Process-Induced Defects

In the above-wavelength or near-wavelength lithography processes (cf. Figure 1.5), most defects in semiconductor manufacturing were attributed to particulates or other contamination in clean-room facilities. However, improved clean-room technology has led to a decrease in the particulate-induced defect rate. Clean-room standards improved dramatically with the advent of high-volume semiconductor manufacturing. As summarized in Table 5.2,[2] clean rooms are classified based on the number of particles of a particular size present within a square area. Today, large fabrication facilities producing high-end chips maintain ISO 4 or higher standards to minimize particle-induced defects and ensure high functional yield.

Given the close-to-vacuum environment for semiconductor manufacturing, process defects today are predominantly caused by equipment and the process itself. Because so many manufacturing steps are now needed to fabricate wafers, there is a large number of defect sources. Most of the processing steps, from wafer slicing to final packaging, are controlled by computerized machinery, so it is imperative to find which steps are increasing the overall defect rate. The number of defects that a manufacturing step may contribute is a function of the precision to which the required process is controlled

	Maximum number of particles per cubic meter				
Class	**≥ 0.1 µm**	**≥ 0.2 µm**	**≥ 0.3 µm**	**≥ 0.5 µm**	**≥ 1 µm**
ISO 1	10	2			
ISO 2	100	24	10	4	
ISO 3	1,000	237	102	35	8
ISO 4	10,000	2,370	1,020	352	83
ISO 5	100,000	23,700	10,200	3,520	832
ISO 6	1,000,000	237,000	102,000	35,200	8,320

TABLE 5.2 Clean-Room Classification in Terms of International Organization for Standardization (ISO) Categories

and executed by automated equipment. It is well known that some process steps cause more particulate defects than others. Chemical vapor deposition, oxidation, and polishing are known to produce a significant number of particulate defects due to the flaking and scattering of large particles. This is where a number of iterative steps among process control, metrology, and failure analysis is most effective in ensuring process performance within specifications to reduce the rate of defects induced by particulate contamination.

5.2.1 Classification of Error Sources

One aspect of failure analysis is that a lab may have several different types of equipment for diagnosing different types of defects. For example, an FA lab may have microprobing stations, laser cutters, microsectioning equipment, a high-resolution x-ray system, an automatic decapsulation system, a reactive ion etcher to strip layers, a scanning electron microcope, light emission microscopes, and spectrometers. Because of differences in cost, not all labs will have all types of equipment; moreover, the expertise of lab personnel in using such equipment may vary widely. As a result, theories concerning a defect's root cause may often be skewed by limitations in available equipment and in the expertise of engineers. This means that troubleshooting results may not accurately reflect the true cause or rate of process defects. It is therefore extremely difficult to establish the precise defect rate for each of the various steps in a manufacturing process.

Another problem is that the debugging equipment itself may operate outside specification, which makes establishing the true yield a noisy learning process. Equipment errors typically depend on the equipment's design and date of manufacture. Control of the process is greater for state-of-the-art equipment, whereas vintage equipment produces a regular supply of particle defects. Equipment manufacturers always upgrade their systems to meet the requirements of current wafer technology—for example, they constantly adapt to changes in the supply chain, material characteristics, and process execution times in addition to accommodating various other specifications that inevitably change from one technology generation to the next.

Wafer mishandling by computerized handlers is a major source of particle defects. Economic factors are driving both changes in wafer thickness and increases in wafer size, so most wafer handling is now performed by robotic arms. Because wafers are so thin, their structure can be damaged even by a minute disturbance. In one study it was observed that undesirable dynamic vibrations caused structural damage to the wafer during handling.[1] Wafers are stored on chamfered slot rails during transfer. Excessive base excitation of equipment at specific frequencies can cause wafers to vibrate and loose contact with the rails. This loss of contact reduces wafer stability in further processing.

Errors that are induced by process parameters can be subdivided into those caused by a material's *chemical state* (i.e., solid, liquid, or gaseous) and those caused by the *process mechanisms* employed during individual stages of manufacturing. Silicon is the most widely used material in the manufacture of semiconductor devices. Other well-known semiconductors used in device manufacturing today include germanium, indium phosphide, gallium arsenide, and indium gallium arsenide. The different semiconductor types produce defects at different rates. The two fabrication steps with the highest defect rates are wafer cleaning and etching, which depend on the chemical properties of etching and cleaning solutions. Wafer cleaning is performed between several different steps of the wafer fabrication process. Impurities associated with the liquids used to clean the wafer can lead to particle defects on the wafer. Typical particulate levels for the etching and photolithography processes are summarized in Table 5.3.[3] It can be seen that the aluminum etch process contributes the highest number of impurities. In plasma-based anisotropic etching, increased sharpness has been accompanied by a reduced level of contaminants. However, plasma generates polymer by-products that can lead to particulate defects. Oxide etch produces a high number of defects due to the hydrophobic nature of oxide, which creates an environment conducive to the settling of polymer by-products.

Chemical properties of gases used in the process stages of oxidation and of deposition of nitride and metal also play a key role in determining the density of particulate defects. Chemical vapor deposition is a well-known process that uses materials in gaseous

Stage	Process	Particles per 5 in^2
Etching	Nitride etch	50 to 2000
	Photo resist strip	10 to 1000
	Presputter clean	50 to 3000
	HF oxide etch	25 to 3000
	Oxide etch	100 to 2000
	Poly etch	50 to 1000
	Aluminum etch	100 to 3000
Photolithography	Resist spin	20 to 500
	Pattern exposure	5 to 2000
	Pattern develop	10 to 200

TABLE 5.3 Particle Counts for Etching and Photolithography Processes (particles > 0.5 μm)

Process	Particles per 5 in^2
LTO deposition	200 to 3000
Poly deposition	100 to 1500
Silicon nitride deposition	100 to 2000
PECVD	300 to 5000
Aluminum sputtering	20 to 1500

Table 5.4 Particle Counts for Vapor Deposition Processes (particles > 0.5 μm)

forms to be deposited on the wafer. This process occurs in a chamber whose temperature and pressure are strictly controlled. Any variation in the chemical properties of the gases or in the environment can lead to the formation of improper chemical bonds at different surface locations of the wafer, which can produce defects. Typical particulate levels for different vapor deposition processes are summarized in Table 5.4.[3]

5.2.2 Defect Interaction and Electrical Effects

Process-induced particulate defects usually lead to catastrophic failures in integrated circuits (see Figure 5.1 for examples). A particle defect that changes circuit behavior is said to have caused a *fault*. Defects that occur in a die are harmless if they do not cause errors in circuit functionality.

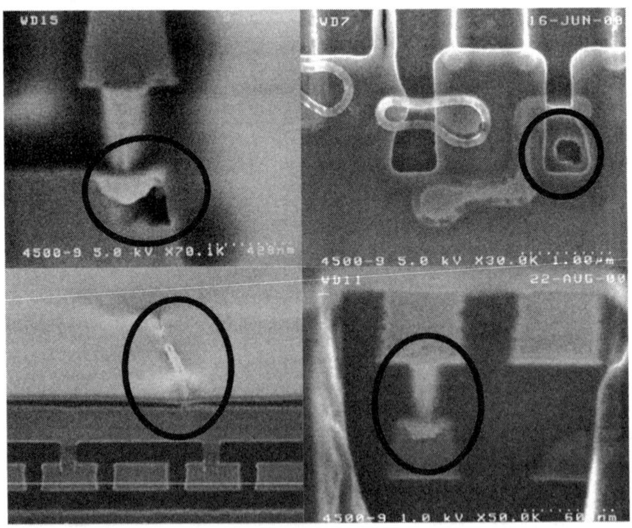

Figure 5.1 Catastrophic failures in ICs. (*Courtesy of Intel Corp.*)

Thus a defect is the cause and a fault is the effect. Not every defect leads to faulty behavior. As shown in Figure 5.2, circuit failures are caused by a void or *open defect* on a line or by a *bridging defect,* which improperly joins two separate lines. A bridging defect becomes a bridging fault when the bridge resistance is low. At higher bridge resistance the defect may not immediately cause a fault, but there remains a latent fault that could manifest itself over time as the bridge resistance changes in response to such reliability effects as electromigration. Similarly, a resistive open defect causes a fault at high defect resistance. A defect that causes faulty behavior leads to yield loss.

Particulate defects can be caused by physical or chemical factors. For example, before the patterning process, a wafer is coated with protective material in order to prevent lower layers from being affected. The coating process may permanently lodge a particulate defect, and the particle may be exposed if the protective coating is removed (as during via formation). This may make the particle electrically active. Similarly, unmasked regions of the wafer are set up to be exposed to UV light and etched away. A defect can cause masking of this region on the wafer; hence there is no exposure of this region, leading to pattern irregularity. Pattern irregularity is an indirect cause of the opens and shorts that can lead to a design's functional failure.

In addition to these physical impacts of defect formation, chemical impacts can also change circuit operation substantially. For example, oxide defects are a major source of circuit failures. An oxide layer is

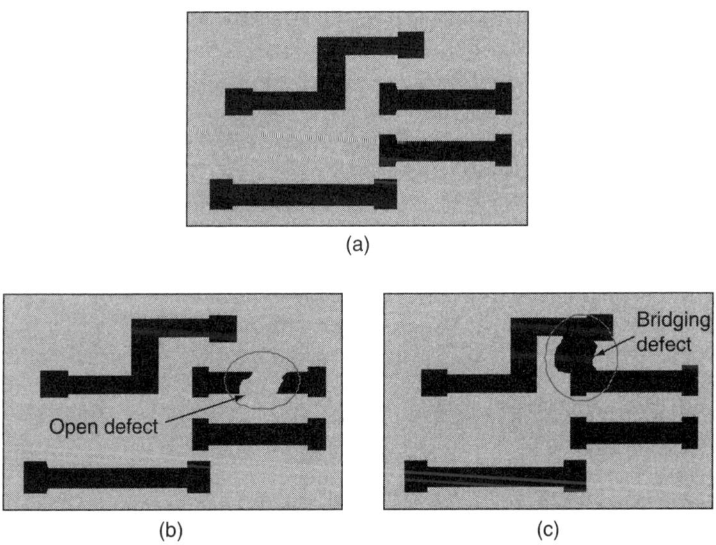

FIGURE 5.2 Defects that can cause circuit failure: (a) original layout; (b) layout with open defect; (c) layout with bridging defect.

used in nano-CMOS fabrication primarily as the transistor's dielectric material and is also used to separate metal layers. Defect formation due to oxide's chemical properties was discussed in the previous section. Defects in transistor gate oxides create dielectric breakdown, which may become evident over time, ultimately leading to device malfunction. Oxide defects also increase the formation of interface traps, which arise from crystal interactions initiated by high-energy electron/hole pairs in the presence of a strong electric field. A defective oxide layer will increase the number of interface traps formed. These traps attach to one another, thereby establishing an unwanted connection between the gate and the channel region and leading to catastrophic failure of the device. Because the CVD process is applied to the entire wafer, any contamination in the deposited substance can also lead to device failures. Chapter 7 addresses long-term device failures attributable to oxide defects.

Physical size of the defect matters when it causes a physical damage to the wafer, whereas chemical properties of substances are important for other types of defects such as contamination. Periodic physical, mechanical, and chemical analyses of individual processes and their environments are an integral part of process control.

5.2.3 Modeling Particle Defects

Particle defects cause various changes in device operation, as described previously. Modeling these defects is necessary in order to improve process control and yield. The salient properties needed to model defects effectively are: (1) defect size, (2) area or region of defect, (3) defect density, and (4) chemical properties of the defect. Of these four properties, defect size and density are the parameters most widely used to model defects.

Quantitative defect models target higher process control, better understanding of a defect's impact on the circuit, and (most importantly) an accurate estimate of process yield. Yield is the primary metric for estimating the overall effectiveness of a manufacturing process. Since yield is a function of the individual process steps, it is important to improve each step by targeting its defect mechanisms. Particle-based defects may cause catastrophic failures. Yield calculation involves computing the functional yield of the process based on particulate defect rate and size. Yield models are classified by whether they address point particle defects or rather gross defects. *Point particle defects* are caused by the physical, mechanical, and/or chemical properties of the process. *Gross defects* affect large areas and are caused by (among other factors) misprocessing, improper mask alignment or usage, or gross handling errors. Defects of this nature are not random, so they can be mitigated with process maturity and effective control. In contrast, the modeling of point particle defects is geared toward random defects, which are not easily controlled by process tuning.

5.2.3.1 Defining Critical Area and Probability of Failure

When estimating the effect of random particle defects on the yield of a process, it is important to identify the layout locations most likely to be affected by such particles. These areas of the layout are known as *critical areas.* Thus, critical area is a measure of design sensitivity to random particle defects of various sizes. The critical area A_c is defined as the region over which the center of the particle must lie in order to cause a catastrophic functional failure. Defects that fall in this region lead to such functional failures as opens and shorts.

Figure 5.3(a) illustrates a critical area related to bridging defects. In this figure we consider a defect of diameter *d,* greater than the separation *s* between the lines. If the center of the defect is close to the center of interline spacing, the result may be bridging; however, if its center is far outside this spacing, then the defect might not cause a fault.

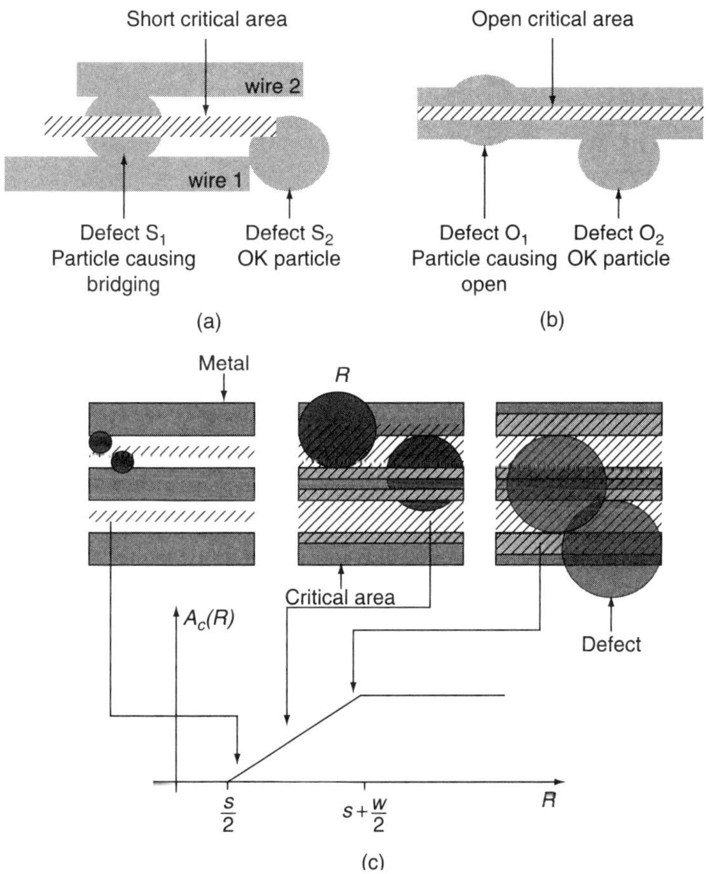

FIGURE 5.3 Critical areas of a layout: (a) short critical area; (b) open critical area; (c) dependence of critical area on the defect radius *R*.

As illustrated in Figure 5.3(a), S_1 causes a bridging defect because its center falls close to the center of the space between lines, whereas S_2 does not result in bridging because its center is away from the center of the space. Critical area is defined as the region in which the center of a circular defect of certain size must lie in order to produce a fault. Referring to Figure 5.3(a), the hatched lines indicate the critical area. The critical area for open faults may be defined analogously. In this case, the defect's center must be nearly along the center of the line. Thus, as shown in Figure 5.3(b), O_1 causes an open defect because it is centered near the line center, whereas the off-center O_2 does not cause an open defect. The critical area is a function of the spacing s between the lines, the width w of the metal line, and the defect diameter d. The size of the critical area depends also on the defect size; see Figure 5.3(c). As the figure shows, defects of size smaller than the spacing s or the width w cannot result in yield loss.

The *probability of failure* (POF) is defined as the fraction of defects that result in faults. The POF depends on defect type, defect size, and circuit geometry. Given a defect of size d, the POF is related to the area in which the center of the defect must lie for the failure to occur. This area (shown by hatched lines in the figure) is the critical area (CA) for open defects and for short or bridging defects. To define the POF for a circular defect with diameter x, let $\theta_i(x)$ be the POF for a defect of type i and diameter x, where θ_i is the POF for type-i defects of varying diameter. The POF θ_i can be obtained by integrating the product of $\theta_i(x)$ and the probability density of x between the maximum (x_{max}) and minimum (x_{min}) defect diameter. Thus,

$$\theta_i = \int_{x_{\min}}^{x_{\max}} \theta_i(x) f_d(x)\,dx \tag{5.1}$$

The probability density $f_d(x)$ is obtained empirically through experiments and can be written as follows:[2]

$$f_d(x) = \begin{cases} \dfrac{u}{x^p} & \text{if } x_{\min} \le x \le x_{\max} \\ 0 & \text{otherwise} \end{cases} \tag{5.2}$$

$$\text{where,}\quad u = \frac{(p-1)x_{\min}^{p-1}x_{\max}^{p-1}}{x_{\max}^{p-1} - x_{\min}^{p-1}}$$

Values for p and x_{max} are also obtained empirically, whereas x_{min} depends on the resolution limit of the lithography system. Let $A_i^{crit}(x)$

be the critical area for defects of type i and diameter x. Then A_i^{crit} is the average over all defect diameters x and is given by

$$A_i^{\text{crit}} = \int_{x_{\min}}^{x_{\max}} A_i^{\text{crit}}(x) f_d(x)\, dx \tag{5.3}$$

Hence POF may be defined as the ratio of critical area to the total chip area:

$$\theta_i = \frac{A_i^{\text{crit}}}{\text{Chip area}} \tag{5.4}$$

5.2.3.2 Critical Area Estimation

If defects are assumed to be circular, then a defect's probability of failure can be defined using the chip's critical area. Critical area analysis (CAA) involves computing the critical area (as defined previously) on a chip for a defect of diameter x. This technique uses a number of factors to predict the POF, including diameter of the defect, width of the metal interconnect, amount of spacing, chip area, and statistics on random process defects. The CAA method is the most widely used technique for estimating design yield for random defects. There are several approaches to performing CAA on a chip: (1) geometry-based; (2) Monte Carlo; (3) grid-based; and (4) stochastic.

The *geometry-based* CAA techniques proposed in the literature use shape expansion, shape overlap, and/or shape intersection techniques to calculate the critical area of a region. For a defect of arbitrary size, a region of length equal to the radius of the defect is drawn around each conduction line as shown in Figure 5.4. The intersection of such regions forms the bridging critical area between multiple conducting lines in the design.

For the case illustrated in Figure 5.4, a circular defect was considered. Some shape expansion techniques don't stop with circular

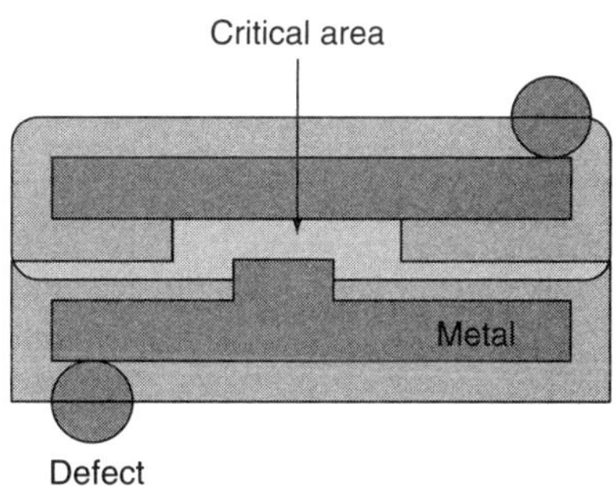

FIGURE 5.4 Critical area estimation based on computing a common region by expanding the conducting polygon by the length of the defect radius.

defects but expand to defects of arbitrary shapes. As shown in Figure 5.5, these shapes are placed at each vertex of the layout pattern and then connected by line segments that are tangential to each feature. These connected segments form a region around each polygon. The critical area associated with the defect shape can be found as the intersection of such expanded regions of electrically disconnected polygons. This intersecting region forms the bridging defect critical area. For open defects, the same procedure is performed but, instead of joining tangential line segments *outside* the polygon, line segments *within* the polygon (and tangential to the defect) are connected to form the critical area. The critical area for a range of defect orientations and sizes can be estimated by scaling and rotating the defect. This shape expansion technique can be used to perform critical area analysis for a particular defect size at a time. A change in defect size warrants a rerun of the critical area estimation procedure.

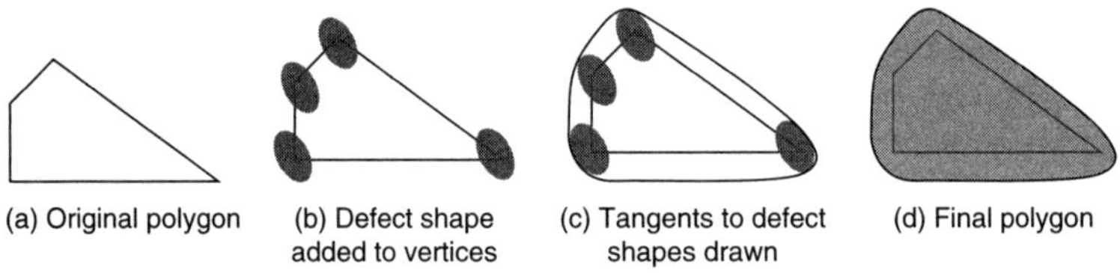

FIGURE 5.5 Shape expansion formulation for noncircular defects.

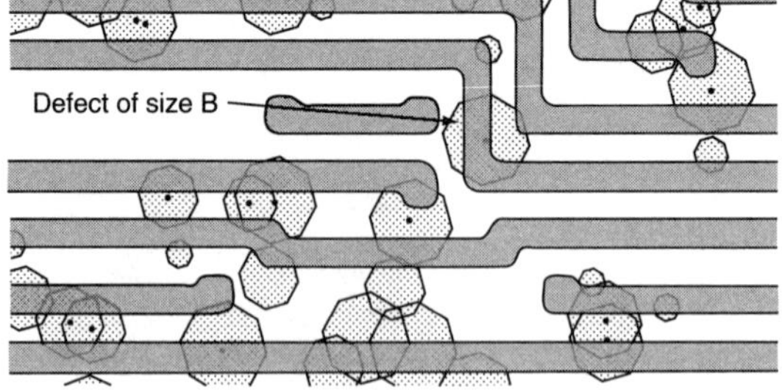

FIGURE 5.6 Monte Carlo–based critical area estimation using defects of various sizes; markers indicate defects that cause faulty circuit behavior.

In contrast, the *Monte Carlo* technique does not restrict itself to a particular defect size. It generates random defect sizes based on the defect distribution in order to estimate the overall chip critical area (see Figure 5.6). The critical area of the chip for a defect of size x is given by the geometric union of the critical area of all the wires in the design:

$$A_{\text{total-CA}}(x) = \int_{x_{\min}}^{x_{\max}} A_c(x)d(x)dx \tag{5.5}$$

Here $A_{\text{total-CA}}$ is the total chip critical area for all defect sizes, $x_{\max}$ and $x_{\min}$ are (respectively) the maximum and minimum defect sizes, and $d(x)$ is the defect size distribution function. If x_0 is the minimum allowable spacing provided in the design rules manual, then a typical defect distribution is given by

$$d(x) = \begin{cases} \dfrac{x}{x_o^2} & \text{if } 0 < x \le x_o \\ \dfrac{x_o^2}{x^3} & \text{if } x_o < x \le x_{\max} \end{cases} \tag{5.6}$$

A large number of defects (of various radii) with distributions as described by Eq. (5.6) are placed randomly over the layout. Each defect is checked for a failure. The POF thus obtained can then be used to estimate yield. This method is not accurate when applied to small sample sizes. The geometry-based and Monte Carlo methods are widely used because they estimate the critical area better. One disadvantage of these techniques is the extent of required computation time.

In the *grid-based* method of critical area estimation, layouts are divided into grids of finite size; then the critical area is estimated based on a defect's grid occupancy, which varies systematically with defect radius.[4] Although this technique is simple, smaller-size grids tend to complicate the algorithm. *Stochastic* and other approximate methods attempt to reduce the computation time of CAA algorithms. To derive an approximate solution, defects are assumed to be rectangular and simple formulas are defined to represent short and open defects. Using these formulas together with layout-specific parameters, it is possible to calculate, with reasonable precision, the critical area for any defect dimension. Stochastic methods combine the known layout parameters with the size and density distribution of defects to derive the layout's sensitivity to open and short defects. The survival probability of each feature is used to estimate the total layout sensitivity and hence the yield.

5.2.3.3 Particulate Yield Models

The earliest yield models for IC manufacturing were based on particle defects, since yield was driven primarily by such defects.[5–12] Random particle defects (aka "spot" defects) are those caused by process discrepancies. The yield of a process is a function of the distribution of spot defects on the wafer. Probabilistic yield models describe the distribution of faults over the chip. If one assumes the defect rate to be constant per unit area, then there is a relationship between the area of a chip and its yield. Let X be the number of faults on a particular chip and let λ be the average number of faults on a chip; that is, λ is the *expected value* of X. The simplest yield model uses λ and the probability of k faults occurring in the chip to define the chip yield. The probability that k faults will occur is given by the Poisson distribution:

$$\mathrm{Prob}(X = k) = \frac{e^{-\lambda}\lambda^k}{k!} \tag{5.7}$$

The average number of faults on a chip is a function of that chip's critical area A_c and defect density D. For a chip without any associated redundancy, the yield is obtained by assuming it has zero defects. If no defects are present in the chip (i.e., if $k = 0$), then the yield of the chip is simply given as

$$Y_{\mathrm{chip}}(k = 0) = e^{-\lambda} = e^{-A_c D} \tag{5.8}$$

Now suppose that the chip area is divided into n statistically independent subareas, each with probability λ/n of having a fault. Then a binomial distribution can be used to describe the probability of k faults occuring, as follows:

$$\mathrm{Prob}(X = k) = \binom{n}{k}\left(\frac{\lambda}{n}\right)^k \left(1 - \frac{\lambda}{n}\right)^{n-k} \tag{5.9}$$

As the subareas become very small and n approaches infinity, Eq. (5.9) reduces to

$$\mathrm{Prob}(X = k) = \binom{n}{k}\left(\frac{\lambda}{n}\right)^k \left(1 - \frac{\lambda}{n}\right)^{n-k} \Rightarrow \frac{e^{-\lambda}\lambda^k}{k!} \tag{5.10}$$

This result motivates the use of a Poisson model to estimate particulate defect yield. The first chip yield model using complex Poisson distribution was given by Murphy.[13] The yield is also a function of the defect density, which may vary across regions of a chip:

$$Y = \int e^{-A_c D} f(D)dD \tag{5.11}$$

A model based on the Poisson distribution is too pessimistic in predicting the yield, because defects typically are not distributed randomly across a chip. Defects occur in clusters on the chip, and this effect must be considered when estimating the yield. Let α be the clustering parameter; then defect clustering can be modeled as a gamma distribution:[9,10,12]

$$\mathrm{Prob}(X = k) = \frac{\Gamma(\alpha + k)}{k!\,\Gamma(\alpha)} \frac{(\lambda/\alpha)^k}{(1 + \lambda/\alpha)^{\alpha+k}} \tag{5.12}$$

Hence the chip yield becomes a negative binomial distribution that depends on defect density. The yield function is given as follows:

$$Y_{\text{chip}} \simeq \left(1 + \frac{A * D}{\alpha}\right)^{-\alpha} \tag{5.13}$$

The defect clustering itself may be distributed unevenly throughout the wafer, and this unequal clustering affects the yield model.[5,6] When this is factored, the preceding yield equation becomes

$$Y = \prod_{i=1}^{W} \left(1 + \frac{(D * A)_i}{\alpha_i}\right)^{-\alpha_i} \tag{5.14}$$

Newer spot defect models also consider defect size when predicting a chip's yield. The different types of random particle-based yield models are summarized in Figure 5.7.[14] The first column illustrates the distribution function used to model the probability of failure. The second column gives the ratio of current yield to the yield with mean defect density D_0, where σ is defined as the standard deviation of the distribution normalized with D_0. The third column lists the number of defects in each case.

5.2.4 Layout Methods to Improve Critical Area

Techniques to mitigate the impact of particle defects on designs are based on minimizing the critical area. Because the yield depends on

DISTRIBUTION OF D	Y/Y_0	$\lambda=$
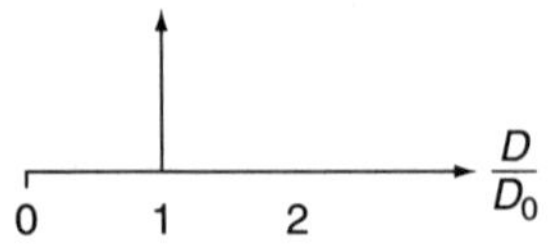	$e^{D_0 A_c}$	0
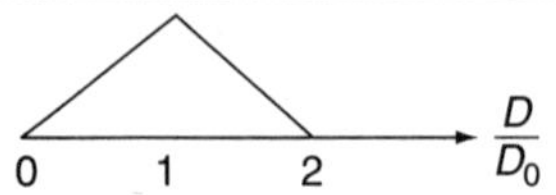	$\left(\frac{1-e^{-D_0A_c}}{D_0A}\right)^2$	0.22(±0.02)
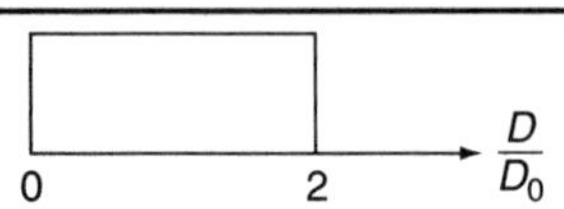	$\frac{1-e^{-2D_0A_c}}{2D_0A}$	0.5(±0.1)
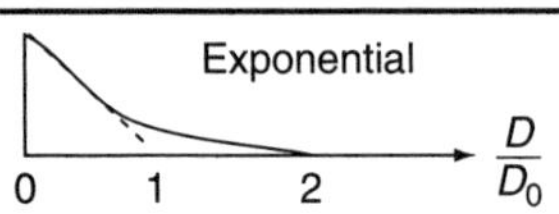	$\frac{1}{1+D_0A}$	1
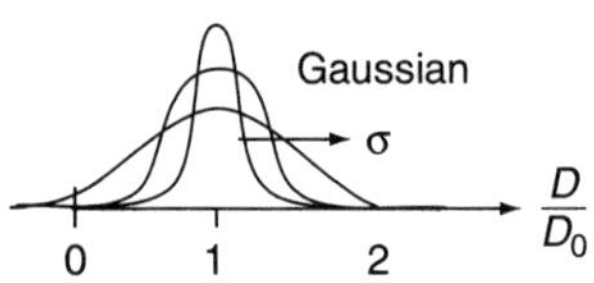 $\frac{1}{\sqrt{2\pi\sigma}}\exp\left[-\frac{1}{2}\left(\frac{\frac{D}{D_0}-1}{\sigma}\right)^2\right]$	$\frac{1}{2}\exp\left(-A_cD_0+\frac{\sigma^2A_cD_0}{2}\right)$ $1+erf\left[\frac{1}{\sqrt{2}}\left(\frac{1}{\sigma}-\sigma D_0A_c\right)\right]$	$\approx\sigma^2$ for small σ
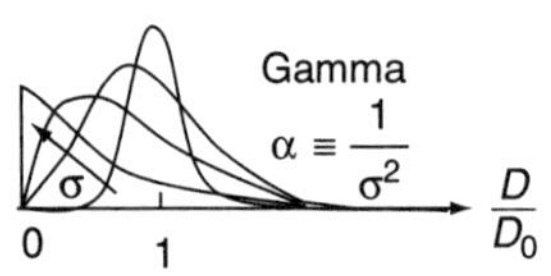 $\frac{\alpha}{\Gamma(\alpha)}\left(\alpha\frac{D}{D_0}\right)^{\alpha-1}\exp\left(-\alpha\frac{D}{D_0}\right)$	$\frac{1}{\left(1+\sigma^2D_0A_c\right)^{1/\sigma^2}}\equiv\frac{1}{\left(1+\lambda D_0A_c\right)^{1/\lambda}}$	σ^2

Figure 5.7 Random particle defect models.

metal width and spacing, these are the two targeted design parameters. Techniques to improve CA include increased spacing, wider lines, and wire pushing or spreading. Increasing the space between lines is a simple way to improve CA-based yield metric. An example suggested in Chapter 4 for better phase assignment in standard cells applies here as well. Poly lines placed at minimum spacing are moved further apart to reduce the possibility of shorts. Similarly, the widening of metal lines within standard cells improves yield, since the probability of defects causing an open is reduced. These two methods are incorporated into standard cell design.

Routing algorithms that are CA-aware have also been suggested. The *allowed spacing* technique scans the layout to compute spacing allowances for all movable wires and then uses this as a measure when routing is performed. The information on allowable spacing is used to help routing algorithms implement wire widening and wire pushing or spreading (see Figure 5.8), thereby reducing the critical area between lines. During the routing process, wire pushing involves finding the optimal route that reduces the overall critical area of the layout.[15] Sensitivity estimation during routing also helps to reduce the critical area. These techniques for wire spreading and estimating allowable space are all derived from the so-called skyline algorithm.[16]

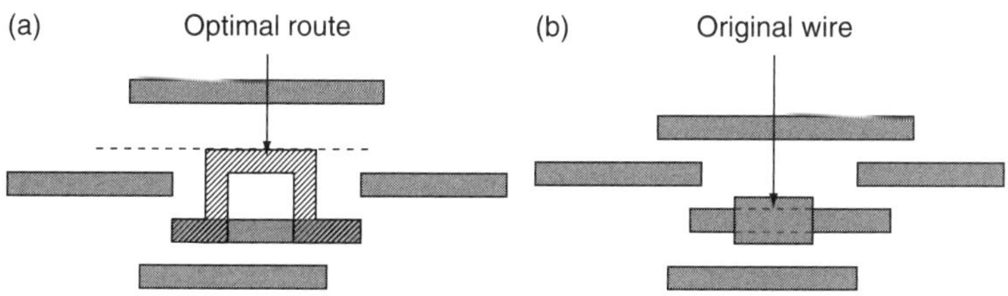

FIGURE 5.8 (a) Wire pushing; (b) wire spreading.

As we have seen, the main methods in critical area improvement are increasing wire width and wire spacing. Such increases tend to increase the chip area. As the chip area increases, the average number of defects that can affect a chip also increases. Thus, any gain from reducing the critical area may be offset by an increasing number of defects. Moreover, increasing the chip area reduces the number of dies per wafer. This means that, even with a lower defect rate, the effective die yield may not improve with increasing chip area. These factors must be carefully weighed in any decision to implement CA-based layout improvement techniques.

5.3 Pattern-Dependent Defects

Photolithography is the process of transferring patterns drawn on a glass mask onto a silicon wafer. As a consequence of Moore's law, designs today are large and highly complex. They also have stringent timing and power constraints. As technology scales to 32-nm devices, the number of rules that control design layouts has increased exponentially—in addition to the proliferation of burdensome design rules and guidelines for effective printability. Pattern-dependent defects are functions of the actual pattern being printed on the silicon. These defects are distinct from particulate defects. Whereas particulate defects tend to decline as the process matures, many pattern-dependent defects cannot be cured by process tuning alone.

5.3.1 Pattern-Dependent Defect Types

Patterning-related problems may occur in vias and interconnects as well as in active devices. The problems related to vias are particularly acute. Because of increased device and interconnect density, there tends to be a large number of vias in today's circuits.

A transistor is formed by patterning of the active area and gate masks. The amount of diffusion under the gate determines the length and width of the device. Rounding of diffusion areas due to proximity effects has been observed experimentally (this effect was shown in Figure 3.24). Diffusion rounding leads to reduction in device width and irregularity of device gate length, both of which affect the amount of current flowing through the channel and the overall circuit performance.

In an IC layout, vias form the connection between wires in successive metal layers. Via density increases with device and interconnect densities. Mask alignment errors during via patterning cause vias to fail either partially or completely. These failures depend on the via's location inside a block. Partial failure of a via increases its contact resistance, and any change in via parameters affects the timing of a circuit. Complete via failures lead to the formation of opens; of course, this impairs the design's functionality. Similarly, improper contact printing due to insufficient active area to contact spacing has been found to reduce yields in 45-nm technology. Improper contact formation leads to resistive and in some cases catastrophic opens. Partial vias may also contribute to long-term device reliability problems.

Apart from the issues discussed so far, patterning problems are also observed when interconnect lines are printed. Resolution enhancement techniques such as OPC and PSM are used to mitigate such printability issues. Even so, post-OPC layouts sometimes have regions that do not produce the required wafer image. The reason is that OPC algorithms cannot guarantee exact reproducibility of an image and therefore result in suboptimal solutions. The algorithms can generate suboptimal solutions also if the procedure is terminated prematurely at certain locations of the layout. Such termination may be caused by iteration constraints, small segment movement area, and/or reduced segment stepping sizes. The maximum iteration count to arrive at the minimum EPE error between modified mask and required image for each simulation point is set as a global constant. If a tool does not produce a suitable solution within this time, the optimization at that point ends prematurely. The high density of layout patterns implies that the allowable space between two neighboring features is typically quite small. Hence, this spacing constraint can prevent certain tools from attaining proper OPC modifications. Mask economics limits the number of segment steppings possible for OPC, which in turn limits the size of features that can be added to quantized dimensions. All these factors compound the spacing constraint for OPC, leading to early termination.

Some "solutions" can miss jogs or features to be added to the mask polygon. Examples of some resultant defects are illustrated in Figure 5.9. Such mistakes in the OPC algorithm can lead to catastrophic failure.

The Alternating PSM technique requires that the spacing between two minimally spaced metal lines be assigned a different phase in order to improve the depth of field and contrast of the main feature. Therefore, the inability to assign unique phases to alternating regions can lead to reduced DOF, causing opens and/or shorts. Some drawn masks have regions that are not phase assignable (see Figure 5.10), but tools are available for checking that all layout regions are phase compliant. In response to this problem, a top priority of today's layout engineers has become the creation of phase-assignable layouts. A "correct by approach" technique was suggested by Kahng[17] to produce phase-compliant layouts. (Examples of the "correct by construction" approach were shown in Figure 4.18(b).) Other RET-based catastrophic faults include SRAF placement failure for 2-D

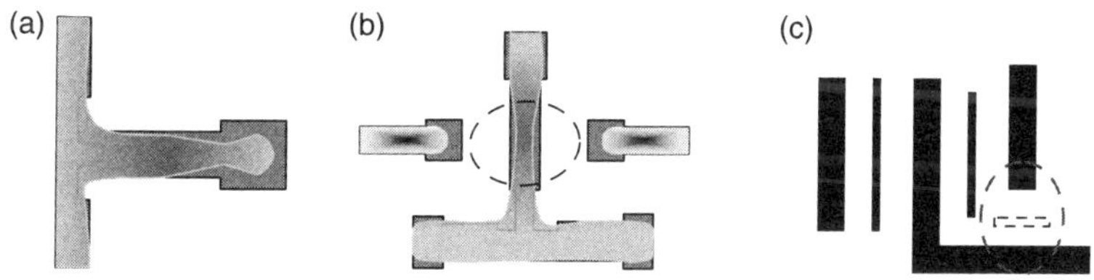

FIGURE 5.9 Patterning defects: (a) necking of small hammerheads; (b) not enough OPC at nonnominal process corners; (c) SRAF placement error.

FIGURE 5.10 Layout phase assignment problem.

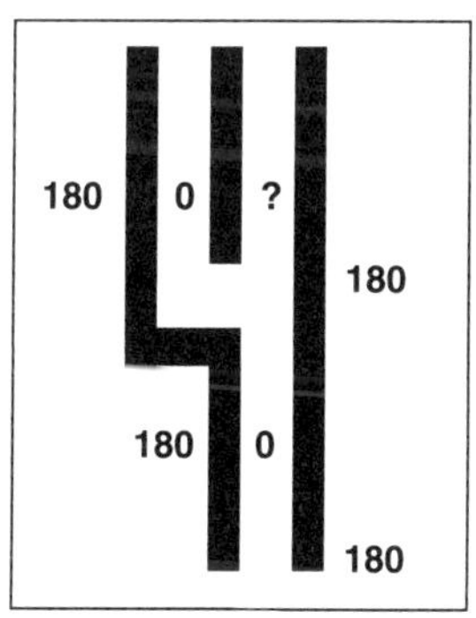

masks. Improper SRAF placement can cause isolated patterns to be printed off-shape, leading to catastrophic defects.

5.3.2 Pattern Density Problems

Chemical-mechanical polishing (CMP) is a technique used to planarize wafers. The wafer is polished on a station using a polishing pad and an abrasive fluid called slurry (see Figure 3.28 and Figure 3.29). The effectiveness of the planarization obviously depends on physical parameters of the polishing pad and chemical properties of the slurry. Apart from these, it has been found experimentally that the patterns on the wafer also have a role to play in determining the planarity of the resulting wafer surface. The peaks and troughs on a wafer's surface depend on the pattern density of the underlying layer(s). Variation in CMP for a particular layer can result in dishing and erosion, as shown in Figure 3.30. Chemical-mechanical polishing variation can also occur as a result of thickness variation in a multilayer dielectric. In short, variation in pattern density can cause improper planarization, leading to DOF issues. Defocus is responsible not only for parametric defects (e.g., gate CD variation) but also for such catastrophic defects as opens and shorts. Insufficient removal of metal can cause shorts and erosion, whereas dishing leads to open defects. Because the failures induced by CMP-related defects vary systematically with the pattern density, the yield can be estimated based on density correlations across the die. The density of patterns also has an affect on the etching process. When minimum-sized patterns are placed close to large patterns, the etchant solution does not adjust to the spacing and ends up removing some portion of the minimum-sized feature.

The CMP process is designed to maintain an upper and a lower bound for copper and dielectric thickness levels in order to ensure circuit operation and yield within specifications. Upper thickness level (UTL) and lower thickness level (LTL) define the minimum and maximum allowable thickness of material remaining on a wafer after planarization. These thickness levels are determined based on focus specifications. A technique described by Luo and colleagues[18] seeks to attain the CMP-based yield of a die by calculating the probability that all post-CMP material thickness across the die lie within the LTL-UTL range. Let n denote the number of locations at which the material thickness is monitored in the yield prediction process, and let Φ denote the joint distribution of the thickness variation at n different locations. Then the yield is given by

$$Y_{\text{CMP}} = \int_{\text{LTL}}^{\text{UTL}} \int_{\text{LTL}}^{\text{UTL}} \cdots \int_{\text{LTL}}^{\text{UTL}} \cdots \Phi(p)\, dp_1\, dp_2 \cdots dp_n \tag{5.15}$$

For p, an n-dimensional probability density vector and $|C_d|$ the determinant of the density covariance matrix,[19] the joint distribution Φ is given as follows:

$$\Phi(p) = \frac{\exp\left\{-(p-\mu)^T \Sigma^{-1}(p-\mu)\right\}}{\sqrt{(2\pi)^n |C_d|}} \tag{5.16}$$

As explained in Sec. 3.5, the Preston equation is used to estimate the thickness of a planarized layer. The thickness of underlying layers is also considered when estimating the thickness of the current layer. In order to obtain an accurate estimate of the yield, a large number of locations must be tracked for thickness variation. Numerical integration of thickness at a number of locations is performed using Genz's algorithm.[19] The number of such locations is typically of the order of 10^6. Populating the correlation matrix for large dies is done with the aid of manufacturing data based on density variation readings from different test structures.[20]

5.3.3 Statistical Approach to Modeling Patterning Defects

Statistical variation in manufacturing parameters affects the printability of layout patterns. For example, linewidths may vary significantly with variations in focus or exposure dose. In extreme cases, when linewidth or interline spacing goes to zero, a defect is created. Such defects are related to general lithography parameters as well as to particular layout patterns. Predicting yield based on variations in such manufacturing parameters is a complex but crucial task.

This section describes two methods for obtaining the lithographic yield of a metal layer based on the mask's critical dimension (CD). The yield estimated with these methods is known as *CD-limited* or *linewidth-based* yield.

5.3.3.1 CASE STUDY: Yield Modeling and Enhancement for Optical Lithography

Charrier and Mack presented a yield model based on the CD distribution obtained under various combinations of input parameters.[21] This prediction technique consists of a four-step process, as summarized in Figure 5.11.[21]

First, the error distribution of each input variable is obtained. The input variables used to characterize lithographic variations include focus, exposure dose, and resist thickness as well as a parameter for development. Next, using a lithographic simulator, a multivariable process response space is generated in order to model CD under input variations. In the third stage, a final CD distribution is generated by mapping the input parameter distributions on the process response space. Finally, the output CD distribution is used to predict CD-limited yield by using a CD acceptance criterion.

Critical dimension error as a total derivative of a number of uncorrelated input errors Δp can be calculated as

$$\Delta CD = \frac{\partial \text{CD}}{\partial p_1}\Delta p_1 + \frac{\partial \text{CD}}{\partial p_2}\Delta p_2 + \frac{\partial \text{CD}}{\partial p_3}\Delta p_3 + \cdots \tag{5.17}$$

where the partial derivatives represent the process response of CD to the input variable p_i. Equation (5.17) holds for small input error when the parameters are not correlated. In the paper by Charrier and Mack, variation in (one-dimensional) input parameter errors is assumed to be Gaussian, as shown in Figure 5.12.[21] This distribution for the

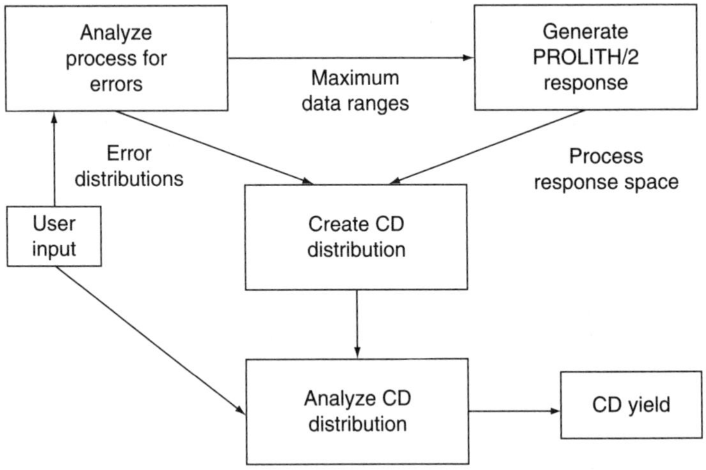

Figure 5.11 Methodology flow for estimating CD-limited yield based on lithography variation.

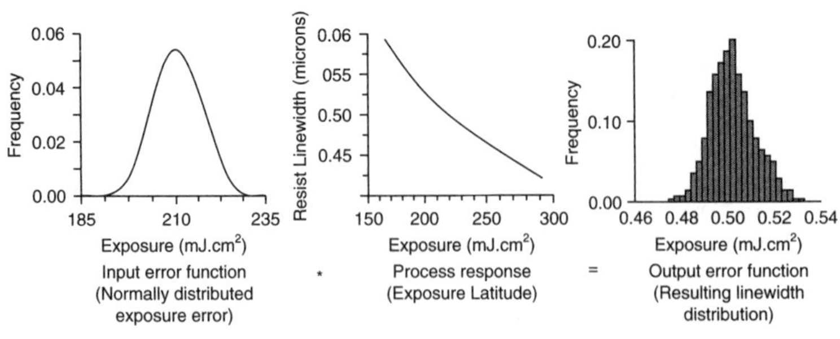

Figure 5.12 Calculation of output error function (i.e., CD distribution).

parameter variation is then convolved with the process response space. The process response curve for the exposure dose case of Figure 5.12 is simply the exposure latitude obtained through process window analysis.

After obtaining the CD distribution for a multivariable error distribution with a complete process response, the calculation of CD-limited yield is fairly simple. Given a CD specification, the frequencies of all CDs within this specification are added and normalized against cumulative frequencies to predict the yield. For this model, Figure 5.13 plots yield against increasing feature size based on this model. The CD-limited yield is observed to be increasing in feature width (and spacing).

5.3.3.2 CASE STUDY: Linewidth-Based Yield When Considering Lithographic Variations

The yield model just described considers manufacturing parameter variations but does not consider actual layout patterns. Sreedhar and Kundu presented a yield analysis method based on mask layout; see Figure 5.14 for an overview of this technique.[22]

The proposed yield modeling methodology is based on lithography simulation, which uses mask patterns to predict the printed shape on silicon. This technique requires electromagnetic field modeling for optical diffraction analysis, vector modeling of

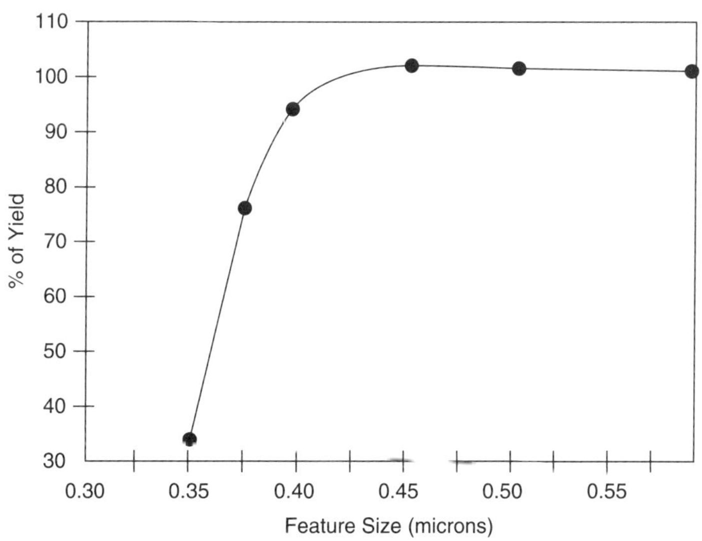

FIGURE 5.13 Yield as a function of feature size for a 0.4-μm i-line process with dense lines and spaces.

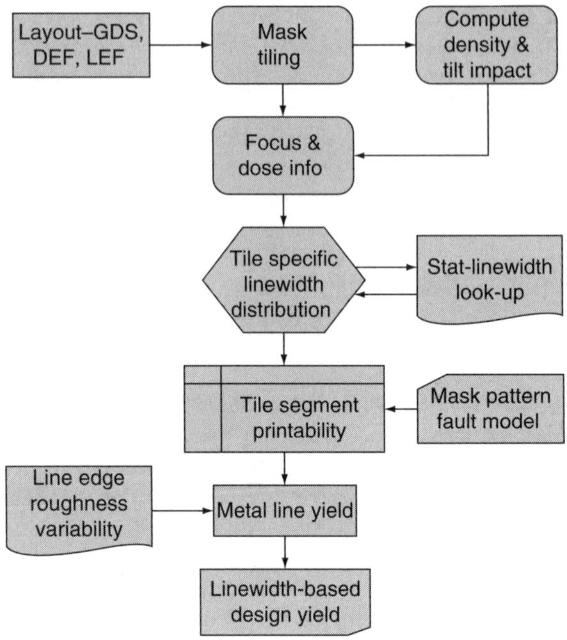

FIGURE 5.14 Linewidth-based yield estimation methodology.

partial coherent imaging, polarized illuminations, Zernike aberrations, Jones pupils, flare analysis, and numerical modeling of photoresist development. The printed shapes are sensitive to many process parameters, including focus, exposure, bake, photoresist development, and etch rate.[23] A small change in any of these parameters—even within the error range specified by the manufacturer—may distort the printed shape, cause line opens and shorts, and reduce yield. It is therefore possible, at least in principle, to predict yield based on statistical lithography simulation. However, it is difficult to translate this principle into practice. In the first place, there is no established relationship between yield and the probability of a line shape. Second, lithographic process simulation is capacity limited and consequently better suited for a library cell than a design layout. Third, the process is slow; and fourth, the interconnect layers cannot be simulated independently. Despite these limitations, the main contributions of this technique are modeling the connection between line shape and yield and obtaining line shape probabilities from the statistical simulation of lithography.

Yield from Line Shape Consider the metal lines shown in Figure 5.15.[22] Because of proximity effects, the postlitho linewidth ($LW_{postetch}$) is either smaller or larger than the expected linewidth (LW_{ideal}). If

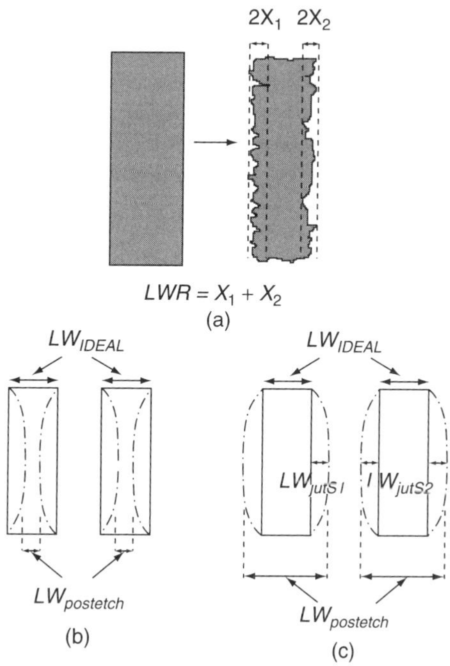

FIGURE 5.15 Mask pattern fault model: (a) worst-case LWR; (b) line open; (c) line short.

$LW_{postetch}$ goes to zero, the result is an open defect on the line. Similarly, if the postetch interline spacing goes to zero then the result is a short defect on the line.

An additional parameter related to line-edge roughness increases the probability of opens and shorts. A narrow line of nominal width greater than zero but less than two times the LER amplitude may become open. Likewise, if the nominal postetch interline spacing exceeds zero but is less than twice the LER amplitude, a short may result. Let LWR denote the LER-adjusted, postetch, worst-case linewidth for each defect case just described. Then the conditions for defects may be written as follows:

$$\begin{aligned} &\mathrm{LW_{postetch}} - \mathrm{LW_{LWR}} \le 0.3\ \mathrm{LW_{ideal}} \\ &\mathrm{LW}_{\mathrm{jut}\ S_1,S_2} = 0.5(\mathrm{LW_{postetch}} - \mathrm{LW_{ideal}}) \\ &\mathrm{LW}_{\mathrm{jut}\ S_1} + \mathrm{LW}_{\mathrm{jut}\ S_2} - \mathrm{LW_{LWR}} \le \text{spacing} \end{aligned} \tag{5.18}$$

Here LW_{ideal} is the ideal (expected) linewidth for the target lithography process, otherwise known as mask CD, and $LW_{postetch}$ is the obtained linewidth after the etch process with variations. The M_{layer} term is the metal layer number, and LW_{jutS1} and LW_{jutS2} denote the protrusions on either side of a metal line. Finally, "spacing" is the edge-to-edge distance between two adjacent metal lines. In this case, "spacing" denotes the limit to which two adjacent lines can expand and not bridge. The expressions displayed above can be used to obtain the linewidth distribution from statistical mask simulation; this distribution is then used to calculate the probabilities of a short or open for a given segment. (For more details, see Sreedhar and Kundu.[22])

Reducing Simulation Points The yield analysis technique just described hinges on lithography simulation with statistical variation of input parameters. Because linewidth is not a linear function of the input variables and is pattern dependent, Monte Carlo simulation is used. However, it is well known that aerial imaging simulation of layout features is computationally intensive and extremely slow.[24] To address this problem, the number of simulation points must be reduced. A *stratified* sampling strategy samples the input parameter space for focus and dose variation while keeping the number of simulation points low. However, this leads to oversampling around edges, so the results need to be weighted in order to compensate. With stratified sampling, the data are split into smaller disjoint sets upon which random sampling is performed. This technique has found various applications in the domain of statistical analysis.[21]

5.3.4 Layout Methods That Mitigate Patterning Defects

Diffusion rounding errors can be mitigated by modifying the layout. Typical candidates for diffusion rounding are edges that connect to the power supply through the source side of the device. Restricted design rules that establish a minimum spacing between such corners from the gate region can help mitigate irregularity in gate width and length. Problems in via patterning are solved by extending the metal contact region surrounding a via to accommodate another via and thereby increase metal-to-metal contact. This process known as *double via insertion,* and it has been incorporated into most design tools. Errors that remain uncorrected after OPC and PSM require manual layout modifications.

Errors in CMP and in etchant-flow-induced overetches can be mitigated by controlling global and local pattern densities, respectively. Pattern density uniformity can be achieved by employing regular layout structures. Layout regularity, in turn, can be enhanced by the placement of unconnected, nonfunctional dummy features between metal lines; these are called *dummy fills.* With fills, isolated features resemble dense patterns for planarization, limiting thickness variation. However, a problem with such fills is that they add capacitance to existing signal lines. Hence capacitive cross talk increases, which leads to reduced circuit performance. In order to avoid performance penalties, filling techniques that are "chip timing aware" aim to improve post-CMP material thickness while satisfying timing constraints. Such techniques often assign dummy fills to power lines by connecting them.

Erosion in planarized layouts is caused by highly dense patterns that are of minimum width and separated by a minimum spacing. The neighborhood around isolated features are filled by dummy fills to normalize global pattern density, which leads to uniform erosion (see Figure 5.16). Dishing is most pronounced in patterns that are large in width. This is typically seen in higher metal layers that feature thick and wide metal lines. Dishing reduces thickness over the metal line. *Slotting* is a technique used with wide features to minimize dishing; it involves placing squared dielectric features within the metal, as shown in Figure 5.17. Together, dummy features and slotting mitigate defocus problems with designs, and they have been observed to produce dramatic effects on dielectric and metal thickness.

Various algorithms for dummy feature placement have been suggested in the literature.[25,26] Effective dummy feature placement reduces thickness-dependent variations in focus. Simple, rule-based dummy fill techniques aim to fill in dummy features between patterns at all location of the die. Predefined dummy feature templates are used to fill in regions for which spacing rules and capacitive coupling

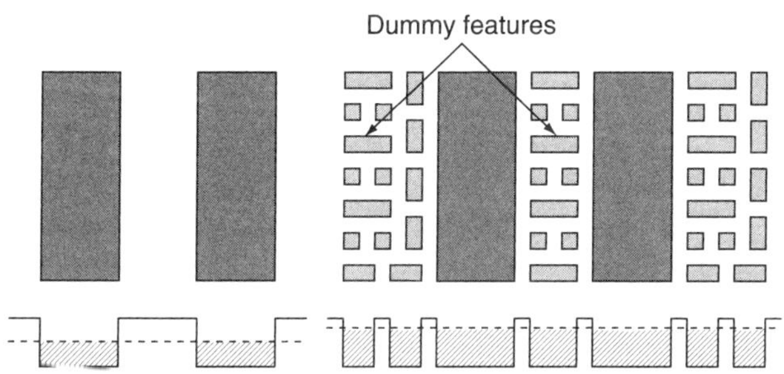

Figure 5.16 Placement of dummy features to increase pattern density and thereby regularize erosion.

Figure 5.17 Slotting in wide metal lines to reduce dishing.

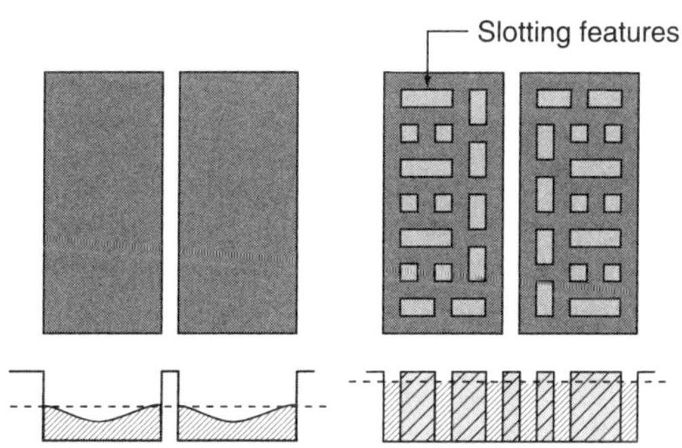

constraints must be taken into account. There are two disadvantages to using template-based dummy fills: (1) they are not entirely flexible with respect to feature size and allowable spacing; and (2) dummy feature placement occurs even for regions where pattern density is already high. As a result, today it is more common for model-based techniques of placing dummy features to be based on pattern density estimates. Methods for measuring pattern density use a moving window of fixed size over the chip to estimate density of overlapping regions of the die; see Figure 5.18. Measured density values suggest regions to be filled with dummies. For slotting, wide metal lines are first chosen and then square oxide slots of fixed width and spacing are created to minimize dishing. (See Kahng and Samadi[27] for a more detailed treatment of various density estimation and dummy fill techniques.) Disadvantages of dummy fills and slots include (1) an increase in the number of patterns in a mask, which leads to increased mask cost; and (2) an increase in the complexity of the RC extraction process, which leads to larger extracted circuits.

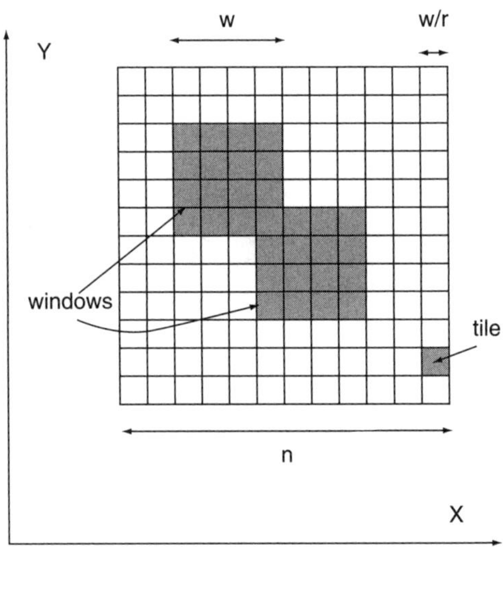

FIGURE 5.18 Overlapping moving window used for estimating pattern density.

5.4 Metrology

Metrology is a part of semiconductor manufacturing that involves data measurement within and outside of the clean room. Figure 5.19 summarizes the classification of metrology. Metrology within the clean room can be categorized as being either in-line or in-situ. *In-line* metrology involves data measurement on test structures that have been fabricated on the wafer; it includes measurement requirements for the process control of fabricating transistors and on-chip

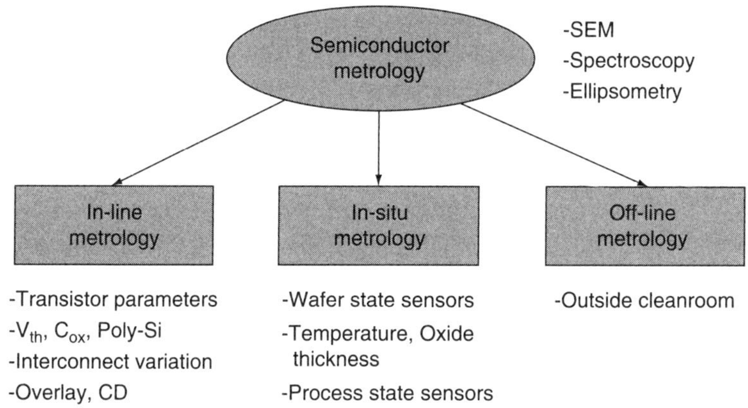

FIGURE 5.19 Semiconductor metrology classification.

interconnect.[3] Typical in-line measurements are CD variation, overlay error, material thickness, wafer resistivity, and etch control. These measurements must be fed back to the designer for incorporation into better circuit design solutions.

In-situ metrology refers to measurement and process control performed using sensors in the test bed and process chamber. Typical in-situ measurements are based on chip-level sensors for temperature variability, power droop, and (most importantly) process state. *Off-line* measurements are those that are performed outside the clean-room facility.

5.4.1 Precision and Tolerance in Measurement

Measurement *precision* is a combined function of the measuring tool's short-term reliability and its long-term reproducibility. Process *tolerance* refers to the variation in parameters and tool properties that can be tolerated by the process. In general, precision can be estimated by repeating measurements of a specific type on reference data for a prolonged observation period. A measurement tool's *repeatability* is the extent to which the measurements it takes do not vary under identical process conditions; a tool or measurement's *reproducibility* is the extent to which the measurements taken are time-invariant. The ratio of measurement precision to process tolerance is a well-accepted metric for evaluating the capacity of automated metrology equipment to produce useful statistical data on process control.[28] Measurement accuracy or precision can be defined as the square root of the sum of squares of repeatability and reproducibility: $\sigma_{\text{precision}} = \sqrt{\sigma^2_{\text{repeat}} + \sigma^2_{\text{reprod}}}$. Process tolerance is defined as the range of variation allowed by the process: $\text{Tolerance}_{\text{process}} = \overline{\lim}_{\text{process}} - \underline{\lim}_{\text{process}}$, where overbar and underbar indicate upper and lower limit (respectively) of process tolerance. The precision-to-tolerance ratio is then given by

$$\frac{P}{T} = \frac{6\sigma_{\text{precision}}}{\overline{\lim}_{\text{process}} - \underline{\lim}_{\text{process}}} \tag{5.19}$$

A P/T value of 30 percent is typically allowed for a measurement tool. However, owing to the increased variability across all manufacturing steps, a P/T value of less than 10 percent is preferred. High precision, lower tolerance, and high resolution form a good measurement framework for reliable back-to-design feedback schemes to help control process parameter variation.

5.4.2 CD Metrology

Critical dimension metrology involves the measurement of linewidth, spaces, and via or contact holes patterned on the wafer. There are three principal techniques for performing linewidth measurement: (1) scanning electron microscopy (SEM); (2) electrical metrology; and (3) scatterometry. Each technique is based on fundamentally different concept of measurement. Scanning electron microscopy, as the name implies, uses electron flow to take measurements; electrical techniques use test structures; and scatterometry is an optical technique.

5.4.2.1 Scanning Electron Microscopy

The technique most commonly used today is SEM, which uses an electron beam to scan a particular region of the wafer. A simple SEM

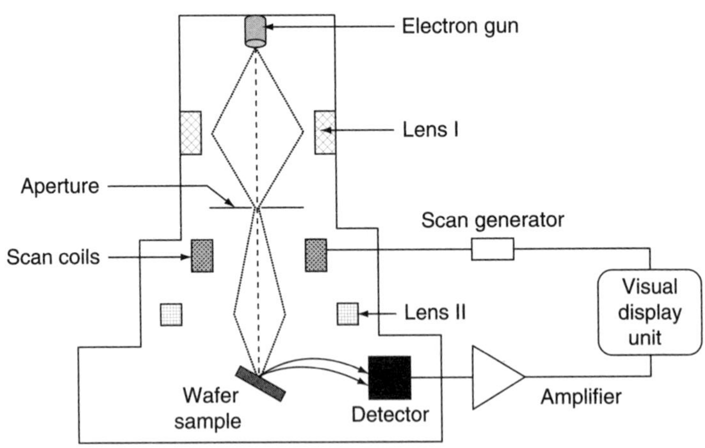

FIGURE 5.20 Simplified schematic for scanning electron microscopy (SEM) setup.

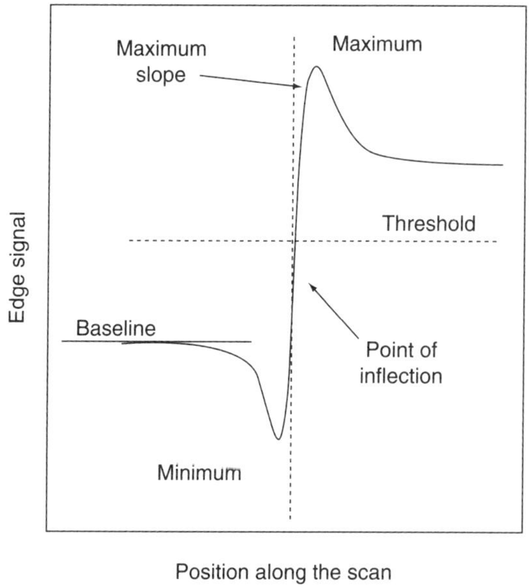

FIGURE 5.21 Electron signal corresponding to one edge feature as measured by SEM; the signal's attributes help pinpoint the edge's location.

setup is shown in Figure 5.20. The voltage of electron beams used to measure changes in the resist ranges between 300 and 1000 volts. The incident beam is scattered and moves in directions that depend on the wafer's material composition and feature topography. This scanning beam is collected and amplified to produce an image. Topography induces variation in the intensity of the detected signal, which is visible in the resulting image. Edges are located based on analysis of the intensity profiles; see Figure 5.21. The change in intensity is at a maximum on the edges and at a minimum on the flat surfaces.

Various edge detection techniques—such as maximum slopes, linear approximation, and inflection-point techniques—have been proposed in the literature.[29] A simple expression for measuring signal threshold (one that is similar to that used for edge detection in aerial imaging simulation) is:

$$I_{\mathrm{th}} = (1 - P)\mathrm{ES}_{\mathrm{min}} + (P)\mathrm{ES}_{\mathrm{max}} \tag{5.20}$$

where ES is the edge signal and the probability P takes a value between 0 and 1. Apart from the incident beam, secondary and back-scattered electrons also cause emission onto the wafer surface. All emissions are proportional to the slope of the feature being scanned.[30,31] Figure 5.22 shows that the number of electrons is proportional to

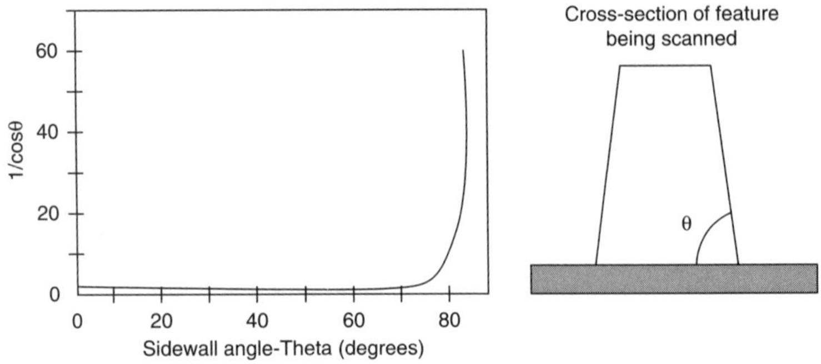

FIGURE 5.22 Secondary emissions depend on the slope of the resist profile.

$1/(\cos\theta)$, where θ is the sidewall angle of the profile being scanned. Hence, the resist profile image can be created by measuring the linewidth and height of the feature (from the incident beam) and the sidewall angle (from the secondary emissions). Extreme care must be taken when using SEM images to predict edge location and slope, because the profile variation of such images is highly sensitive to errors.

Wafer images created by SEM also have application to defect identification and other device measurements. However, a major problem with SEM imaging is charging of the sample being imaged. Electrons from the incident beam induce charging of the substrate, which can have a significant effect on the measurement. The extent of charging depends on the voltage of the incident beam and on the composition of the substrate material, so any change in either of these factors can lead to measurement errors. At low voltage, energy beams have high numbers of primary electrons and and low numbers of secondary and back-scattered electrons. This balance changes at higher voltages. The measurement is error-free only when the material being imaged remains electrically neutral, yet the material being exposed may accumulate a net charge.[32,33] Negatively charged material deflects electrons, which causes measurements to be narrower than the actual linewidth (see Figure 5.23). The opposite happens when the material is positively charged. Recent work has shown that error magnitudes can be reduced by taking a 2-D Fourier transform of the resulting image.[34]

5.4.2.2 Electrical CD Measurement

Linewidth can be measured electrically as a supplement to SEM measurement. For a particular CD specification, SEM measures the material surface for best focus, the exposure dose, and the wafer tilt. However, it is not practically feasible to perform line edge measurements for all lines on a wafer. Electrical measurements

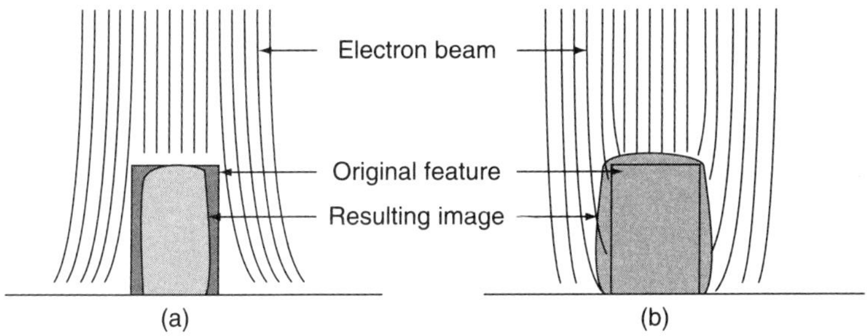

Figure 5.23 Measurement error caused by behavior of scanning beam electrons: (a) electrons are deflected by negatively charged material, so the measured linewidth is narrower than actual; (b) electrons are absorbed by positively charged material, so the measured linewidth is wider than actual.

complement SEM by reading CD values at different regions of the wafer.

There are two ways to perform electrical CD measurement. One technique for obtaining gate CD value is by calculating the transconductance g_m of the transistor. This transconductance is defined as the ratio of the derivative of current I_d and the derivative of the gate-to-source voltage V_{GS} at a constant value of drain-to-source voltage:

$$g_m = \left.\frac{dI_d}{dV_{gs}}\right|_{V_{ds}} \tag{5.21}$$

The transconductance changes as follows for different regions of transistor operation:

$$g_m = \begin{cases} \beta V_{ds} & \text{for } V_{ds} < V_{dsat} \text{ (active region)} \\ \beta V_{gt} & \text{for } V_{ds} \geq V_{dsat} \text{ (saturation region)} \end{cases} \tag{5.22}$$

where $\beta = L_{eff}^{-1}$ is used to estimate the effective gate delay and leakage during frequency binning and reliability tests.

The second well-known procedure is to use test structures (like the one depicted in Figure 5.24) to estimate printed CD for the mask under preset imaging conditions.[35,36] This method estimates W_{TS}, the electrical linewidth from the test structure, in terms of the sheet resistance R_{sh} and the bridge resistance R_b, both obtained from potential values at the probe pads. That is,

$$W_{TS} = \frac{R_{sh}}{R_b} L_{TS} \tag{5.23}$$

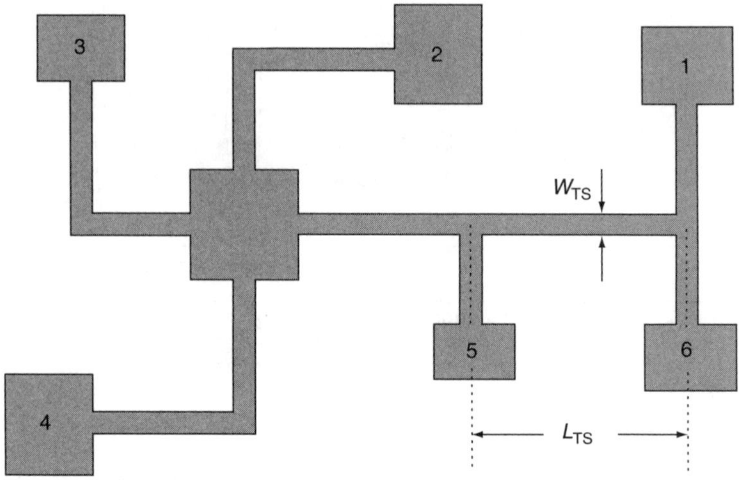

Figure 5.24 Resist test structure for electrical CD measurement.

The sheet resistance is obtained as a function of voltages observed at the probe pads of the test structure:

$$R_{sh} = \frac{\pi\left(|V_{2,5}| + |V_{5,2}| + |V_{4,5}| + |V_{5,4}|\right)}{8I_{sh}} \tag{5.24}$$

Thus, R_{sh} is obtained by first passing a current I_{sh} between pads 3 and 4 while measuring the voltage $V_{2,5}$ between pads 2 and 5; the current is then reversed to obtain $V_{5,2}$. The same procedure is adopted for measuring the voltage between pads 4 and 5 (and vice versa) as a current I_{sh} flows though pads 2 and 3. Likewise, the bridge resistance R_b is determined by passing a current I_b through pads 1and 3 while measuring the voltage change between pads 5 and 6. Thus,

$$R_b = \frac{|V_{5,6}| + |V_{6,5}|}{2I_b} \tag{5.25}$$

Line end effects are avoided by ensuring that the test structure has length significantly greater than the width of an interconnect (i.e., $W_{TS} >> L_{TS}$). Electrical linewidth measurement is also used to measure contact and via dimensions.[37] Let $L_{v\text{-}c}$ and $W_{v\text{-}c}$ denote (respectively) the width and length of vias or contact holes in the test structure, and let N be the number of these features. Then the effective diameter of via-contact holes is given by

$$D_{\text{v-c}} = \frac{W_{\text{ref}}}{2\sqrt{12}\pi}\left[\sqrt{1+\frac{48}{N}\left(\frac{L_{\text{v-c}}}{W_{\text{v-c}}}-\frac{L_{\text{v-c}}}{W_{\text{ref}}}\right)}-1\right] \tag{5.26}$$

where W_{ref} is the reference width of a similar test structure. Electrical measurements can be used to verify stability of results from SEM images.[38]

5.4.2.3 Scatterometry

Scatterometry is a method that complements the SEM technique. As with SEM, this procedure requires a large enough area to deduce CD and resist profile information. Scatterometry involves the use of a beam of light that is incident on a grating printed on the wafer. The reflectance of the light is measured as a function of the wavelength in order to obtain a profile. A schematic of a scatterometry setup is shown in Figure 5.25.

There are two types of scatterometry: one that changes the wavelength to obtain the reflectance of images, and one that uses varying incidence angles. When it's the wavelength of the light that is changed, the method is known as *spectroscopic ellipsometry*.[39] The plot

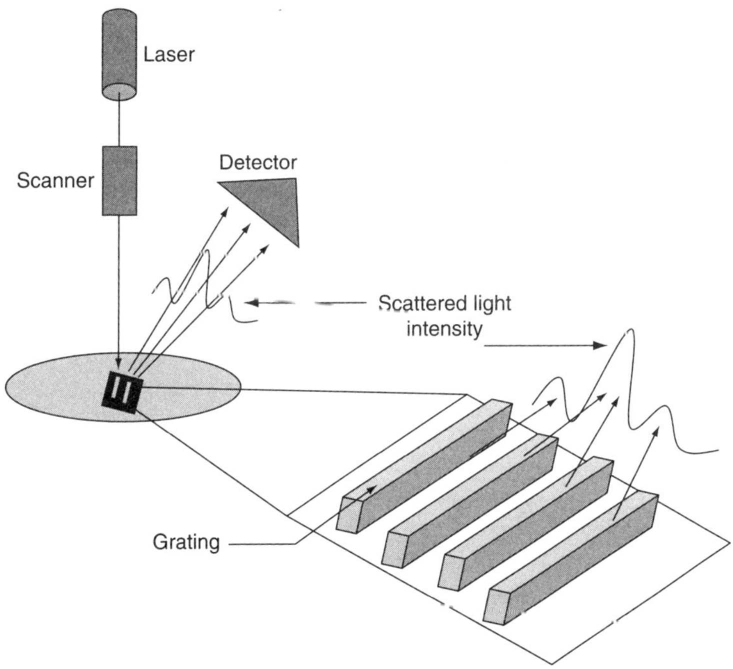

Figure 5.25 Scatterometry setup.

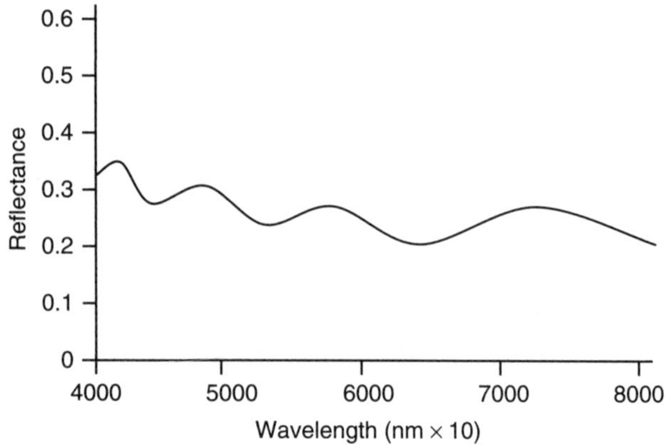

Figure 5.26 Reflectance of SiO_2 on Si in water.

of reflectance versus wavelength (see Figure 5.26) is compared to a precharacterized library for information on CD value, resist profile, and thickness. When a good match for this comparison is obtained, the profile is finalized. The method is known as *angle-resolved spectrometry* if it's the angle of the incident beam that is varied.[40]

The ellipsometer measures two parameters, referred to as "Del" (Δ) and "Psi" (Ψ), from which reflectance can be estimated.[41] Del is the phase difference $\varphi_1 - \varphi_2$ after reflection, and Psi is defined as follows:

$$\tan \psi = \frac{|R^p|}{|R^s|} \tag{5.27}$$

Note that the reflectance ρ can be described as the ratio of parameters R^p and R^s:

$$\rho = \frac{R^p}{R^s} = \tan \psi \times e^{i\Delta} \tag{5.28}$$

Unlike SEM, scatterometry is capable of generating the complete resist profile; hence it can be used more effectively than SEM to obtain CD information regarding variation in dose and focus.[41] Scatterometry is best suited for one-dimensional measurement of the resist profile. It cannot be used to measure post-OPC layout or contact and via features because the analysis of two-dimensional wavelength is too complex.

5.4.3 Overlay Metrology

Overlay refers to the registration of a mask pattern with the patterned structure of the underlying layer present on the wafer.[3] Most overlay tools use simple optical measurements to automatically evaluate how far the current feature's center is from the center of the pattern on the wafer. Overlay metrology is required for all features on the mask. Control of gate CD variation is critical in integrated circuit manufacturing, so overlay metrology for the smallest of features is especially important. Figure 5.27 shows examples of overlay errors.

Overlay is measured at specific points on the wafer by using special features (see Figure 5.28). Consider now the feature shown in Figure 5.29; the measurements L_{X0} and L_{X1} along the X axis are used to calculate the overlay for this particular wafer orientation as follows:

$$(\Delta X)_{0^\circ} = \frac{L_{X1} - L_{X0}}{2} \tag{5.29}$$

Tool-induced misalignment can occur because layer 0 and layer 1 of the exposure step use different materials. This misalignment causes asymmetric patterns on the wafer, as shown in the figure. Equipment problems that can lead to asymmetric imprints include nonuniform illumination, wafer tilt, lens aberrations, decentered lenses, and

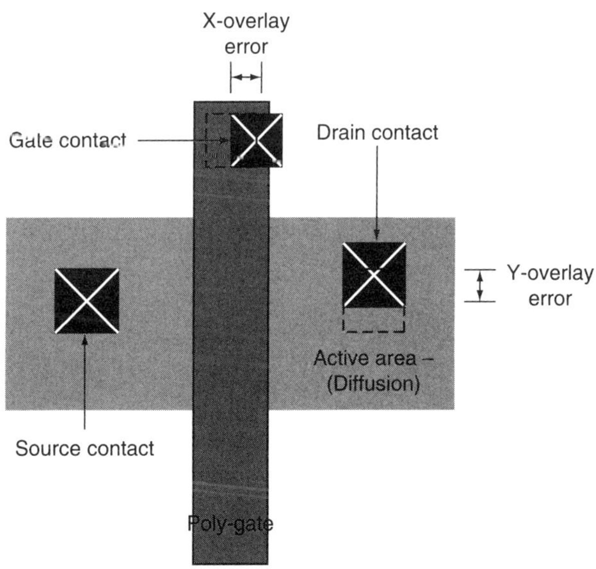

FIGURE 5.27 Overlay errors.

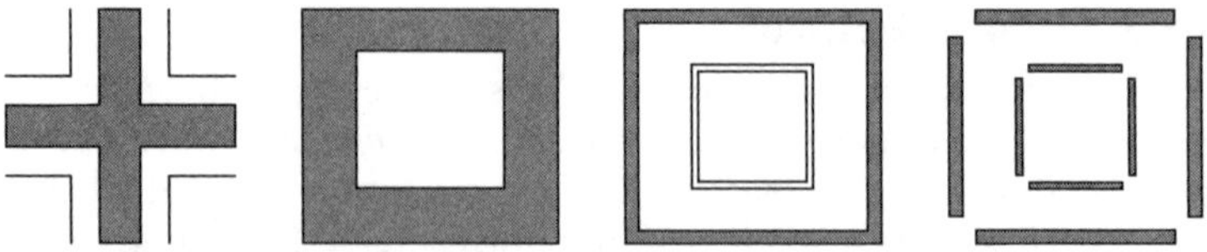

FIGURE 5.28 Typical overlay patterns.

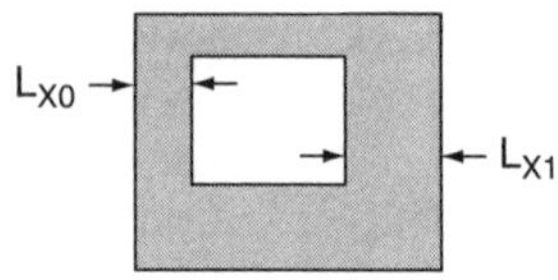

FIGURE 5.29 Misaligned overlay: the inner box is the underlying (substrate) pattern on wafer; the outer box is the second (polysilicon) pattern.

nonuniform detector response.[42,43,44] A simple verification of wafer overlay due to asymmetry can be performed by rotating the wafer by 180° and then recalculating the ΔX value:

$$(\Delta X)_{180^\circ} = \frac{L_{X0} - L_{X1}}{2} = -(\Delta X)_{0^\circ} \tag{5.30}$$

The sum of ΔX at the two orientations must be zero in order for the pattern imprint to be symmetrical. This method of wafer overlay measurement was the first to suggest appropriate shifts in mask position to improve patterning symmetry.[45] Improvements to this technique have been used to identify overlay errors such as translation errors, scaling, orthogonality and wafer rotation. Overlay errors can also result if the reticle is not flat and square (see Figure 5.30). To monitor wafer process errors that frequently occur with a particular equipment setup, a check process consisting of a dry wafer run is performed to keep track of within-die and die-to-die overlay errors.

As pattern density has increased with scaling, CMP causes material thickness undulations. This leads to problems with acquisition of alignment patterns and also creates overlay measurement issues (see Figure 5.31). Overlay measurement structures that compensate for material thickness variations have been suggested, as shown in Figure 5.28. The box-and-frame structures shown are typically less prone to CMP-induced measurement variation than are the cross and box-in-box structures. Better overlay measurements result when the features that help alignment are printed at linewidths greater than the minimum feature width. When feature sizes shrink to less than 45 nm, lens aberration and variations in line edge roughness can induce linewidth changes that exceed 10 percent. Feature placement errors for actual interconnect and gate patterns

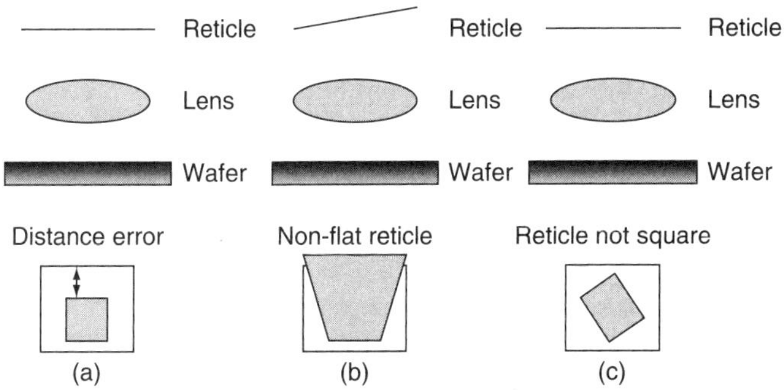

FIGURE 5.30 Lithographic stepper diagrams illustrating the relationship between reticle position and overlay errors.

FIGURE 5.31 Distinguishing overlay error from measurement error is complicated by CMP-induced edge irregularity.

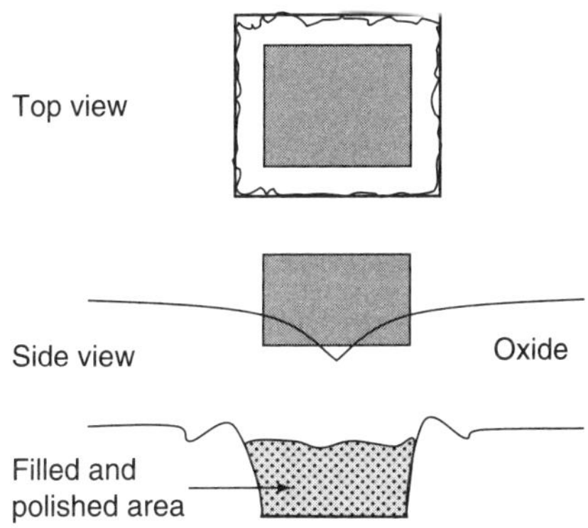

will differ from overlay errors. Hence, overlay patterns are printed at minimum feature widths not only to improve alignment but also to help with other measurements.

5.4.4 Other In-Line Measurements

In addition to measurements of CD and overlay, metrology is also required for measurements of gate dielectric thickness and other device parameters. Thus, metrology plays a vital role in establishing the proper specification of device and oxide variations for the SPICE modeling used in circuit simulation by design engineers. Scatterometry is used to measure gate dielectric thickness on both patterned and unpatterned wafers. The dielectric thickness is estimated using the optical parameters of the dielectric film structure, and the observed measurement depends on the size of the area being analyzed. Multiple-wavelength reflectance techniques similar to those of CD metrology are used to measure the exact thickness of the dielectric on the wafer. The ellipsometer measurement is affected not only by the process-dependent refractive index of the medium but also by the interaction at the oxide-silicon boundary (interface layer). Since the measurement is sensitive to both parameters, computationally robust optical models for the substrate-oxide interface are required in order to attain precise measurements.

An electrical technique for estimating effective dielectric thickness uses capacitance-voltage (c-v) data obtained from test structures on the wafer. The *effective* thickness refers to those regions in which the material acts either as a dielectric in a capacitor or as a transistor between two conducting surfaces. Thickness values measured electrically can differ from those measured optically because doping concentrations of polysilicon differ above and below the oxide. Given the current levels of dopant fluctuations and dielectric thicknesses approaching 2 nm, the thickness measurements used to estimate threshold voltages of devices on the wafer must be extremely accurate. Better process control will require the development of new methods that minimize errors in translating between optical to electrical information.

Because dopants control a transistor's operation, accurate measurement is crucial for an effective analysis of their implantation. Ion implantation is the predominant method used in doping semiconductors today. The types and concentration of implants vary with the terminal of a device. Implant steps are classified into three types: (1) high-energy dose for retrograde doping to form well structures; (2) medium-energy dose to form source, drain, and drain extension regions; and (3) low-energy dose for the threshold voltage implant in the channel. These doping types have different tolerance levels, so the metrology and its required precision varies. Well-known techniques for dopant measurement include the four-point probe, depth profiling with "secondary ion mass" spectrometry, and optically modulated reflections. For more details on the concepts of device dopant metrology, the reader is encouraged to consult the work of Current, Larson, and Yarling.[46,47]

5.4.5 In-Situ Metrology

There are two major limitations of the statistical process control used by manufacturing houses. First, only one or a few parameters are considered to be varying; and second, all measurements are delayed versions of the past state of the process or tool. In-situ metrology incorporates sensors to make a wide range of measurements throughout the manufacturing processes. In-situ metrology has three principal aims: (1) find process anomalies by considering a wide domain of parameters; (2) detect process variation within a short time frame; and (3) decrease the variance of wafer state parameters, not the number of checkpoints. Advanced process control is a reactive engine that involves the use of in-situ sensors. The control system starts with processing information from sets of in-situ sensors and other metrology tools to arrive at a performance metric for the tool. When measurements indicate that some predefined specifications are not being met, the process is terminated or model tuners are used to reoptimize the tool settings so that performance returns to being within specifications. Thus in-situ metrology is part of the feedback process control engine.

In-situ sensors used during semiconductor manufacturing are either wafer-state sensors or process-state sensors. The most critical parameters concern the wafer state because they are directly tied to effective monitoring and control of the manufacturing process. *Wafer-state* sensors are used to assess film and resist thickness, uniformity of thickness, and resist profile. The sensors use optical techniques (e.g., scatterometry, interferometry, and reflectometry) to measure the required parameters. Optical models are then used to relate the phase change of incident light or other electromagnetic beams to the process parameter. However, certain process stages are not amenable to wafer-state sensors—either because the appropriate sensor technology does not exist or because sensors are poorly integrated with the processing tools. In this case, *process-state* sensors are used to monitor the manufacturing tool. In many cases these sensors are less expensive and easier to control. The most important application of process-state sensors is to determining the endpoint, which is accomplished by continuous measurement of a particular signal during the processing of the wafer. The primary function of these sensors is to identify when parameter attributes change. Typical measurement applications include temperature, gas phase composition, and plasma properties. There are currently no rigorous models that relate process-state measurements to actual wafer parameters, so these sensors are mainly employed in fault detection. The use of process-state sensors for this specific purpose has been found to increase process yield.[3]

5.5 Failure Analysis Techniques

Semiconductor failure analysis is the process of determining why and how a particular device failed and how future occurrences can be prevented. *Device failure* refers not only to catastrophic failures but also to a device's failure to conform with electrical, mechanical, chemical, or visual specifications. Failures can be functional or parametric in nature. A flowchart for semiconductor failure analysis is shown in Figure 5.32. The first stage is failure verification, which confirms that a failure has indeed occurred. This verification stage also characterizes the *failure mode* of the system (i.e., the conditions of system failure). It is critical to validate the reproducibility of failure and to characterize the failure mode before engaging in further analysis.

Given the characterized system and information on its failure modes, the next stage is failure analysis. Different FA techniques are

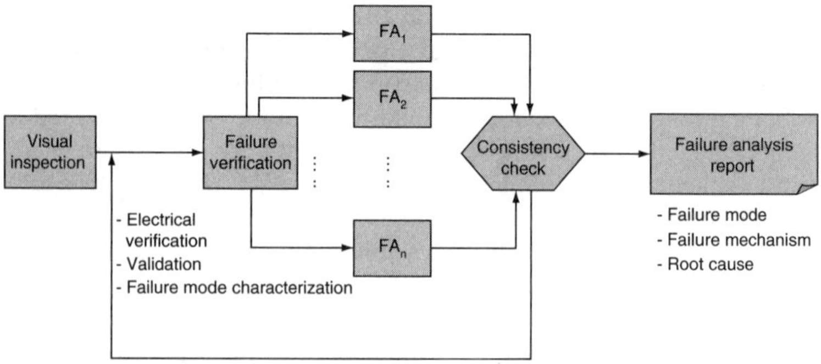

FIGURE 5.32 Flowchart for analyzing semiconductor failure.

used depending on failure mode characterization. Nondestructive FA techniques are used before destructive ones. During each step of the analysis, attributes of the device and its response to inputs is observed periodically. Consistency among various FA techniques is important for finding the correct failure mode and targeting the exact failure location. If two FA techniques are used to analyze a particular failure, then both must provide the same type of result (positive or negative) regarding the failure. For example, two separate FA techniques cannot report the presence *and* absence of a bridging defect at the same location. In order to ensure accurate information on failure mode, consistency checks are repeated by rerunning the failure verification procedures. If consistency has been achieved, then the results of the FA will point to the true failure site. Failure analysis is complete once the location of the failure, mode of occurrence, and mechanism have been identified (see Figure 5.32). That is, FA yields a report on the following information.

1. *Failure mode:* describes how the device actually failed, by how much it deviated from specifications, etc.
2. *Failure mechanism:* details the possible mechanism of failure occurrence (e.g., radiation, corrosion, ESD, thermal stress).
3. *Failure cause:* lists the input events or conditions that triggered the failure.

Failure analysis techniques are individual approaches to discovering the cause, mechanism, and location of failures. Each FA technique employs a unique procedure to analyze the device, thereby providing specialized information on the failure.

Destructive testing is necessary for a large portion of the defects. Of course, failure verification and nondestructive testing must be performed *before* destructive testing, as the latter process results in irreversible damage. If a component is already severely damanged, then electrical testing and verification may not be possible. Care must be taken when removing electrical components so that secondary damage is not introduced. In such cases, optical examination may be the only suitable alternative.

5.5.1 Nondestructive Techniques

5.5.1.1 Failure Verification

Failure verification is essential for establishing the existence of a failure and also for characterizing the suspect device. The first step is to perform electrical testing using automatic test equipment (ATE). The target device is tested under varying input conditions to detect which patterns lead to failure. Results from ATE-based verification are correlated with production standards, which enhances the validity of the result. The FA process uses the report generated by ATE equipment to facilitate finding the error site, failure mechanisms, and input conditions.

Characterizing the failure more completely involves additional testing of the component, which may include I-V (current-voltage) analysis and simulation of the failure condition. Plotted I-V characteristics reveal the behavior of various attributes of the device under test. Selected nodes are connected using microprobes, after which voltage is applied to the component. The resulting current characteristic is measured and compared to an ideal model to identify suspect failure modes during device operation. Another type of testing is to excite the component and then measure its responses to a given input. One drawback of this approach is that the testing circuit setup changes when a new set of parameters are to be analyzed. To overcome this limitation, device operation can be simulated while attempting to replicate the failure. The simulation is set up to modify various input parameters and characteristics of the device, observe the response, and identify the failure.

5.5.1.2 Optical Microscopy

With optical microscopy, failures are located and analyzed by inspecting the component with a high-power optical microscope. The setup consists of the sample being placed perpendicular to the direction of light. Light incident on the sample reflects off the surface and back to the lens, thus providing an enhanced, magnified image. The actual image captured by the microscope depends on the light wavelength, the lens system, and the sample material. Three types of illumination are used to detect varying types of device failure. *Light-field* illumination consists of uniformly focused light over the sample under test. The reflectance information from this illumination setup detects topographical modulation of the sample surface, which can be used to detect thickness variations in the component. *Dark-field* illumination eliminates the central light cone and only allows light through the periphery, forming a ring. This light impinges on the surface at an angle, which causes scattered light from rough edges and other irregularities in the surface. Dark-field illumination is used to detect surface scratches and other contamination. *Interference contrast illumination* uses polarized light to displace light rays that take different paths along the sample surface. The reflected waves produce interference fringes in the image plane, which are used to detect surface defects such as etch errors and cracks. This method is especially effective for examining fractures, chemical damage, and other small defects not visible under normal illumination.

5.5.1.3 X-Ray Radiography

X-ray radiography is a nondestructive method of analysis that is well suited to detecting many types of interior package defects observed in semiconductors. X-ray radiography is based on the phenomenon that x-ray transmission through different materials varies systematically with material density. Different regions of the package transmit x rays with differing contrast levels, which can be imaged onto a film. X-ray signals transmitted through the system are collected by a detector and amplified to produce a good image. Low-density regions of the package appear bright in the x-ray image. X-ray radiography is commonly used to inspect cracks, wire bonding problems, and voids in the die or package.

5.5.1.4 Hermeticity Testing

Hermeticity testing is often used to assess the integrity of device encapsulation. This analysis is typically used to detect package seal cracks, incomplete sealing, and wetting within the package. The purpose of the encapsulation is to prevent gases or fluids from leaking into the package. Moisture intrusion and gas attack on a device's internal surfaces will result in corrosion and failure. In semiconductor devices, moisture attacks cause physical corrosion, electrical leakage, and shorts. To analyze package integrity, two types of leak testing are performed: the gross leak test and the fine leak test. The first step in each test type involves a vacuum cycle to remove trapped gases and moisture within the package. In the gross leak test, the package is soaked in fluorocarbon liquid under pressure; this is followed by visual inspection of the package for signs of bubble emission, which would indicate leak failure. In the fine leak test, the package is soaked in pressurized helium gas, which drives helium atoms into accessible places in the package; then the leakage rate of helium atoms is measured in a vacuum. Alternative techniques have been suggested that use the same principle steps of vacuuming, immersion, and inspection but with various other fluids and dyes. Hermeticity testing is a nondestructive secondary FA technique that is commonly used to aid in postulating the probable cause of failure.

5.5.1.5 Particle Impact Noise Detection

Particle impact noise detection (PIND) systems are employed to detect loose particles within unfilled cavities present in the device.[48]

When applied to a device, noise signals can excite loose particles, which are then detected by a transducer. This testing procedure is performed in the presence of high leakage, intermittent behavior, and/or shorts. Devices that fail PIND tests must be further analyzed to find the nature of the particle under stress. The PIND test is also a secondary nondestructive FA technique.

5.5.2 Destructive Techniques

5.5.2.1 Microthermography

Microthermography is a well-known semiconductor FA technique used to locate areas on the die surface that exhibit high thermal gradients (aka hotspots). Excessive heat indicates high current flow, which could be caused by circuit abnormalities, high current density, dielectric breakdowns, and open or short junctions. Hotspots are detected by dropping a ball of liquid crystal onto the die surface while maintaining the die in a biased (powered-up) state to create a temperature gradient. At low temperatures, the liquid crystals remain solid; but as the temperature rises, the crystal liquefies and so its visual appearance on the die changes.

Liquid crystal can appear in one of two different phases. In the *isotropic* phase, the liquefied crystal is highly homogenous; thus it appears completely black under an optical microscope when polarized light is shone upon the die. This phase is unsuitable for detecting temperature gradients. In the *nemetic* phase, however, the light reflected back passes through the analyzer to form a distinct, prismatic pattern on the die. When a die is coated with nemetic films, hotspots appear black under the microscope, and this is how they are detected.

5.5.2.2 Decapsulation

Decapsulation is an FA technique used to reveal internal construction and to uncover device failure. The plastic package is opened without altering the failure mode. Decapsulation techniques can be either mechanical or chemical. The *mechanical decapsulation* process involves the application of opposing forces to the top and bottom of the package in order to remove the seal glass or pry the lid of the ceramic package. *Chemical decapsulation* techniques include chemical, jet, and plasma etching that employ external etchant materials to perform chemical decapsulation. Acid-based chemical etching involves the use of fuming acids such as sulfuric and nitric acids. These acids do not etch selectively but instead attack materials indiscriminately while performing decapsulation.

5.5.2.3 Surface Analysis

Surface analysis using x-rays is a useful supplement to SEM metrology. When a sample is bombarded with a high-energy beam of electrons, the x rays emitted from the surface can be captured with a detector. The penetration depth of the electron beam within the silicon is a function of the energy of the emitted x ray. X rays interact with silicon atoms to generate electron-hole pairs, thereby generating currents. These currents are sampled to find magnitudes that are correlated with the x-ray peaks, signaling the presence of various elements in the specimen. A specialized surface analysis known as *auger electron spectroscopy* (AES) involves ion etching of the surface followed by analysis of the resulting depth profile of the contamination. Other techniques for surface analysis include *secondary ion mass spectrometry* (SIMS), which is used to measure directly the dopant profiles in the semiconductor, and *energy spectroscopy chemical analysis* (ESCA), which utilizes information on the valence state of the material to analyze material composition on the device surface.

5.6 Summary

This chapter began with a brief discussion of the semiconductor manufacturing processes. We provided an overview of process-induced defects, of their sources and electrical impact, and of defect models. The various proposed particle defect models were explained, along with their use in CA-based yield analysis. We discussed patterning problems that can lead to pattern-dependent catastrophic device failures due to errors in diffusion, vias, contacts, and interconnect. Variations in thickness due to CMP can cause defocus errors, contributing to defect formation. We described how pattern density can be correlated to CMP-related thickness variations and also to local etch problems. We then proceeded to examine various layout engineering techniques to mitigate both particulate- and pattern-induced errors. In addition, various metrology techniques and their applications to semiconductor measurement for process control were described. Finally, we introduced semiconductor failure analysis by describing the various destructive and nondestructive techniques in use today.

References

1. *International Technology Roadmap for Semiconductors Report*, http://www.itrs.net (2007).
2. A. V. Ferris-Prabhu, "Role of Defect Size Distributions in Defect Modeling," *Transactions of Electron Devices* **32**(9): 1727–1736, 1985.
3. R. Doering and Y. Nishi, *Handbook of Semiconductor Manufacturing Technology*, CRC Press, Boca Raton, FL, 2007.
4. B. R. Mandava, "Critical Area for Yield Models," IBM Technical Report no. TR22.2436, 1992.
5. T. Okabe, M. Nagata, and S. Shimada, "Analysis of Yield of Integrated Circuits and a New Expression for the Yield," *Proceedings of Electrical Engineering Japan* **92**: 135–141, 1972.
6. C. H. Stapper, "Defect Density Distribution for LSI Yield Calculations," *IEEE Transactions on Electron Devices* **20**: 655–657, 1973.
7. I. Koren, Z. Koren, and C. H. Stapper, "A Unified Negative Binomial Distribution for Yield Analysis of Defect Teolerant Circuits," *IEEE Transactions on Computers* **42**: 724–737, 1993.

8. I. Koren, Z. Koren, and C. H. Stapper, "A Statistical Study of Defect Maps of Large Area VLSI ICs," *IEEE Transactions on VLSI Systems* **2**: 249–256, 1994.
9. C. H. Stapper, "One Yield, Fault Distributions and Clustering of Particles," *IBM Journal of Research and Development* **30**: 326–338, 1986.
10. C. H. Stapper, "Small-Area Fault Clusters and Falut-Tolerance in VLSI Circuits," *IBM Journal of Research and Development* **33**: 174–177, 1989.
11. I. Koren and C. H. Stapper, "Yield Models for Defect Tolerant VLSI Circuits: A Review," in *Proceedings of Workshop on Defect and Fault Tolerance in VLSI Systems*, IEEE Computer Society Press, Los Alamitos, 1989, vol. 1, pp. 1–21.
12. O. Paz and T. R. Lawson, Jr., "Modification of Poisson Statistics: Modeling Defects Induced by Diffusion," *IEEE Journal of Solid-State Circuits* **12**: 540–546, 1977.
13. B. Murphy, "Cost-Size Optima of Monolithic Intergrated Circuits," *Proceedings of IEEE* **52**: 1537–1545, 1964.
14. W. E. Beadle, R. D. Plummer, and J. C. Tsai, *Quick Reference Manual for Silicon Integrated Circuit Technology*, Wiley, New York, 1985.
15. V. K. R. Chiluvuri and I. Koren, "New Routing and Compaction Strategies for Yield Enhancement," in *Proceedings of IEEE International Workshop on Defect and Fault Tolerance in VLSI Systems*, IEEE Computer Society, Los Alamitos, 1992, pp. 325–334.
16. J. Fang, J. S. K. Wong, K. Zhang, and P. Tang, "A New Fast Constraint Graph Generation Algorithm for VLSI Layout Compaction," in *Proceedings of IEEE International Symposium on Circuits and Systems*, IEEE, New York, 1991, pp. 2858–2861.
17. A. B. Kahng, "Alternating Phase Shift Mask Compliant Design," U.S. Patent no. 7,124,396 (2006).
18. J. Luo, S. Sinha, Q. Su, J. Kawa, and C. Chiang, "An IC Manufacturing Yield Model Considering Intra-Die Variations," in *Proceedings of the Design Automation Conference*, IEEE/ACM, New York, 2006, pp. 749–754.
19. A. Genz, "Numerical Computation of Multivariate Normal Probabilities," *Journal of Computational and Graphical Studies* **1**: 141–149, 1992.
20. J. P. Cain and C. J. Spanos, "Electrical Linewidth Metrology for Systematic CD Variation Characterization and Causal Analysis," in *Proceedings of SPIE Optical Microlithography*, SPIE, Bellingham, WA, 2003, pp. 350–361.
21. E. W. Charrier and C. A. Mack, "Yield Modeling and Enhancement for Optical Lithography," *Proceedings of SPIE* **2440**: 435–447, 1995.
22. A. Sreedhar and S. Kundu, "On Linewidth-Based Yield Analysis for Nanometer Lithography," in *Proceedings of Design Automation and Test in Europe*, IEEE/ACM, New York, 2009, pp. 381–386.
23. C. A. Mack, *Fundamental Principles of Optical Lithography*, Wiley, New York, 2008.
24. R. Datta, J. A. Abraham, A. U. Diril, A. Chatterjee, and K. Nowka, "Adaptive Design for Performance-Optimized Robustness," in *Proceedings of IEEE International Symposium on Defect and Fault-Tolerance in VLSI Systems*, IEEE Computer Society, Washington DC, 2006, pp. 3–11.
25. D. Ouma, D. Boning, J. Chung, G. Shinn, L. Olsen, and J. Clark, "An Integrated Characterization and Modeling Methodology for CMP Dielectric Planarization," in *Proceedings of IEEE International Interconnect Technology Conference*, IEEE Press, Piscataway NJ, 1989, pp. 67–69.
26. R. Tian, D. F. Wong and R. Boone, "Model-Based Dummy Feature Placement for Oxide Chemical-Mechanical Polishing Manufacturbility," in *Proceedings of Design Automation Conference*, IEEE/ACM, New York, 2000, pp. 902–910.
27. A. B. Kahng and K. Samadi, "CMP Fill Synthesis: A Survey of Recent Studies," *IEEE Transactions on Computer-Aided Design of Integrated Circuits and Systems* **27**(1): 3–19, 2008.
28. D. H. Stamatis, *TQM Engineering Handbook*, CRC Press, Boca Raton, FL, 1997.
29. R. R. Hershey and M. B. Weller, "Nonlinearity in Scanning Electron Microscope Critical Dimension Measurements Introduced by the Edge Detection Algorithm," *Proceedings of SPIE* **1926**: 287–294, 1993.
30. J. I. Goldstein, D.E. Newbury, P. Echlin, D. C. Joy, C. Fiori, and E. Lifshin, *Scanning Electron Microscopy and X-Ray Microanalysis*, 2d ed., Plenum Press, New York, 1984.

31. J. Finders, K. Ronse, L. Van den Hove, V. Van Driessche, and P. Tzviatkov, "Impact of SEM Accuracy on the CD-Control during Gate Patterning Process of 0.25-µm Generations," in *Proceedings of the Olin Microlithography Seminar*, Olin Microelectronic Materials, Norwalk CT, 1997, pp. 17–30.
32. M. Davidson and N. T. Sullivan, "An Investigation of the Effects of Charging in SEM Based CD Metrology," *Proceedings of SPIE* **3050**: 226–242, 1997.
33. C. M. Cork, P. Canestrari, P. DeNatale, and M. Vascone, "Near and Sub-Half Micron Geometry SEM Metrology Requirements for Good Process Control," *Proceedings of SPIE* **2439**: 106–113, 1995.
34. M. T. Postek, A. E. Vladar, and M. P. Davidson, "Fourier Transform Feedback Tool for Scanning Electron Microscopes Used in Semiconductor Metrology," *Proceedings of SPIE* **3050**: 68–79, 1997.
35. L. J. Zynch, G. Spadini, T. F. Hassan, and B. A. Arden, "Electrical Methods for Precision Stepper Column Optimization," *Proceedings of SPIE* **633**: 98–105, 1986.
36. L. W. Linholm, R. A. Allen, and M. W. Cresswell, "Microelectronic Test Structures for Feature Placement and Electrical Linewidth Metrology," in K. M. Monahan (ed.), *Proceedings of Handbook of Critical Dimension Metrology and Process Control*, SPIE Press, Bellingham, WA, 1993.
37. B. J. Lin, J. A. Underhill, D. Sundling, and B. Peck, "Electrical Measurement of Submicrometer Contact Holes," in *Proceedings of SPIE* **921**: 164–169, 1988.
38. E. E. Chain and M. Griswold, "In-Line Electrical Probe for CD Metrology," *Proceedings of SPIE* **2876**: 135–146, 1996.
39. N. Jakatdar, X. Niu, J. Bao, C. Spanos, S. Yedur, and A. Deleporte, "Phase Profilometry for the 193nm Lithography Gate Stack," *Proceedings of SPIE* **3998**: 116–124, 2000.
40. P. C. Logafătu and J. R. Mcneil, "Measurement Precision of Optical Scatterometry," *Proceedings of SPIE* **4344**: 447–453, 2001.
41. J. Allgair, D. Beniot, R. Hershey, L. C. Litt, I. Abdulhalim, B. Braymer, M. Faeyrman, et al., "Manufacturing Considerations for Implementation of Scatterometry for Process Monitoring," *Proceedings of SPIE* **3998**: 125–134, 2000.
42. R. M. Silver, J. Potzick, and R. D. Larrabee, "Overlay Measurements and Standards," *Proceedings of SPIE* **3429**: 262–272, 1995.
43. D. J. Coleman, P. J. Larson, A. D. Lopata, W. A. Muth, and A. Starikov, "On the Accuracy of Overlay Measurements: Tool and Mask Asymmetry Effects," *Proceedings of SPIE* **1261**: 139–161, 1990.
44. A. Starikov, D. J. Coleman, P. J. Larson, A. D. Lopata, and W. A. Muth, "Accuracy of Overlay Measurement Tool and Mask Asymmetry Effects," *Optical Engineering* **31**: 1298–1310, 1992.
45. M. E. Preil, B. Plambecj, Y. Uziel, H. Zhou, and M. W. Melvin, "Improving the Accuracy of Overlay Measurements through Reduction in Tool and Wafer Induced Shifts," *Proceedings of SPIE* **3050**: 123–134, 1997.
46. C. B. Yarling and M. I. Current, "Ion Implantation Process Measurement, Characterization and Control," in J. F. Zeigle (ed.), *Ion Implantation Science and Technology*, Academic Press, Maryland Heights, MO, 1996, pp. 674–721.
47. L. L. Larson and M. I. Current, "Doping Process Technology and Metrology," in D. G. Seiler et al. (eds.), *Characterization and Metrology for ULSI Technology*, AIP Press, New York, 1998.
48. P. L. Martin, *Electronic Failure Analysis Handbook*, McGraw-Hill, New York, 1999.

CHAPTER 6

Defect Impact Modeling and Yield Improvement Techniques

6.1 Introduction

With increasing device density, manufacturing defects and large process variations lead to higher rates of device failure. The previous chapter dealt with two types of defects that occur in semiconductor manufacturing: particle (process-induced) defects and lithography (pattern-dependent) defects. With large process variations, circuits suffer from parametric failures. Metrology and failure analysis techniques aim to identify the root cause of a failure induced by a defect, its failure modes, and its input conditions. A defect is said to cause circuit irregularity only if it manifests itself as a fault. A fault can cause either logic failure or parametric failure. Catastrophic failures are usually tied to logic faults, whereas parametric variations occur because of changes in some attributes of a device. For example, a defective oxide layer can lead to threshold voltage variation, which may manifest as a parametric fault.

A designer's role is to hypothesize about such potential defects and to generate test patterns that can effectively screen for defective parts. In a designer's world, fault models are pivots to generating test patterns. Without effective test patterns, defective chips cannot be screened at a manufacturing site. When a defective part ends up in a board, the downstream cost of testing it and replacing the faulty chip is typically far greater than the cost of a manufacturing test at the factory. This explains the importance of fault models. Section 6.2 describes defect and fault models, whose purposes are to hypothesize about a defect's location and its behavior in faulty mode.

In contrast, the purpose of a yield model is to predict the number of good chips produced per wafer or per lot; this prediction is based on defect and fault distribution statistics. Yield may be classified as functional or parametric. Any chip that works correctly under *some* voltage, frequency, and temperature conditions is included in functional yield. Parametric yield is a proper subset of functional yield that consists of chips that work under *stipulated* voltage, frequency, and temperature conditions. Parametric yield may differ significantly from functional yield if there are large parameter variations in a design. In the previous chapter we saw how defect probability may be translated into a probability of failure, which then translates into a yield model. The process-induced yield model calculates the yield of a design based on the statistical distribution of defect size and the critical area of the layout.

Manufactured devices vary in performance, so optimizing parametric yield is a design responsibility. Designers need tools to predict parametric yield at design time. Yield predictions are based on statistical timing analysis, which requires information about the range of process variation. A process variation that is large creates a design optimization problem as well as parametric yield problems. Assuming the worst-case scenario may lead to unreasonable device sizing during design optimization, increasing power dissipation and creating design convergence problems. On the other hand, basing a design only on "nominal" process variation may lead to parametric yield loss. Thus, assumptions made during the design phase affect the area, power characteristics, and parametric yield of a design. These factors must be traded off carefully.

Failure analysis leads to better understanding of defect profiles. For example, some defects may be related to specific layout structures that can be avoided by excluding those structures. Many random defects occur in clusters and can be isolated and replaced by spare functions. It is thus possible to minimize defect formation with better design and also to survive any defects that do arise with planned redundancy. *Fault avoidance* refers to a collection of design practices that reduce the possibility of a defect. *Fault tolerance* employs hardware design and software techniques that compel circuits to behave according to specification even when a defect may be present. Designs that incorporate fault avoidance and tolerance lead to better yield.

Many spot defects occur as clusters. This knowledge has been exploited to improve the yield of several design structures at a low overhead. Today's SRAM and DRAM chips incorporate spare rows, spare columns, or spare blocks to take advantage of defect clustering; this implementation of low-overhead redundancies allows these memories to survive defects. Such designs use fuse programming to activate and deactivate spares. Redundancy-based, fault-tolerant design techniques—together with the use of programmable fuses—are discussed in Sec. 6.3. The properties of transistors and interconnect wires also vary when defects are introduced by the patterning process. New techniques have been devised that help harden the circuit against such defects. A few of the techniques described here involve widening of wires, transistor sizing, and gate biasing.

6.2 Modeling the Impact of Defects on Circuit Behavior

Errors can occur during the circuit realization process and also during the manufacturing process. Circuits are "realized" through successive steps of refinement. Each step involves manipulating the circuit representation, a process that is prone to errors. Such errors are addressed by design verification at every step of the circuit transformation process. Errors that occur during circuit realization are typically referred to as presilicon errors. Such errors include:

1. *Logic errors*—improper functionality of the circuit
2. *Timing errors*—not meeting required performance goals (often due to insufficient tolerance to process variations)
3. *Physical design errors*—resulting from DRC and/or DFM guidelines not being met, inadequate OPC, etc.

These presilicon errors must be detected, analyzed, and corrected *before* the product is sent to the foundry, where downstream repairs

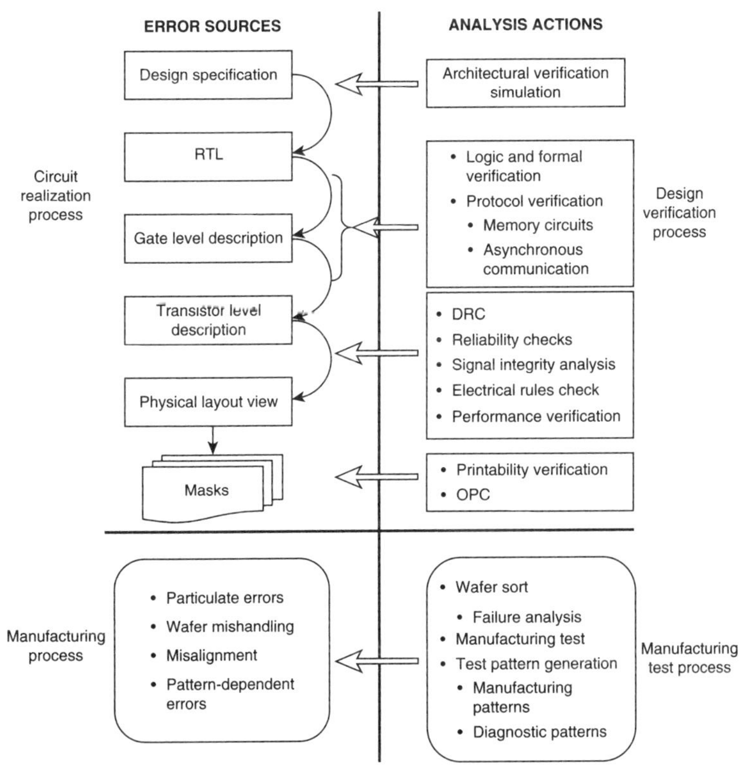

FIGURE 6.1 Error sources and analysis actions.

are much more expensive. Figure 6.1 depicts the sources of manufacturing and circuit errors and the associated responses of analysis and verification. A typical verification suite includes architectural pattern simulation, formal verification, logical equivalency checks, design rule checks, signal integrity analysis, printability verification, electrical rules check, timing analysis, and reliability analysis.

Defects that occur during the manufacturing process are referred to as *postsilicon defects*. These defects include random or spot defects as well as systematic defects, which are usually lithography and/or pattern dependent. The sources of such defects and their mechanisms were detailed in Chapter 5. The first analysis action performed after manufacturing is a wafer test that distinguishes good from bad dies. Good dies then continue through the process, undergoing further tests and quality assurance checks; defective dies may be targeted for failure analysis or simply discarded. As explained in Sec. 5.5, failure analysis is used to obtain detailed reports on the cause, failure mode, and mechanism of various postsilicon defects. Once a failure mechanism's origin is identified, steps may be taken during subsequent design work to prevent such defects at the outset. These considerations underscore the importance of modeling defect mechanisms.

6.2.1 Defect-Fault Relationship

A manufacturing defect becomes a fault when it manifests itself as an observable design failure. Manufacturing defects such as spot defects and systematic defects may result in functional and parametric design failures. The relationship between manufacturing defects (aka deformations) and integrated circuit failures is illustrated in Figure 6.2.[1] The lower part of the figure classifies defects by type and extent; the upper part classifies circuits in terms of their operating condition. A functional failure is attributed to structural fault, whereas performance-related failures are attributed to parametric faults. Solid lines indicate a direct relationship between defect and failure, while the dashed lines indicate an indirect relationship[1]. This type of representation for a classification of faults and error sources has been derived from the defect-fault relationship described in Maly *et.al.*[1]

Each deformation may manifest as a number of different faults whose underlying causes are defects arising from particulate- or pattern-induced errors. Particle defects can be classified based on the geometric and electrical effects. Pattern-dependent defects originate from the etching process, diffraction during image transfer through the projection system, or chemical-mechanical polishing. A particular defect's region of influence determines whether it is global or local. Global defects lead to malfunction of multiple devices or interconnects present in an integrated circuit, and they are relatively easy to detect. Local defects affect a smaller region of the IC, so they are harder to spot without targeted tests. Hard and soft performance failures are classified as a function of defect-induced electrical effects, which

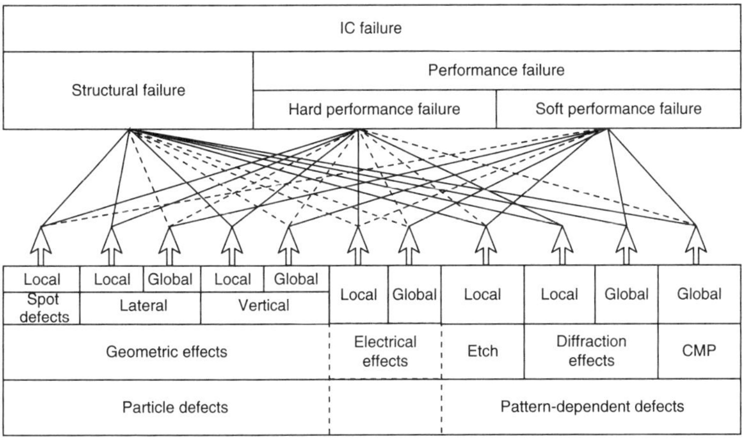

Figure 6.2 IC manufacturing process deformations and their relationship to IC faults.

rarely affect the IC's structural operation. Global defects are more likely to cause soft performance failure, and they are typically controlled by effective process control and process maturity.

6.2.2 Role of Defect-Fault Models

There are three principal motivations for the modeling of defects and faults: avoiding defects, estimating yield, and generating test patterns. Defect avoidance requires understanding the nature of defects and how they relate to circuit structures. Yield estimation requires understanding the spatial distribution of defects in terms of their size and frequency. Test patterns are usually generated based on fault models in which defects are characterized by their impact on circuit behavior rather than by their frequency, physical size, or extent of clustering.

Although these three applications are quite different, the unifying theme in modeling is to understand how defects relate to physical structures, the scope of a defect's influence, and the impact of defects on circuit behavior. Fault modeling for the purpose of generating test patterns can similarly be classified into three categories: (1) *defect-based* fault models, which use defect location and size to estimate impact on circuit behavior; (2) *abstract* fault models, which rely solely on abstract models of failure; and (3) *hybrid* fault models, which are rooted in defect structures but ultimately map to abstract fault models.

Defect-based modeling is the most direct technique for modeling faults. This method aims to model the exact behavior of the defect, the failure mode, and the input/output behaviors. Defects are often related to physical structures, in which case their likelihood and location can be predicted reasonably well. Such predictions are central to defect-based fault models. A complex fault model may incorporate the elements and classification structure illustrated in Figure 6.3.[2]

Fault models are used in automatic test pattern generation (ATPG) to create test vectors or in pattern simulation to estimate coverage. The patterns are then applied to the circuit in order to screen for defects. The role of fault models is to provide a logical basis for test pattern generation. If the derived test patterns are effective in screening defects, then the underlying fault model is considered valid regardless of its type.

6.2.2.1 Defect-Based Fault Models

Defect-based fault modeling (DBFM) relies on circuit structures to predict where a defect might occur as well as its severity and modes of failure. Fault models that are based on realistic defects are extremely precise and hence provide effective fault coverage if used for pattern generation. Yet even though DBFM is comprehensive in capturing fault behavior, it is not ATPG-friendly. This is because ATPG algorithms are effective only when dealing with a limited set of constraints. Given the elaborate constraints and timing relations between signals, ATPG often fails to provide a solution. These ATPG problems deter more extensive use of DBFM. However, the simulations run by defect-based fault models are usually not complex and often provide useful diagnostic information. Examples of DBFM include bridging faults, stuck-open faults, IDDQ faults, and analog faults.[3] These models attempt to predict the behavior of a circuit under the defect condition. It is tempting to continue improving the accuracy with which fault models mimic defect behavior. But just as in other areas of engineering, there is always a trade-off between accuracy and computational efficiency.

Defect-based fault modeling is predominantly used to model open or short lines and stuck-on or stuck-open transistors. Recall

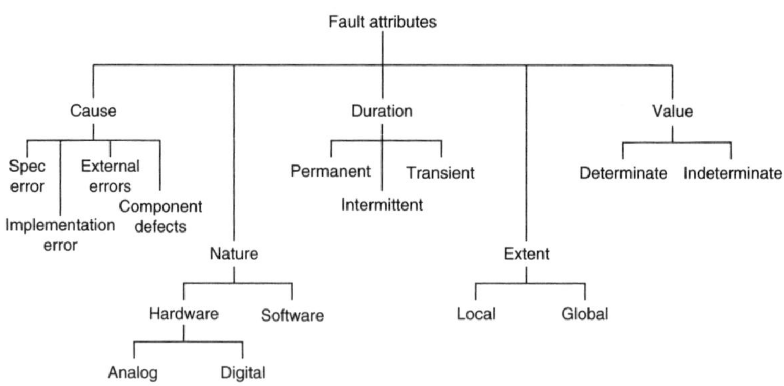

Figure 6.3 Classification of fault (failure) attributes.

that, when two or more lines are bridged as a result of particulate or printability defects, a line short fault (aka bridging fault) is created; in contrast, a line open fault is created by particulate- or lithographic patterning-induced electrical discontinuity. Similarly, if defects in linewidth, alignment, or doping induce "punch through" in a transistor, the result is a stuck-on fault. A transistor that cannot be turned on is manifesting a stuck-open fault. The fault models may further be refined when resistances are considered. For example, a transistor in the stuck-on state may behave like a resistor, and resistive line opens and shorts may be described similarly. In CMOS circuits, a fault that creates a floating condition may also induce memory effects. Two or more consecutive test patterns may be required before such faults are detected.

Consider the circuit sketched in Figure 6.4. It shows a CMOS NAND gate with inputs A and B connected to pMOS transistors P1 and P2 and nMOS transistors N1 and N2. Let us assume that a CMOS transistor may be idealized as switch that turns ON (switch closed) and OFF (switch open) depending on its input. The stuck-open fault in transistor P1 results in a condition where this transistor never conducts. During the normal operation of the gate, if inputs AB = 11 then the N1 and N2 switches are closed while the P1 and P2 switches are open, thus connecting the output node OUT to ground (GND). Under input condition AB = 01, P1 should establish a path from output to power supply (V_{DD}). However, in a faulty circuit where P1 is stuck open, there is no conducting path to any of the power terminals from the output node. Because there is no conducting path to either V_{DD} or GND, the output node is in a floating state. In this state, the voltage at the node OUT is determined by the stored charge in the capacitor. If the inputs 01 were immediately preceded by the inputs 11, then the capacitor would store a value of 0 even though the fault-free output would be 1. This discrepancy leads to detection of the fault. However, if the capacitance were charged to logic 1 prior to the inputs AB = 01, then the output logic value would remain at 1; under this condition, the fault is not detected. Therefore, the detection

FIGURE 6.4 Stuck-open fault in a CMOS NAND gate.

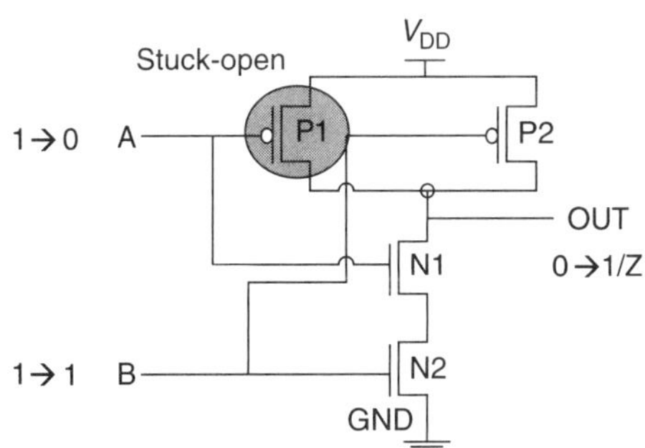

of such faults requires a *two-pattern* test. In this example, the fault cannot be detected unless the capacitor is precharged to a value 0, which is possible only if the inputs are 11. The faulty gate may be embedded inside a larger circuit. In the circuit illustrated here, inputs AB to the gate are determined by primary inputs of the circuit. As the primary inputs switch, inputs AB to the gate may transition from 11 to X0 to 01; in this case, the intermediate value may destroy the precharge condition and thus invalidate the test. A test that can guarantee no such invalidation will occur is called a *robust* test. In practice, robust tests are difficult to create. (For more on this subject, see Jha and Kundu.[4]) Whenever an output node is not driven, it is in a floating state whose value is denoted by Z.

Transistor stuck-on faults may be detected with a single test pattern. The stuck-on transistor causes a short circuit between the power rails (V_{DD} and GND), which leads to an increased quiescent current that can be detected by I_{DDQ}-based tests.[5] Faults of this type are detected by comparing the quiescent current of a faulty and a fault-free circuit. In large circuits, the quiescent current may itself be several orders of magnitude larger than the faulty current caused by a stuck-on transistor. For example, in 45-nm technology, I_{ON} for a pMOS transistor may be of the order of 300 μA per micron of transistor width and the total quiescent current may be in the tens of amperes. Hence, I_{DDQ}-based fault detection is usually impractical with large circuits. Fortunately, two-pattern tests may work well in such cases. If transistor P1 had been stuck ON in our example, we could have applied the two-pattern transition test $0X \rightarrow 11$. The first pattern initializes the output node to a value 1, and the second pattern is presumed to discharge the capacitor and thus take its output value to 0. Yet the presence of a stuck-on fault will significantly delay the discharge; hence, even if the final logic value settles at or near 0, it will take much longer for the current to flow through P1. Consequently, this fault may be detected as a delay fault.

Line open faults and bridge faults may be further classified as resistive or nonresistive. Nonresistive line faults may be detected by bridge tests and stuck-at tests, as described in Sec. 6.2.2.3. Resistive line opens or shorts are typically detected as transition faults or small delay faults. A two-pattern excitation may be required in order to test for resistive opens and shorts. When the delay δ introduced by such a fault is small and quantifiable, the fault is classified as a *small delay fault*. Low-voltage testing effectively increases the value of δ in CMOS circuits, which enhances fault detectability. A resistive defect that causes infinite delay (i.e., $\delta \rightarrow \infty$) is modeled as a *transition fault* and tested using transition fault tests. Examples of transition faults include slow-to-rise (STR) and slow-to-fall (STF) faults.

6.2.2.2 Defect-Based Bridging Fault Model

In order to find defect-based bridging faults, information on circuit voltage, bridge resistance, transistor sizing, and transistor technology (e.g., TTL, CMOS, ECL) is used to simulate the circuit. The result is an approximated truth table that represents the boolean function corresponding to the bridging fault. In a more comprehensive model, the entire output cone (field) of a bridge may be simulated at circuit level to determine the propagation of intermediate voltage values. As noted previously, the engineering trade-off between accuracy and computational efficiency has led to the development of a range of solutions for modeling defect-based bridge faults. Some approaches employ analog simulation for bridge sites only, whereas others simulate the fault's entire output cone. Still others perform analog simulation at the cell level but then combine this with precomputed information to model propagation of the intermediate signal value.

In CMOS, each node in a circuit represents a capacitive load that is charged or discharged by the driver gate; the current supplied by this gate is based on its input voltage trigger. The voltages on the bridged nodes of a circuit are a function of the driver strengths, and the voltage at each node is determined by the driver providing the greatest current. The drive strength of a gate is a function of its size. A popular defect-based bridging fault model known as the "voting" model determines the logic value at the shorted node based on relative strengths of its drivers. In this model, the gate providing the largest drive current determines the logic state of the driven node. An alternative model considers the intermediate voltage levels that arise from such bridging faults. Under this approach, downstream logic gates with varying input voltage thresholds may interpret the logic state of the bridge node differently. This ambiguity is referred to as the "Byzantine general's problem."[6]

A transistor can be represented as an equivalent resistance between the source and the drain when the device is turned on. The greatest drive current is provided by the path of least equivalent resistance to either V_{DD} or ground. The equivalent resistance can be computed both statically (based on the type, size, and number of transistors connected to the node) and dynamically (through simulation). The resistance-based fault model and the so-called ladder model use the equivalent resistance information to arrive at a suitable logic value at the shorted node.

Another model assumes that the circuit's fault-induced analog behavior extends beyond the fault site and is interpreted differently by gates fanning out from the shorted node. The EPROOFs simulator implements this technique by performing SPICE-like analog simulation at each shortened node to assign an exact value to the node during logic simulation.[7] However, the computational complexity of this simulation, and hence the time required to perform it, is high.

6.2.2.3 Abstract Fault Models

Abstract fault modeling (AbsFM) is an alternative to DBFM. The fault model in this case offers a surrogate that represents defect behavior in the circuit. Typical faults that are modeled with AbsFM are stuck-at, transition, and path delay faults. The ATPG process using AbsFM tends to be much faster than a realistic defect model, and AbsFM is typically a stripped-down version of the defect-based fault model.

Abstract fault modeling targets the fault attributes classified in Figure 6.3. The most important attributes that must be considered when modeling a defect are the technology (e.g., TTL, CMOS, ECL), the defect source, its duration, and its value. We have discussed fault sources previously. A fault's *duration* is a measure of its effect on circuit operation. Permanent faults are those that make the interconnect line hold the same state throughout the circuit lifetime. Transition faults are activated when a line or a gate changes value—for example, with a *slow-to-rise* (STR) transition fault the observed (faulty) value is 0. Small delay faults are transition faults that occur within a finite duration (delay) of δ. Intermittent faults are caused by radiation-induced soft errors; these are modeled as transition faults or as stuck-at faults that occur only at some clock cycles of operation. Intermittent faults are not repeatable, and they occur randomly. Although transient faults are repeatable, they may not occur during each clock cycle. Errors of this type are typically caused by problems related to signal integrity and may be modeled as constrained transition faults. Path delay faults are activated when a specified signal transition takes place along a specified path. Example includes STR and STF output delays at the end of a path.

Stuck-at Fault Model The most commonly used fault model is the stuck-at fault model. A stuck-at fault is a proxy for such design defects as metallization shorts, oxide opens, missing features, and source-drain shorts. Whereas defect-based faults may exhibit complex errors in logic behavior, a stuck-at fault model simplifies matters by associating a constant value at a line or a node. This makes the ATPG process considerably simpler. Studies have shown that test pattern sets generated by considering stuck-at faults are almost always able to detect the defects described here.

In stuck-at fault models, the interconnect line can take only one of two values and so test generation is computationally light. The single stuck-at fault model is the simplest of them all (see Figure 6.5). This model proceeds under the assumption that a circuit contains only one fault at a time. Hence there are only $2n$ faults to be tested, where n is the number of nodes. A more sophisticated version of the stuck-at fault model is the multiple stuck-at (MSA) fault model. This model assumes the existence of two (or more) faults at a time in the circuit, so the number of potential faults increases to 3^n-1. This model improves the overall coverage of physical defects, but its major drawback is the exponential fault count and the resulting huge number of required tests. (In manufacturing, a test's compactness is important because larger test sets increase test time and product cost

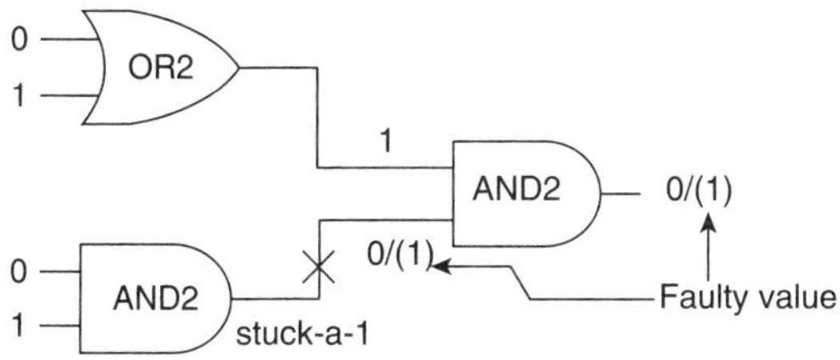

FIGURE 6.5 Stuck-at fault in a simple circuit.

significantly.) Shorts in CMOS designs can also be modeled as constrained stuck-at faults. The resulting logic value at the two conducting regions bridged together by the defect is obtained by assigning values at each node. Today's ATPG tools can handle such constrained abstract fault models with relative ease.

Bridging Fault Model Bridging fault models predict the behavior of a circuit when two nodes of the circuit are shorted together. A bridging fault, which is classified as a permanent fault, can occur within any of the following contexts: within a logic element, such as source and drain terminals of the transistor; between two logic nodes without any feedback in the circuit; or between two logic nodes or circuit elements with feedback. Various models have been proposed to model circuit bridging defects.[8-11] Wired-OR and wired-AND are the simplest bridging fault models. In the AND bridging case, if two nodes A and B are bridged then the resulting behavior is AB for both nodes. Thus, for A = 0 and B = 1 the resulting behavior is that both nodes have value 0. In this case only, node B has a faulty value. The OR bridges are defined similarly. One of the more interesting bridging fault models is the *dominant* bridge model. Here, the value of A prevails over the value of B in an "A dom B" bridge fault. So if B is 1 (resp. 0) and A is 0 (resp. 1), then B gets a (faulty) value of 0 (resp. 1). However, the "wired" logic does not accurately reflect the actual behavior of bridging faults in static CMOS circuits.[9,10,12] This divergence is caused by the existence of intermediate values that do not clearly correspond to 0 or 1 and thus may be interpreted differently under static versus transition conditions. Wired-logic-based fault models may propagate invalid logic states through the circuit. Figure 6.6 illustrates the bridging fault between two logical nodes when a feedback path is involved. The feedback loop can convert a combinational circuit into an asynchronous sequential circuit.

Typically, abstract fault models are used to generate test patterns during the circuit realization phase. These test patterns are typically derived from models of stuck-at, transition, path delay, and bridge faults. As mentioned before, defect-based fault models are usually unsuitable for ATPG. However, the simulation of defect-based fault

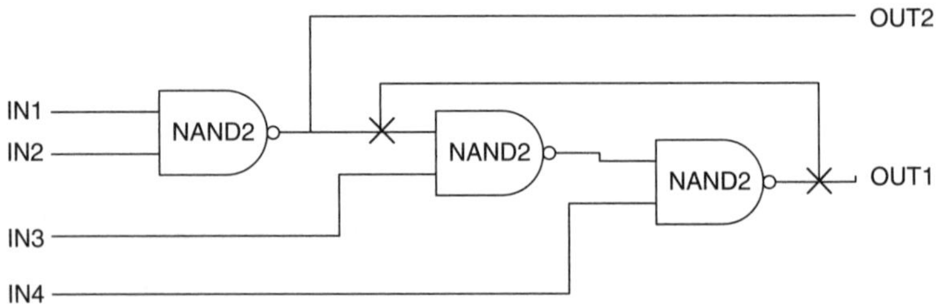

FIGURE 6.6 Bridging fault with feedback.

models is relatively straightforward. For defect-based test coverage, test patterns derived from abstract fault models may be simulated. The relative success of AbsFM-derived test patterns in detecting complex faults is due to the inherent controllability and observability of such tests.

6.2.2.4 Hybrid Fault Models

We know that defect-based fault models are constructed using data on actual defects, their probability of occurrence, and their effect's intensity. A major drawback of these models is the prohibitive expense of using them to generate test patterns. Test pattern generation using abstract models is simpler and less expensive but not as accurate or comprehensive. Hybrid fault modeling (HybFM) aims to approach the accuracy of defect-based fault models without sacrificing the ease of test pattern generation based on abstract fault models.

A hybrid fault model uses a combination of abstract fault models and constraint information from defect-based fault models. An example that illustrates this effect is shown in Figure 6.7. Consider the two nets A and B, which are joined by a resistive bridge. If a stuck-at-0 fault is assigned to the point X in net B, then the abstract model will have only one condition capable of exciting the stuck-at fault; this is not sufficient for detecting the bridge fault. Yet if we add the constraint that B be stuck at 1 subject to A held at 0, the condition for detecting the bridge fault is described entirely at the logic level. Additional logic constraints of this nature make it easy for ATPG to produce a test pattern that can detect such bridging faults. Similarly, for the slow-to-rise transition at B subject to constraint A=0, a transition ATPG tool can use an abstract fault model to produce a test pattern that will detect this resistive bridge fault.

In essence, ATPG relies on logic constraints rather than defect information to arrive at a pattern. Its success is a function of the extent to which reality is accurately portrayed by a simple fault model employing only boolean (either-or) constraints and errors.

6.2.3 Test Flow

The manufacturing test flow comprises four basic steps designed to facilitate and optimize the following procedures: defect screening,

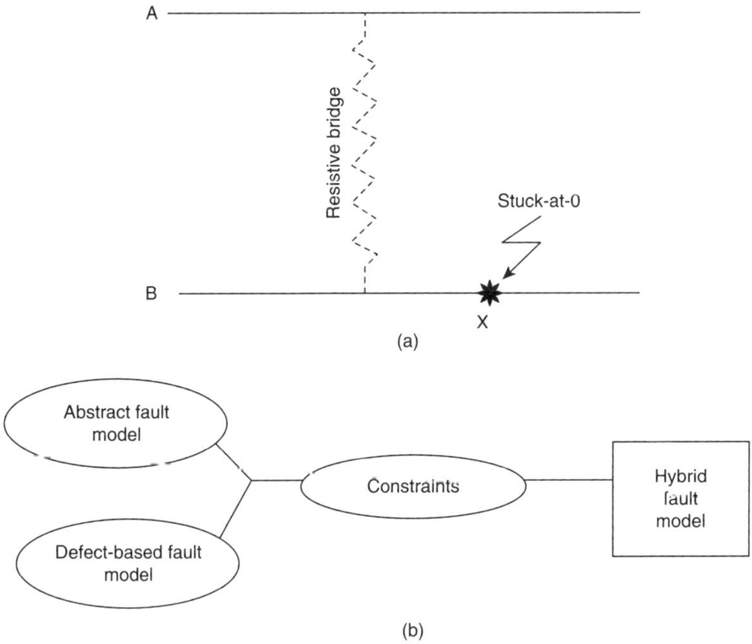

FIGURE 6.7 Hybrid fault models: (a) example context; (b) elements.

performance binning, lifetime acceleration (to screen for aging defects), product quality assurance, and failure analysis. The components of a typical test flow are illustrated in Figure 6.8. The entire test suite is usually not applied at every testing stage, which reduces the cost of testing but does require that test *scheduling* involve minimal overlap between stages. During the process of circuit realization, test patterns are generated by designers using ATPG, fault simulation, and manual pattern writing that are based on a set of fault models appropriate for the particular design objectives and process technology.

The *wafer sort test* (aka probe test) is the first step in the manufacturing test flow. The main goal here is to separate the good chips from the bad ones in order to reduce downstream packaging costs. The dies are then cut from the wafer, and the defect-free chips are packaged. Representative samples from the defective bins are sent for failure analysis. The main objective of wafer sort test is gross defect coverage.

The next step is *burn-in*, which is performed on packaged dies. Packaged dies are subject to high voltage and high temperature stress to accelerate the aging process. High mechanical stress and strong vibrations are used to test package rigidity. Test patterns are applied during burn-in, but typically the response is not observed. This is because the burn-in environment is usually outside of product specifications and so the circuit may not operate correctly.

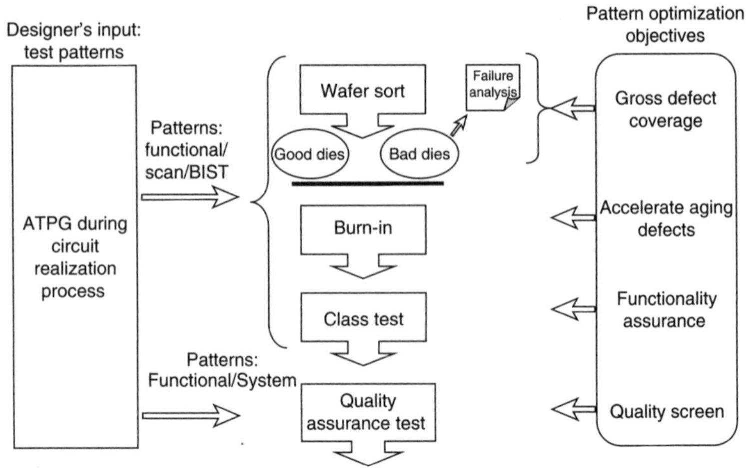

FIGURE 6.8 Typical test flow, with designer's ATPG input and pattern optimization objectives.

After burn-in, the next step in manufacturing test flow is *class test*. This is the final defect screen, so extensive fault coverage is necessary. Because frequency binning is performed in this step, at-speed tests are applied here. The class test also includes parametric tests, which include testing for quiescent current, input/output voltage level, and slew rates. Built-in self-tests are frequently used at this stage in order to reduce test application time or to avoid reliance on high-performance testers.

Finally, system vendors perform a series of inspections to test incoming chip quality. Such inspections are typically performed not on every chip but rather on a statistically representative sample of chips. These inspections are known as *quality assurance* checks. The chip manufacturer may also perform quality assurance tests on a sample of chips in order to ensure the quality of shipped product. In each of the four steps just described, the set of patterns used is consistent with the main objective for that particular step.

The scheduling of tests is based on the manufacturing process, the parametric and measurement environment, and cost issues. For example, suppose that a fault A can be detected by using either functional or scan-based test patterns. (This fact may have been discovered through logic fault simulation during the circuit realization process.) Suppose further that, in the wafer sort test, scan-based test patterns were applied and the chip passed this step. It would then be preferable to apply a high-speed functional test during the class test. Similarly, a type-X test may be used during the wafer sort whereas a type-Y test is used during the class test. These choices simply reflect the strengths and limitations of the particular tests. Choosing appropriate test patterns in each step to improve the effective fault coverage is sometimes referred to as the *test pattern scheduling and optimization* process.

In sum, fault models have two aims: (1) *modeling* defects at the highest level of abstraction to facilitate fault simulation and test pattern generation; and (2) *classifying* the large number of potential individual defects into groups of defects that exhibit common failure mechanisms and have similar effects on circuit behavior.

6.3 Yield Improvement

Every new manufacturing technology goes through a "maturing" process: the chip yield may be low at first but it rises gradually with time as the technology matures. However, an increasingly competitive market dictates that the best profit opportunities occur early in the product cycle and then decrease with time. This conflict has motivated the search for design techniques that can produce higher yields with a shorter maturity process.

The manufacturing failure rate for the current technology generation ranges between 10^{-16} and 10^{-15} (i.e., one defective structure or polygon per 10^{16} structures).[13] Given the increasingly complex architectures that use nanometer devices to execute up to a billion instructions per second, fault avoidance and tolerance techniques for reliable information processing and storage must be able to operate within a regime where some devices are unreliable.

Fault tolerance is the process of exploiting and managing architectural and software resources to reduce the vulnerability to failures. The core components of fault tolerance are redundancy and reconfiguration.[14,15] *Reconfiguration* is addressed in Sec. 6.3.1. Examples of *redundancies* are multiple copies of devices, circuits, or even architectural blocks that serve as spares. Error-correcting codes (ECCs) are perhaps the best-known instance of information redundancy. One example of time redundancy is the recomputation and reevaluation employed by "Razor."[16] Software redundancies include redundant software threads, checkpointing, and rollback recovery. (For a detailed treatment of this subject, see Koren and Mani Krishna.[17]) A brief overview of nonstructural redundancy techniques is provided in Sec. 6.3.1.2.

We have previously discussed how manufacturing defects can be either permanent (catastrophic) or transient (parametric). The aim of redundancy-based techniques just described is to achieve fault tolerance against specific defect types. Some techniques are more efficient at reducing the probability of failure in the presence of permanent errors (e.g., clustered spot defects); other techniques excel at making the design resilient to transient errors (e.g., radiation from radioactive contaminants or cosmic rays). Thus, optimal fault-tolerant design amounts to choosing the lowest-cost solution to guarding against a specific defect type.

In Sec. 6.3.2 we shall examine layout techniques that avoid potential faults by reducing printability errors. Similar protection may also be obtained through transistor and interconnect sizing and other circuit modifications. Defect avoidance techniques of this type are achieved at the cost of suboptimal area, performance, and power characteristics.

Fault-tolerant design techniques date back to the days of vacuum tubes. In 1952, von Neumann proposed a multiplexing technique for effecting redundancy in system architectures and components to obtain higher reliability.[18] He proved that, by managing redundancies, high reliability could be achieved from unreliable logic and storage units. In his experiments, von Neumann mainly considered three methods: voting scheme, standby scheme, and NAND multiplexing scheme. He demonstrated that redundancy schemes can improve system-level reliability.

6.3.1 Fault Tolerance

As noted earlier, fault tolerance can be achieved through several means. These include structural redundancy, nonstructural redundancy (e.g., time, software, and information redundancy schemes), multiplexing, and reconfiguration.

Before venturing into further discussion of fault tolerance techniques, we define systems that are configured in series or in parallel. Knowledge about system architecture is required because the techniques used to achieve fault tolerance change with the structure of the system. A model of a *series* system is shown in Figure 6.9(a). This system consists of different blocks $U_1, U_2, \dots, U_n$ that are connected in series. The system is operable (and has nonzero yield) only if all its components function correctly. Complete redundancy in such systems is achieved by providing replicas of each component. A *parallel* system is illustrated in Figure 6.9(b). In this system, the blocks $U_1, U_2, \dots, U_n$ are connected in parallel; therefore, the system will be remain operable if at least one of the blocks is functioning. Here the redundancy, that is characteristic of parallel system may be provided for a particular set of frequently used blocks. Clearly, a parallel system is less likely to fail that an otherwise comparable serial system.

6.3.1.1 Traditional Structural Redundancy Techniques

Fault tolerance techniques based on structural redundancy have been popular in memory arrays since the 1970s. Early in that decade, semiconductor companies faced problems with low yield in DRAMs, problems that stemmed from clustered manufacturing defects. One early solution incorporated error-correcting codes, whereby additional parity bits were added to an information word. This technique

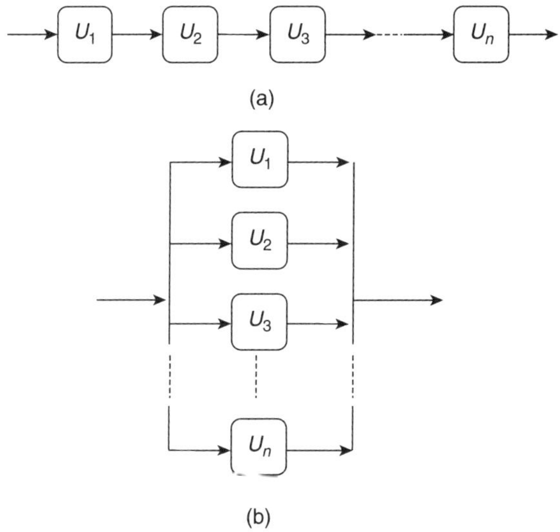

Figure 6.9 System configurations: (a) series; (b) parallel.

did not scale well beyond single or double errors, since the large number of such parity bits required negated any benefit from scaling. The solution that was eventually found required the invention of fuse technology that allowed "substations" of rows, columns, and blocks. In this scheme, a spare row, column, or block is added to the memory array. Fuses can be burnt to swap in, say, a row that replaces the defective one (see Sec. 6.3.1.6). These techniques have been successfully deployed in the semiconductor industry for many decades, improving yields by a factor of 3.[17] The spare elements are good examples of structural redundancy.

Figure 6.10 illustrates structural redundancy for memory ICs. Memory integrated circuits have cells arranged in rows and columns, so redundancy here is achieved by adding more columns and rows than are strictly necessary. During manufacturing test, if a set of defective cells are identified then the corresponding defective row and/or column can be disconnected by blowing a fusible link or fuse.[19] The disconnected element is now replaced by a spare element that uses a programmable decoder with fusible links that are burnt during the same process. The success of spare rows and columns is rooted in the clustering of defects. If defects occurred in random locations, then so many spares would be needed that the probability of defect would actually increase. Yet because defects are usually clustered, a single row or column is often sufficient.

With transistor scaling, memory units became larger. Larger units require more spare elements. Large memory units are divided up into

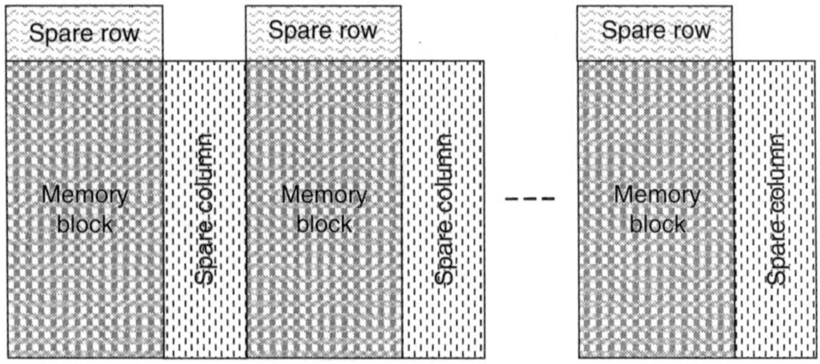

Figure 6.10 Spare rows and columns, a structural redundancy technique for memory blocks.

smaller banks for ease of access and reduced access penalty. In this case, each bank will have separate spare rows and columns so that the overall defect tolerance remains under control. However, this constraint means that some banks cannot get the required number of spare rows or columns and hence will tend to have reduced yield compared to those banks that can. The problem could be mitigated by a more efficient method of allocating redundancy resources. One solution is to share spares between banks; this way, a particular block of memory does not become a bottleneck as long as there are unused redundant blocks in other areas of the memory.

Triple Modular Redundancy The triple modular redundancy (TMR) approach to defect tolerance uses three identical blocks to perform the same operation. To ensure reliable operation and integrity of the result, a voting mechanism is used to select the proper output. The setup is illustrated schematically in Figure 6.11.

Triple modular redundancy is typically employed for improving the tolerance to transient defects that occur in a device. A majority vote is taken through the voting block; this ensures correct output if one can safely assume that errors are confined to a single block. With a voting mechanism in place, a failure in one of the blocks still ensures a correct output. If an odd number of blocks is used, the voting results will be unambiguous.

Figure 6.11 Triple modular redundancy (TMR) using a reliable voting mechanism.

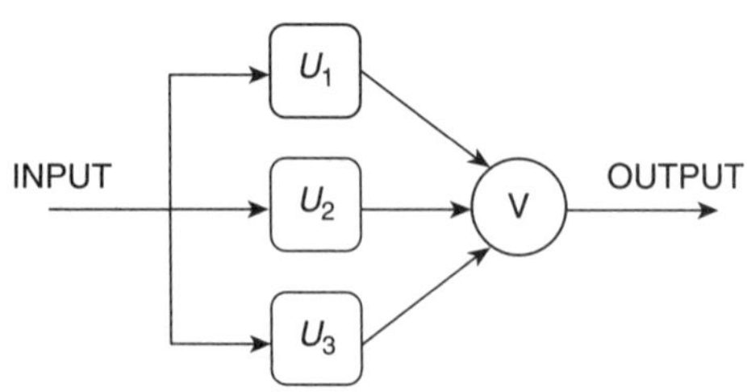

The TMR technique has proved to be a effective defect-tolerance mechanism that increases overall yield. However, TMR improves reliability only when the reliability of the original block is greater than 0.5.[13] System reliability increases rapidly in response to higher reliability of each element. In this configuration it is assumed that the voting block is completely reliable.

The reliability of a TMR system is gated by the reliability of the voting block. If the voting block is thought to be unreliable, then voting block redundancy is required; this is illustrated in Figure 6.12. The basic idea here is to use two or more voting blocks to overcome the intrinsic unreliability of voting circuits.

In most practical scenarios, each constituent block has a different level of reliability. In this case, the overall TMR reliability is gated by the most unreliable unit. The implication is that maximizing TMR reliability requires that a system be subdivided into nearly equal and independent blocks.

N-Modular Redundancy The approach known as *N*-modular redundancy is a generalized version of the TMR technique where, instead of three blocks, there are now *N* blocks in parallel (see Figure 6.13). An *N*-bit voting block is used to obtain the correct output. Each block

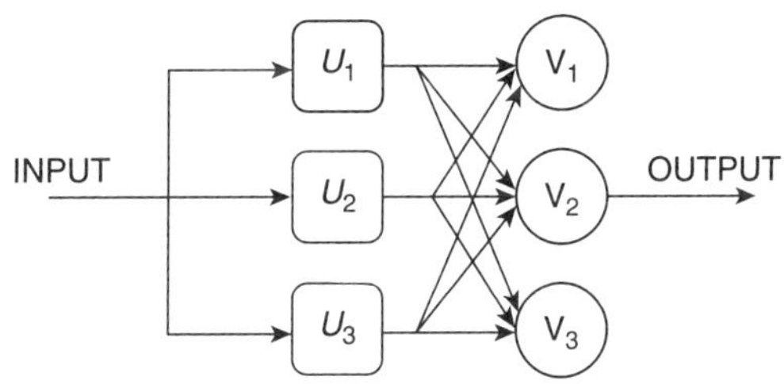

FIGURE 6.12 Triple modular redundancy using an unreliable voting mechanism with redundant voters.

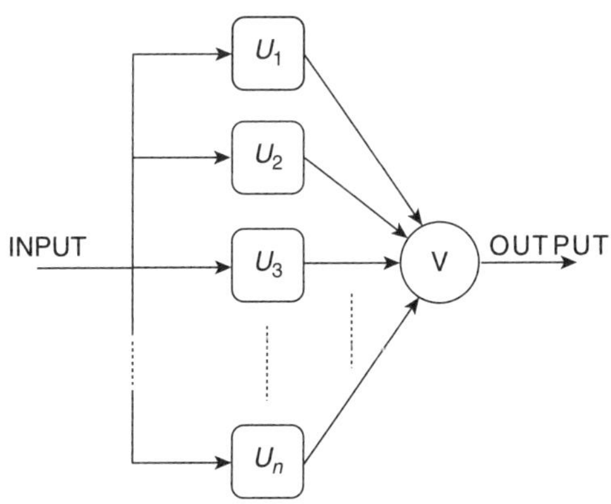

FIGURE 6.13 N-modular redundancy.

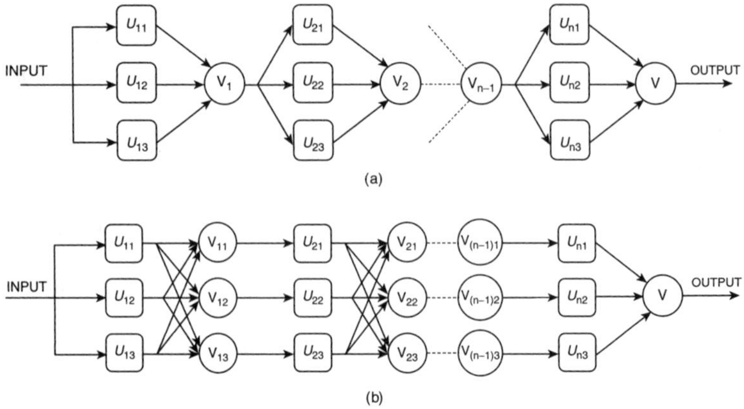

FIGURE 6.14 Cascaded triple modular redundancy: (a) simple TMR with simplex voters; (b) complex TMR with redundant voters.

in the system may consist of *N* units. The reliability of each block is assumed to be uncorrelated, which would preclude any common mode failures from occuring in a very large-scale integrated (VLSI) system.[13]

Cascaded Triple Modular Redundancy The TMR process can be repeated by combining three of the TMR blocks with another majority voter to form a second-order (and so on, up to an *n*th-order) TMR block with even higher reliability. This is the *cascaded* triple modular redundancy (CTMR) technique, depicted in Figure 6.14(a); Figure 6.14(b) shows an extension with redundant voters. The reliability improvement associated with CTMR is observed only when the number of units within each block is high.

Standby Redundancy The standby redundancy technique consists of an arrangement whereby several copies of each block are added to a parallel system; see Figure 6.15 for a basic version. Unlike TMR, a switch is used to choose a copy of the block when the original has a defect. The copies of the block are called spare blocks, and they can be either *cold* or *hot* spares. Cold spares are are powered off until they are utilized, whereas hot spares are powered on and are ready to be used at any time. Standby redundancy with hot spares strongly resembles TMR.

Two versions of the basic standby technique that have been used in designs are *duplexing* and *pair and spare*; see Figure 6.16 and Figure 6.17, respectively. These two methods use comparators to verify the original block's performance and assign the correct switch.[2] A hybrid technique that uses both standby-based block switching and TMR-based voting mechanisms is illustrated in Figure 6.18.

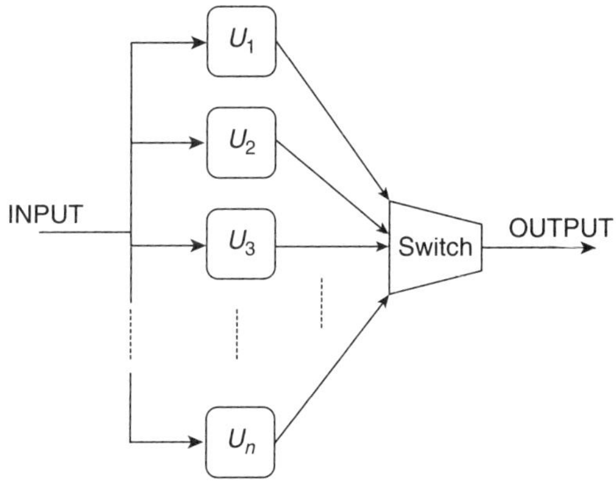

Figure 6.15 Basic standby redundancy setup.

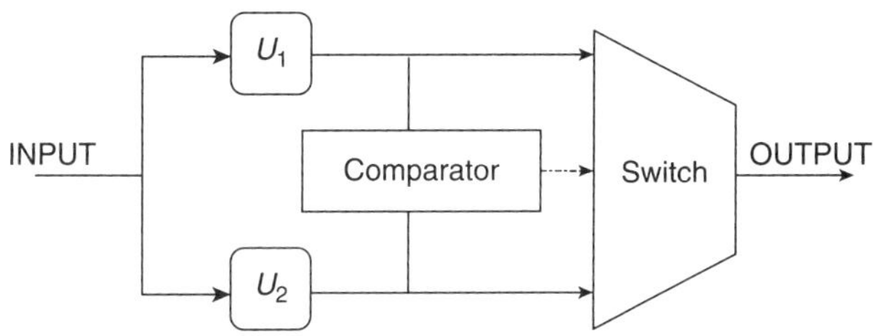

Figure 6.16 Configuration for duplexing version of standby redundancy.

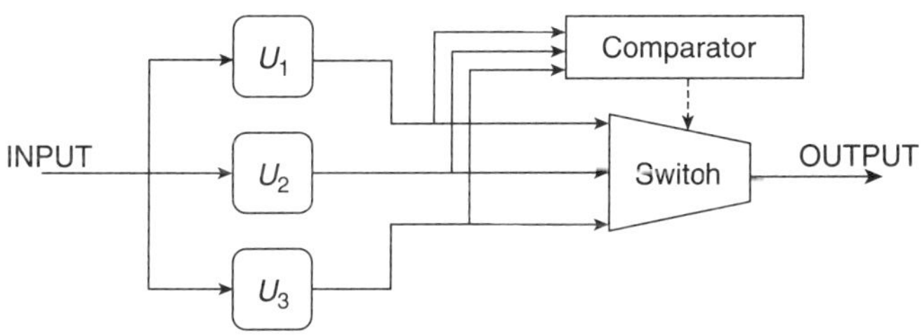

Figure 6.17 Configuration for "pair and spare" version of standby redundancy.

6.3.1.2 Nonstructural Redundancy Techniques

Nonstructural redundancy encompasses information, time, and software redundancies. For example, memory banks may be protected by error-correction code, which is an example of *information redundancy*. The ECC technique works extremely well for detecting

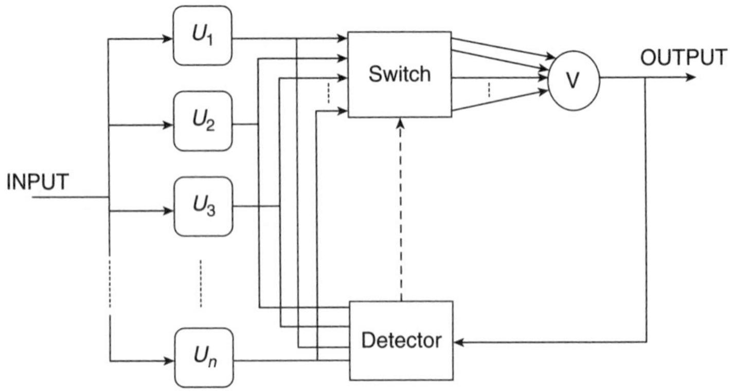

FIGURE 6.18 Configuration for hybrid redundancy scheme.

and correcting *t*-memory cell failures, where *t* is typically a small number. Most of the commonly deployed memory coding techniques are based on linear block codes. Hamming code is one example of linear block code for correcting single errors. A Hamming code of length *n* consists of $n - \lceil \log_2 n + 1 \rceil$ information bits. For example, a Hamming code of length 12 consists of $12 - \log_2 13 = 8$ information bits. Stated differently, eight information bits require four redundant or check bits for single error correction. Multibit errors may also be corrected using ECC, but the number of check bits needed to protect information bits grows rapidly with the number of faulty bits that are correctable.

Time redundancy techniques include recomputation and reevaluation. These techniques are used to detect transient errors modeled by single-event upsets (SEUs) and single event transitions (SETs). Single-event upsets can be caused by soft errors and other transient errors. (Soft errors and their impacts on circuits are discussed further in Chapter 7.) The idea here is to perform detection based on the fact that transient errors occur only for a short duration. A circuit may be reevaluated in separate clock cycles, or its output may be double-sampled based on the relation between cycle time and duration of the soft error. For example, if a transient error persists for 50 ps or less and if the clock period is much longer, then the output of a combinational circuit may be double-sampled at intervals that exceed 50 ps. However, if the cycle time is short then such evaluations must be made during a separate cycle.[4,20] The double-sampling method, which was proposed by Nicolaidis, is illustrated in Figure 6.19.[21] (For a modified version of this approach known as "Razor," see Ernst et al.[16])

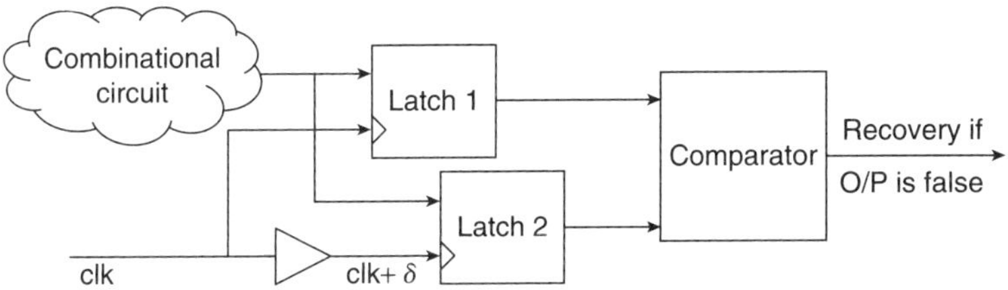

Figure 6.19 Time redundancy technique for detecting transient errors and timing failures.

Time redundancy schemes have also been used at the resource scheduling level. A microarchitectural modification suggested by Pan et al.[22] uses time-scheduled resource sharing to generate yield improvements. This solution centers on exploiting natural redundancy among multicore systems. In homogeneous chip multiprocessor systems, faulty cores use the services of good cores to execute instructions that the former can no longer execute correctly. This procedure improves reliability and yield but with some loss of performance. A special intercore queue or buffer is maintained between the faulty and helper cores, as shown in Figure 6.20.[22] Instructions that require the services of a faulty unit are automatically transferred to a helper core.

Software redundancy schemes for fault tolerance include redundant multithreading (RMT), checkpointing, and rollback or recovery mechanisms. In multithreaded environments, defects can be detected by running two copies of the same program as separate threads and

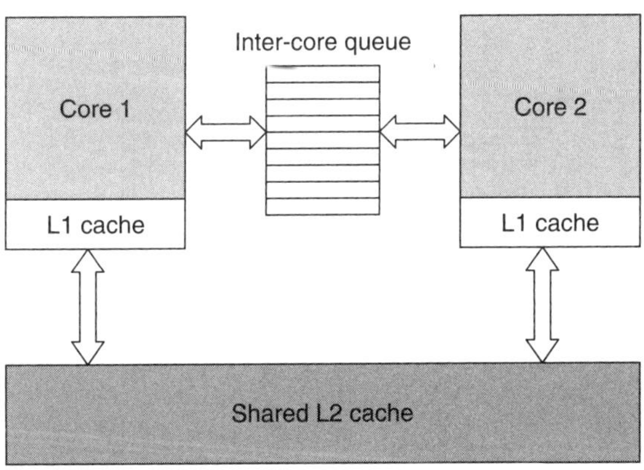

Figure 6.20 Microarchitectural scheduling: resource sharing between cores can improve reliability and yield.

then comparing the outputs. A mismatch between the two threads is flagged as an error, which initiates a recovery mechanism. This RMT approach has been used in simultaneous multithreaded processors.[23] Another software approach to improving fault tolerance is through checkpointing. A checkpoint is a snapshot of the process state at a given moment in time. The checkpoint represents all pertinent information that would be required to restart the process from that point. When an error is detected, checkpointed information is used to return the program to a stable state. Each checkpoint may store a large amount of information about the process state, so checkpointing imposes a time overhead. If the number of checkpoints for a process is large, then overhead due to checkpointing may be excessive. On the other hand, if checkpoints are few and far between then a program may be rolled further back, leading to an undesirably high execution time. Therefore, the optimal number of checkpoints is a function of checkpointing overhead and the intrinsic failure rate.

6.3.1.3 NAND Multiplexing

In the mid-1950s, the NAND-based multiplexing technique was proposed by von Neumann to improve the reliability of design modules in computing systems. This technique has received renewed attention for its application to mitigating the effects of transient faults and also (though to a lesser extent) spot defects in manufacturing. Consider the NAND gate depicted in Figure 6.21, and replace each input and the output of the NAND gate with *N* signal lines, as shown in the figure. Also, duplicate the NAND gate *N* times. Let A and B be

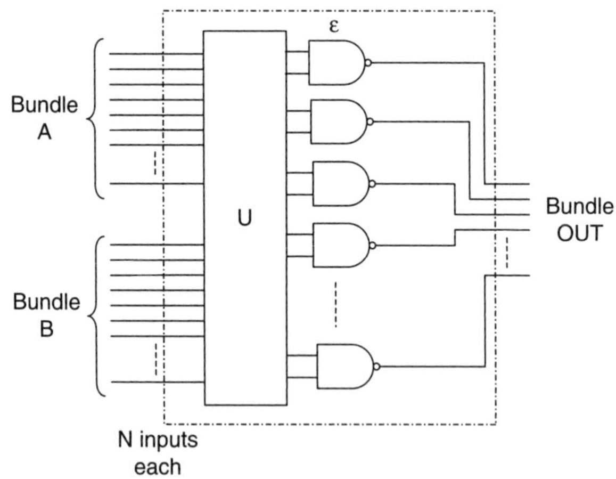

FIGURE 6.21 Redundancy based on NAND multiplexing.

two bundles of input lines, and let OUT be the output bundle. The rectangular region marked by the dashed lines performs random permutation of the input signals to be provided to the *N* NAND gates. The first input is selected from bundle A, is paired with any signal from bundle B, and is then connected to a NAND gate. According to von Neumann's theory, this NAND multiplexing scheme is effective against single faults only when the number of lines in each bundle is high. This fact renders the scheme relatively impracticable, so the technique is not very popular. Figure 6.22 depicts a multistage version. (See the original work of von Neumann, published in 1955,[18] for additional details on this method.)

6.3.1.4 Reconfiguration

A *reconfigurable* architecture is one that can be programmed after fabrication to perform a given functionality. With this technique, faulty components are detected during the testing phase and excluded during reconfiguration. Reconfigurable architectures have been explored as possible means of improving tolerance to manufacturing defects. A good example of this technique is provided by the Teramac,[14] which was created by HP labs as an efficient, defect-tolerant, reconfigurable system. Programmable switches and redundant interconnects form the Teramac's backbone. It was observed that, in the presence of large number of defects, the Teramac was able to produce results a hundred times faster than conventional computing engines.

The reconfigurable computing system for defect tolerance relies on the same concept as field programmable gate arrays (FPGAs).[14,15] The FPGAs contain a regular array of logic units, called configurable logic blocks (CLBs) or look-up tables (LUTs). Each of these blocks can take the form of any logic function with a given set of inputs and outputs. Two CLBs capable of implementing different logic functions with a given set of inputs and outputs are diagrammed in Figure 6.23. Each CLB can communicate with any other CLB through a regular

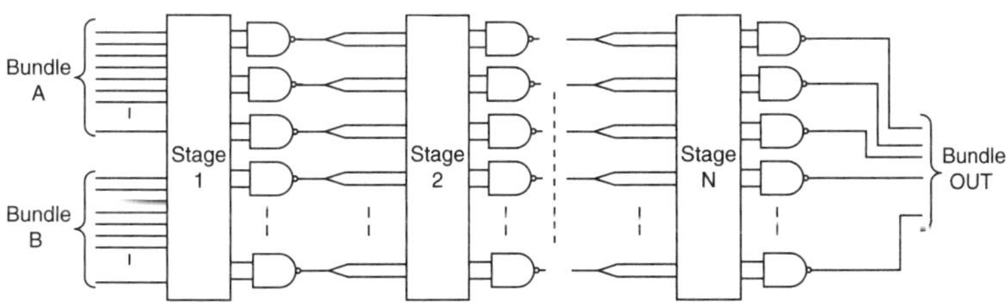

FIGURE 6.22 Multistage NAND multiplexing.

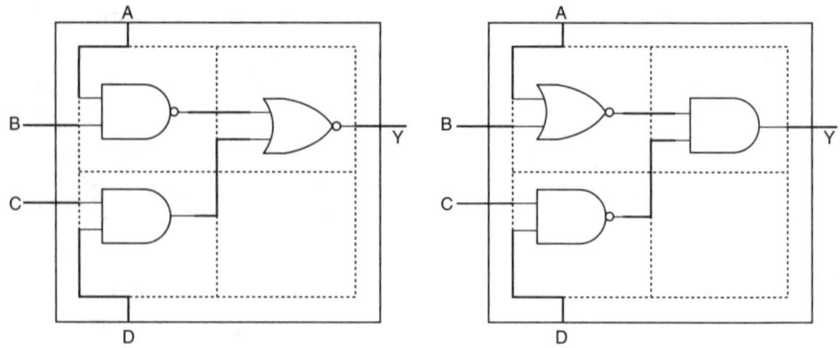

FIGURE 6.23 Configurable logic blocks (CLBs), the basic element of reconfiguration theory.

structure of interconnect wires and crossbar switches. Multiple CLBs form blocks, which in turn form clusters. The logic and memory mapping of each CLB is done in the field. The main advantage of reconfiguration is the ability to detect manufacturing defects, locate the CLBs that are faulty, and then work around them during configuration. The Teramac performs all three steps by using self-diagnostic software to create a database of defective CLBs before configuring the CLBs for a given function. Thus, instead of relying on defect-free circuits, the logic is implemented with available fault-free CLBs—provided the mapping can be satisfied. During configuration, an important part is played by the probability of finding x clusters from the available N good clusters that can be used to map the required logic and memory elements. This same approach is also used for the reliability analysis of VLSI systems.

6.3.1.5 Comparison of Redundancy-Based Fault Tolerance Techniques

The efficiency of multiplexing, redundancy, and reconfiguration schemes is plotted against intrinsic device failure rates in Figure 6.24. The probability of failure of each device in a chip that has approximately 10^{12} devices should be smaller than 10^{-10}. At this failure rate, reconfiguration schemes have much less overhead compared to the N-modular and NAND multiplexing schemes. As the figure shows, reconfiguration is appropriate for high defect rates, though at an expense of large redundancy overhead. The N-modular technique provides good coverage for chips with a large number (~10^{12}) of devices for failure levels of 10^{-9} or lower. The NAND multiplexing scheme is appropriate for a design with 10^{12} devices when the device failure probability is ~10^{-3}. Given the device failure rates characteristic of today's technologies, such large overheads are usually not justifiable. In atomic-scale devices, however, such redundancies may play a larger role in implementing designs with large number of devices.

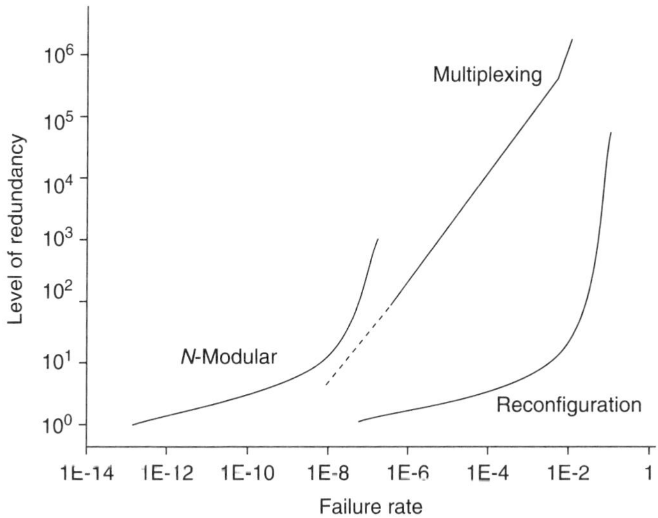

FIGURE 6.24 Failure rate versus level of reduncancy for three redundancy-based fault tolerance techniques: *N*-modular redundancy, multiplexing, and reconfiguration.

6.3.1.6 Fuses

Even when the memory layout is highly optimized, DRAM memories are known to be susceptible to process defects. The redundancy techniques described so far have been used extensively to protect different parts of the memory units, including cells, sense amplifiers, word line drivers, and decoders.[2] Detection of defects typically occurs in a postfabrication environment and is followed by repair and redundancy allotment. Detected faulty parts of the memory are disabled from the actual working portion by burning laser-programmable fuses. A laser source physically "blows" fuses placed in different regions of the wafer, thereby disconnecting the defective portions of the chip and replacing them with spare rows and/or columns, drivers, and decoder circuitry. The laser fuses are made of either polysilicon or metal, and they are built in such a way that just a temporary exposure to a repair laser will blow the fuses accurately. The fabrication of laser fuses must be precise in both location and dimensions so that they can be effectively blown out and also make the required connections/disconnections. Laser fuse patterning also must obey design layout rules and, of course, satisfy the requirement that the laser not cause defects in other functionally nondefective regions surrounding the fuse. To help minimize defects during the blowing of a laser fuse, the laser fuse heads are carefully placed end-to-end at a constant spacing on the wafer. Minimizing the number of fuse rows can also help improve the accuracy and consistency of such laser repair. Finally, special alignment markings (aka keys) for each fuse row are used to align the laser repair machine head for each exposure.

The exposure of a polysilicon link results in less debris than does the exposure of a metal fuse. Moreover, the polysilicon link ensures a reliable separation. A laser fuse array is illustrated in Figure 6.25(a). Figure 6.25(b) shows the blown fuse creating a void (of diameter *D*) that is roughly equal to twice the laser wavelength. The diameter *D* places a limit on the minimum spacing (fuse pitch) between fuses, for if fuses are placed within this pitch then adjacent fuses may be inadvertently blown off. Although shorter wavelengths have better precision, they increase the probability of damage to the underlying substrate.[24] This may increase the total number of defects in the wafer. For this application, lasers of shorter wavelengths are avoided. Decreasing feature widths of fuses will require improvements in focusing and alignment of laser. The main disadvantage of laser fuses is the high capital cost of laser repair equipment. Because these tools cannot be employed in any other process step, the cost of IC production increases drastically with the use of laser fuses. Fuse devices are typically used in CMOS chips for implementing redundancy; trimming capacitors, resistors, and other analog components; and holding permanent information such as chip-id, decryption keys, and the like.

The electrical fuse or eFuse is another type of programmable memory unit. Unlike the laser fuse, the eFuse typically uses large transistors that are blown to program the fuse. A cross section of this transistor is shown in Figure 6.26.[19] There is a layer of thin insulator material (e.g., oxygen-nitrogen-oxygen[25] or amorphous silicon[26]) between polysilicon and the metal. An opening is created in this layer

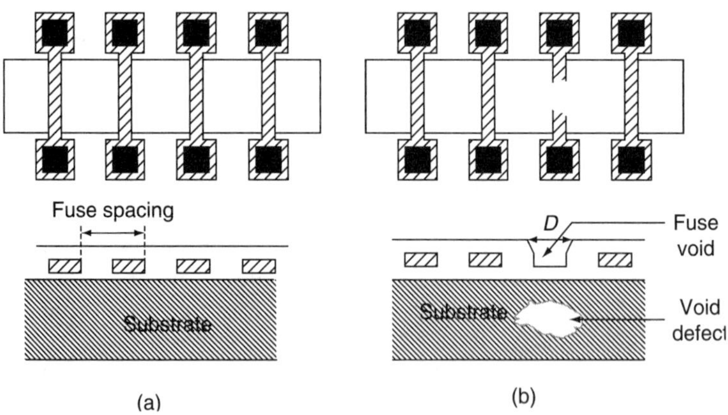

FIGURE 6.25 Bird's-eye view (top) and cross-sectional view (bottom) of (a) laser fuse array and (b) blown fuse array creating a void and possibly a substrate defect.

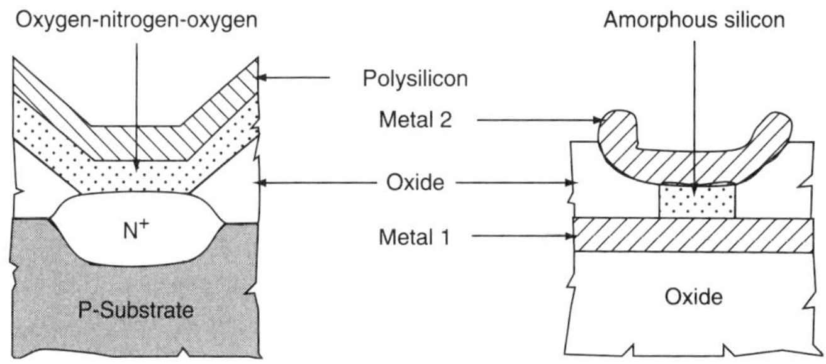

FIGURE 6.26 Electrically programmable fuses (eFuses).

by applying high-joule heat (i.e., electromigration), thus forming a conducting path between the two conductive layers. The eFuses allow an engineer to program the devices after packaging the chip, which is not possible with laser type fuses. The other main advantage of an eFuse is that it can be reprogrammed by running high current in the opposite direction to restore insulation between the connecting layers. The drawback here is that large transistors consume a significant amount of power, which can adversely affect the test throughput. As a result, eFuses are typically used only with small SRAM memories.

Another type of programmable fuse technology is the oxide rupture fuse (aka antifuse). As in the eFuse setup, strong currents are applied to the device, creating a programmed oxide state. Oxide-based antifuses can be produced using standard CMOS process and are small compared to the other two fuse technologies. This small size makes antifuses applicable to larger memory ICs.

All fuse technologies depend on the resistance at the fuse location to ensure reliable disconnection. There is an inherent reliability issue with any fuse, because changes in resistance over time can lead to costly in-field device failures.

6.3.2 Fault Avoidance

The extent of process-induced, lithography-induced, and design-centric defects is on the rise owing to scaling of device and interconnect feature sizes and shrinking of the process variability window. We know that defects may cause catastrophic failures, such as opens and shorts in interconnects and devices, as well as parametric failures that affect performance, noise, and signal integrity. Several circuit and layout techniques have been studied in the past two decades to improve design tolerance to defects. A few of these techniques are discussed in this section.

Spot defects due to process imperfections cause opens and short in metal lines. In Sec. 5.2 we discussed the concept of critical area analysis for yield prediction. The most commonly used layout technique for spot defects is critical area reduction. The CA for opens depends on the width of the metal line, while the CA for shorts depends on the spacing between two adjacent metal lines. Hence, critical area for opens can be reduced by widening metal lines wherever possible, although such metal widening must comply with the design rules. Critical area for shorts is reduced by increasing the metal-to-metal spacing. The projected yield improvement due to layout modifications for critical area reduction is plotted in Figure 6.27.[17] The two CA-improvement techniques just described can also be incorporated into existing OPC algorithms.

Linewidth variation due to pattern dependence was described in Sec. 5.3.2. Such variations result from perturbations or errors in such input parameters as focus, dose, and resist thickness. The main technique for avoiding printability errors due to layout patterns is based on OPC and multipatterning. Newer OPC approaches incorporate statistical variations.

Techniques for improving the parametrics of circuits and layouts play a vital role in the presilicon design optimization phase. Parametric defects include modified threshold voltage V_T, change in circuit path or gate delay, and other deviations from design parameters. Subthreshold leakage constitutes a major portion of the current flowing through a transistor in its OFF state. The amount of this subthreshold leakage increases exponentially with decreases in V_T. Transistor threshold voltage is adjusted through ion implantation, whereby a discrete number of dopant atoms are introduced into the transistor channel region in order to attain an appropriate threshold voltage. For the current generation of technology, the required number of such dopant atoms is of the order of 100. However, the

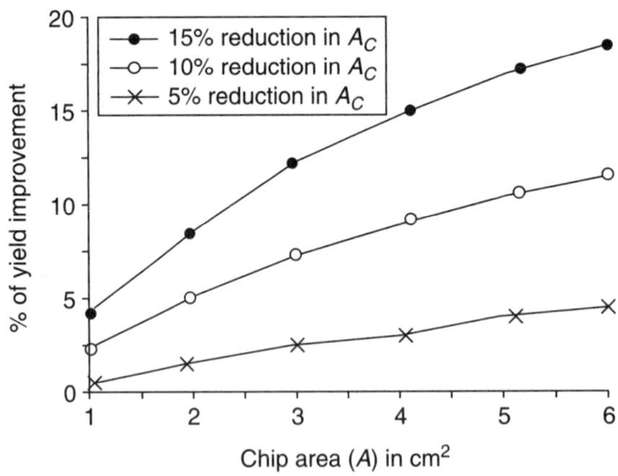

FIGURE 6.27 Effect of reduced critical area on yield improvement.

difficulty of placing a precise count of dopant atoms uniformly over the width of the device leads to random dopant fluctuation (RDF) in the channel. This fluctuation leads to threshold voltage variation that can change such circuit parameters as leakage and propagation delay. Because circuit frequency is determined by the slowest logic path, parametric yield is reduced by large variations in circuit paths.

Gate length biasing may be used to mitigate the effects of performance loss. Gate length biasing involves modifying gate lengths of a selected number of transistors in the design to improve performance and to reduce the dissipation of power in the standby state. This procedure decreases (resp. increases) the gate length of transistors on critical (resp. noncritical) paths. Gate lengths are typically adjusted by the placement of subresolution assist features (SRAFs) to guide optical diffraction during the lithography process.

With transistor scaling, both supply voltage and node capacitances decrease. The reduction in stored charge renders circuit nodes vulnerable to errors due to external (e.g., radioactive or cosmic) radiation. Such errors are known as *soft* errors. The soft error rate (SER) in computing systems has been rising steadily.[27-29] Radiation typically causes single-event upsets (SEU), whose propagation induces faulty values to be "latched" (captured). An observable error that is caused by an SEU is also classified as a soft error. Various techniques to mitigate the effect of SEUs on memory cells, standard cells, and latches have been proposed in the literature. We next look into one such technique that uses adaptive body biasing and voltage division to mitigate SEU impact on circuit operation.

Consider the inverter depicted in Figure 6.28.[28] The inverter consists of a pMOS transistor and an nMOS transistor whose drains are connected to form the output of the gate. The pMOS and nMOS devices have their body terminals connected to V_{DD} and GND rails, respectively. When a logic-1 value is supplied to the inverter input, the output value is a logic-0. A high electric field is generated at the drain-body terminal of the pMOS transistor; this is indicated in the

FIGURE 6.28 Simplified conventional inverter circuit.

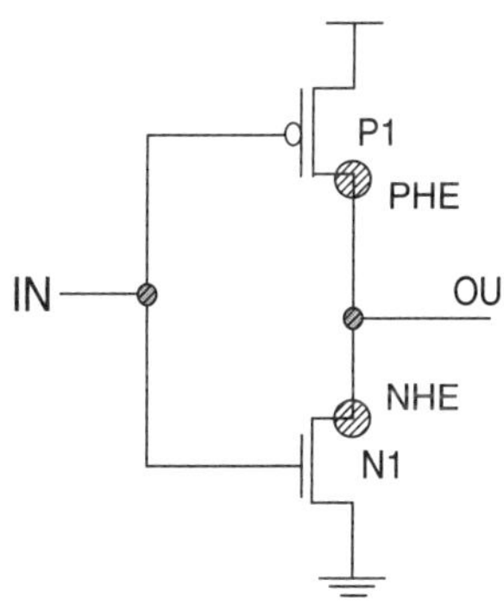

figure by the shaded circle labeled PHE. If a particle strikes near this region then the electric field increases momentarily, causing an increase not only in the drain voltage of the pMOS transistor but also in the charge at the output terminal; this produces an 0–1-0 output value. The effect is clearly illustrated by an attack of cosmic ray particles on a chain of inverters; see Figure 6.29.[28] The plots in panel (b) show the desired output and the momentary signal glitch due to the SEU. Mitigating this effect requires "hardening" the circuit against such intermittent attacks, as we describe next.

A modified inverter circuit is shown in Figure 6.30.[28] This circuit has input ports IN_P and IN_N that feed the same logic value into the

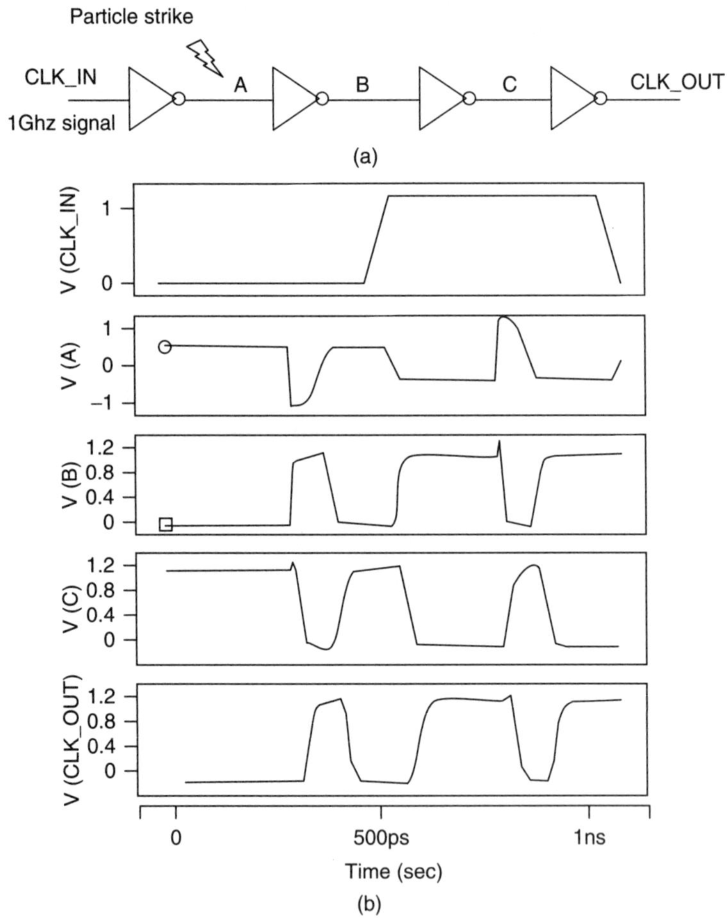

FIGURE 6.29 (a) Conventional inverter chain under attack from cosmic radiation; (b) waveforms illustrating propagation of the resulting soft error.

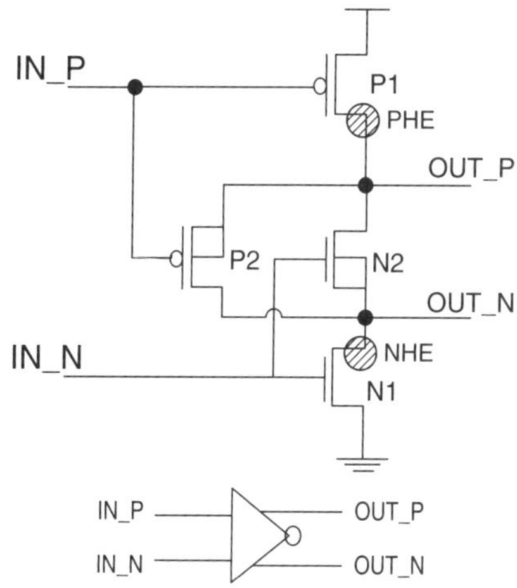

Figure 6.30 Inverter circuit hardened against radiation (top) and its symbolic representation (bottom).

pMOS and nMOS transistors, respectively. Observe that nMOS N1 is outside the isolated well whereas nMOS N2 is inside, so its body terminal is tied to GND. Similarly, the pMOS device P2 is "body biased" to VDD. When a logic value-1 is provided to the input of the inverter, the region susceptible to SEU impact is the shaded circled labeled PHE in the figure. When a particle strikes this modified circuit, the glitch induced at the affected node will not alter circuit operation because that node is connected to the pMOS of the succeeding gate. The mitigation effect is evident in Figure 6.31,[28] where the 0–1-0 glitch does not affect the operation of the inverter chain; this is because the voltage division across the chain of pMOS devices mitigates the intermittent signal change.[28] Various other cells have likewise been designed for such hardening against radiation, since the technique described here can be applied also to standard cells and dynamic circuits.

6.4 Summary

In this chapter we studied various techniques for yield improvement. Yield is the proportion of good chips to all manufactured chips. Defective chips are caused by mask alignment problems, variation in manufacturing parameters, particulate defects related to chemical processes, improper use of equipment, and handling errors. Two important observations made in this chapter are that defect locations are correlated with certain layout patterns and that many manufacturing defects occur in clusters. Solutions to the yield problem are based on fault avoidance, analysis of faulty behavior,

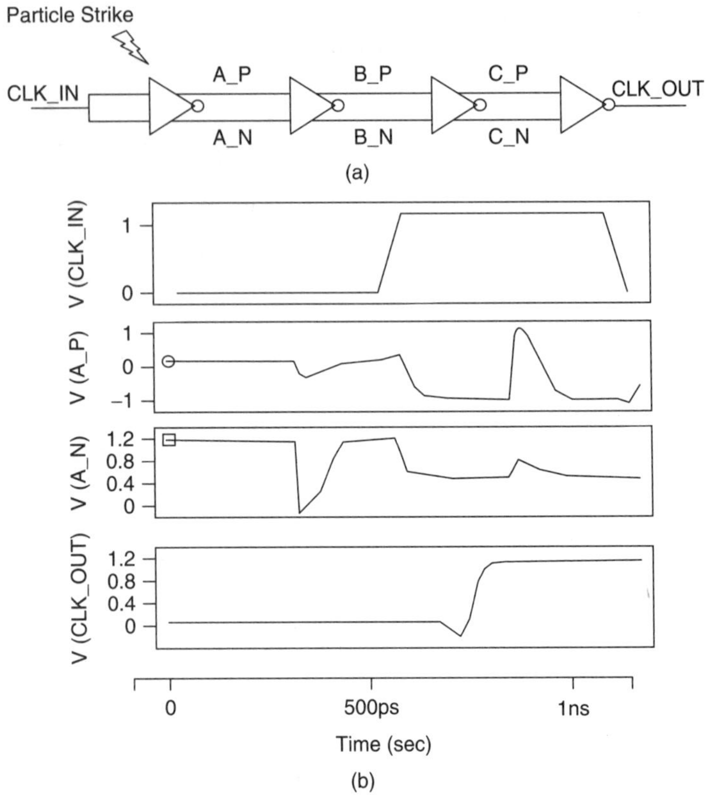

FIGURE 6.31 (a) Radiation-hardened inverter chain under particle attack; and (b) wavesforms illustrating mitigation of the soft error propagation.

and fault tolerance. Faulty behavior analysis is based on fault models, which attempt to capture the logical manifestation of a physical defect. Fault avoidance techniques benefit from two salient defect characteristics: defect locations can often be predicted with high probability, and defects can be reduced or eliminated by layout changes. Fault tolerance techniques involve architectural or software modifications to circumvent faulty circuits. These techniques require redundancies, which may be implemented in the form of time, information, logic, or software. The cost of redundancy is related to logic function and defect clustering: clustered defects are easier to circumvent using logic or hardware solutions, whereas random defects are typically better handled by using information redundancies. Also, software solutions are more suitable for dealing with intermittent errors, whereas time or information redundancies are more suitable for parametric defects.

References

1. W. Maly, A. J. Strojwas, and S. W. Director, "VLSI Yield Prediction and Estimation: A Unified Framework," *IEEE Transactions on Computer Aided Design* **5**(1): 114–130, 1986.
2. V. P. Nelson and B. D. Carrol, *Fault-Tolerant Computing,* IEEE Computer Society Press, Washington DC, 1987.
3. M. L. Bushnell and V. D. Agarwal, *Essentials of Electronic Testing for Digital, Memory, and Mixed-Signal VLSI Circuits,* Springer, New York, 2000.
4. N. K. Jha and S. Kundu, *Testing and Reliable Design of CMOS Circuits,* Kluwer, Dordrecht, 1990.
5. M. Sachdev, *Defect Oriented Testing for CMOS Analog and Digital Circuits,* Kluwer, Boston, 1998.
6. J. M. Acken and S. D. Millman, "Fault Model Evolution for Diagnosis: Accuracy vs. Precision," in *Proceedings of Custom Integrated Circuits Conference*, IEEE, New York, 1992, pp. 13.4.1–13.4.4.
7. G. Greenstein and J. Patel, "EPROOFS: A CMOS Bridging Fault Simulator," in *Proceedings of International Conference on Computer-Aided Design*, IEEE, New York, 1992, pp. 268–271.
8. J. M. Acken, "Testing for Bridging Faults (Shorts) in CMOS Circuits," in *Proceedings of Design Automation Conference,* IEEE, New York, 1983, pp. 717–718.
9. J. M. Acken and S. D. Millman, "Accurate Modeling and Simulation of Bridging Faults," in *Proceedings of Custom Integrated Circuits Conference*, IEEE, New York, 1991, pp. 17.4.1–17.4.4.
10. S. D. Millman and J. P. Garvey, "An Accurate Bridging Fault Test Pattern Generator," in *Proceedings of International Test Conference*, IEEE, New York, 1991, pp. 411–418.
11. J. Rearick and J. Patel, "Fast and Accurate CMOS Bridging Fault Simulation," in *Proceedings of International Test Conference,* IEEE, New York, 1993, pp. 54–62.
12. F. J. Ferguson and T. Larabee, "Test Pattern Generation for Realistic Bridge Faults in CMOS ICs," in *Proceedings of International Test Conference*, IEEE, New York, 1991, pp. 492–499.
13. K. Nikolić, A. Sadek, and M. Forshaw, "Fault-Tolerant Techniques for Nanocomputers," *Nanotechnology* **13**: 357–362, 2002.
14. J. R. Heath, P. J. Kuekes, G. S. Snider, and R. S. Williams, "A Defect-Tolerant Computer Architecture: Opportunities for Nanotechnology, " *Science* **280**: 1716–1721, 1998.
15. J. Lach, W. H. Mangione-Smith, and M. Potkonjak, "Low Overhead Fault-Tolerant FPGA Systems," *IEEE Transactions on Very Large Scale Integrated Systems* **6**: 212–221, 2000.
16. D. Ernst, N. S. Kim, S. Das, S. Pant, T. Pham, R. Rao, C. Ziesler, et al., "Razor: A Low-Power Pipeline Based on Circuit-Level Timing Speculation," in *Proceedings of International Symposium on Microarchitecures*, IEEE/ACM, New York, 2003, pp. 7–18.
17. I. Koren and C. Mani Krishna, *Fault-Tolerant Systems,* Morgan Kaufmann, San Mateo, CA, 2007.
18. J. von Neumann, *Probabilistic Logics and the Synthesis of Reliable Organisms from Unreliable Components,* Princeton, NJ: Princeton University Press, 1955, pp. 43–98.
19. T. P. Haraszti, *CMOS Memory Circuits,* Springer, New York, 2000.
20. M. Goessel, V. Ocheretny, E. Sogomonyan, and D. Marienfeld, *New Methods of Concurrent Checking,* Springer, New York, 2008.
21. M. Nicolaidis, "Time Redundancy Based Soft-Error Tolerance to Rescue Nanometer Technologies," *Proceedings of VLSI Test Symposium*, IEEE, New York, 1999, pp. 86–94.
22. A. Pan, O. Khan, and S. Kundu, "Improving Yield and Reliability in Chip Multiprocessors," in *Proceedings of Design Automation and Test in Europe*, IEEE, New York, 2009.

23. Steven K. Reinhardt and Shubhendu S. Mukherjee, "Transient Fault Detection via Simultaneous Multithreading," in *Proceedings of International Symposium on Computer Architecture*, IEEE, New York, 2000, pp. 490–495.
24. A. M. Palagonia, "Laser Fusible Link," U.S. Patent no. 6,160,302 (2000).
25. E. Hamdy et al., "Dielectric Based Antifuse for Logic and Memory ICs," in *International Electron Devices Meeting* (Technical Digest), IEEE, New York, 1988, pp. 786–789.
26. J. Birkner et al., "A Very High-Speed Field Programmable Gate Array Using Metal-to-Metal Antifuse Programmable Elements," *Microelectronics Journal* **23**: 561–568, 1992.
27. S. Mukherjee, *Architecture Design for Soft Errors*, Morgan Kaufman, San Mateo, CA, 2008.
28. Ming Zhang and Naresh Shanbhag, "A CMOS Design Style for Logic Circuit Hardening," in *Proceedings of IEEE International Reliability Physics Symposium*, IEEE, New York, 2005, pp. 223–229.
29. P. Shivakumar et.al., "Modeling the Effect of Technology Trend on the Soft Error Rate of Combinational Logic," in *Proceedings of International Conference on Dependable Systems and Networks*, IEEE Computer Society, Washington DC, 2002, pp. 389–398.

CHAPTER 7

Physical Design and Reliability

7.1 Introduction

The long-term reliability of integrated circuits has become an important concern as today's devices approach the end of the CMOS technology roadmap. Scaling of feature size coupled with variability in manufacturing process has lead to increased reliability problems. Semiconductor ICs in consumer and business sectors are typically engineered to last for about ten years. In satellite and space systems, or in mission-critical applications, product life expectations may be much longer. In order to improve the reliability of semiconductor products, the underlying failure mechanisms must be clearly understood. Some reliability failures stem solely from manufacturing problems. Physical corrosion due to leaks and moisture, electrical leakage, package encapsulation problems, and/or loose bonding are examples of manufacturing issues that degrade IC reliability. Other reliability failures are rooted in design. High-current density, improper input/output (IO) terminations, and poorly designed heat sinks are examples of chip and package design issues that contribute to reliability failures. Here we are primarily concerned with reliability problems that are related to the design process, so we focus on changes in design that can affect overall chip reliability.

In a typical design system, reliability guidelines are the principal mechanism for guarding against reliability failures. These guidelines are designed to provide enough margin to prevent reliability failures. For example, interconnect electromigration failures are related to current density, which in turn is a function of interconnect width. For a given driver strength, a conductor should thus be sized to minimize current density, thereby averting electromigration failures. Similar guidelines guard against other failure mechanisms. In order to formulate a comprehensive set of guidelines, failure analysis of test chips and actual chips must be conducted so that the design attributes that give rise to the failures are identified. Reliability guidelines are also a function of manufacturing process quality, which changes as the process matures; also, the guidelines will vary from product to product as a function of reliability expectations. Often there are warning signs that reliability problems are imminent. In the case of electromigration, for example, an interconnect may pass through a phase of extremely high resistance before it becomes completely open. Knowledge of such phenomena makes it possible to design an early warning system.

A device is said to be *reliable* if it performs intended operations under a given set of conditions over its scheduled lifetime. Yet variation in manufacturing process parameters, when coupled with designs that are not "margined" adequately, may lead to reliability failures *before* the design's expiration date. Each product from a manufacturing line may fail at a different time. Thus, *mean time to failure* (MTTF) is the reliability measure used to describe useful product life. A related measure is *shipped product quality level* (SPQL), which quantifies the number of bad chips per million at the beginning of and at various points in a product's life. *Mortality rate* is defined as the ratio of the number of chips failing during a specified time interval to the total number of chips. Whereas the MTTF indicates a product's expected lifetime, the mortality rate is a better indicator of the failure pattern.

The reliability failures that occur during the lifetime of an IC can be categorized into three phases: (1) early-lifetime failures (aka infant mortality); (2) normal-lifetime random failures; and (3) end-of-lifetime wear-out failures. When plotted on a graph of time versus mortality, as in Figure 7.1, the IC lifetime phases form a "bathtub" curve. Infant mortality occurs during the early stage of a product's

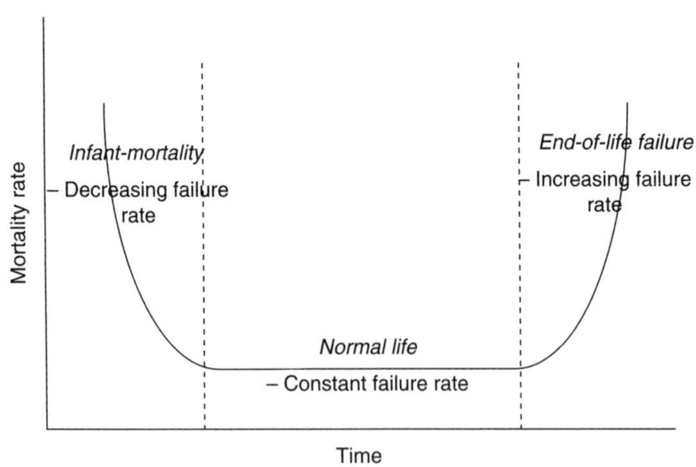

FIGURE 7.1 "Bathtub" curve formed by plotted phases of device reliability.

life. The most frequent causes of infant mortality are manufacturing imperfections. A manufacturing stress test is applied to accelerate infant mortality, so that products leaving the factory do not fail early at the customer's site. This stress test is also known as the *burn-in* test or *shake-and-bake* test, terminologies that are associated with the stress conditions applied. When ICs are subjected to vibrations with high g-force, many of the mechanical failures are accelerated. This is the "shake" part of the shake-and-bake test. The "bake" part consists of applying high voltage and high temperature in a burn-in chamber to accelerate electromigration and gate oxide short failures. The sharper the curve at the infant mortality region, the better the manufacturing stress test. Early failure rate decreases with maturity of the manufacturing process and also with early removal of weak wafers through prudent screening.

A constant failure rate is seen during the IC's normal-lifetime operation. Failures that occur during this period are considered to be random because they depend on external factors of stress and performance overload. The constant failure rate in some cases can extend well beyond the IC's expected useful lifetime. The increasing failure rate in the third and final phase is attributed to the circuit's wearing out, which leads to degradation of performance or complete failure. This IC wear-out, also known simply as aging, results from several failure mechanisms—such as electromigration, oxide breakdown, and negative bias temperature instability—that are discussed in this chapter.

The long-term reliability of a product is defined as the time over which the constant failure rate is maintained and before the wear-out failure mechanisms begin to appear. Integrated circuits of high reliability tend to have a longer than expected lifetimes and are resistant to aging. Device aging mechanisms can be classified as catastrophic mechanisms, gradual degradation mechanisms, and radiation-induced mechanisms (soft errors discussed in Sec. 7.6). *Catastrophic* mechanisms are the flagship of reliability problems because they induce complete failure of the device. Abrupt and catastrophic failures may be caused by electromigration, electrostatic discharge, or oxide breakdown. These failures lead to irreparable catastrophic defects such as metal or device opens and shorts. Electromigration leads to breakage in a metal line due to excessive current flow in the region. Under such high current density, the atoms of the metal line are "blown away" by the constant flow of electrons. Thinner than normal metal lines formed by improper patterning are generally susceptible to electromigration failures, which are also seen in metal contacts to the device and polysilicon gate regions.

Electrostatic discharge is the damage caused by a sudden discharge of static electricity, through a gate terminal of the transistor, that may permanently alter its control. This rapid electrostatic transfer usually occurs in IO devices that come in contact with electrostatic charges.

Gate oxide breakdown is the rupture of the oxide layer, which can lead to reliability problems. An oxide layer (typically silicon dioxide) serves as a dielectric material between the gate and the channel, for shallow trench isolation, and as an interlayer dielectric between metal lines. Oxide breakdowns occur when a high voltage applied across the oxide causes a permanent conduction path through the oxide where the current flows. With technology scaling, the thickness of gate dielectric materials have been reduced to less than 20 Å, which makes the gate oxide layer more susceptible to reliability problems.

Reliability problems can also manifest as errors that cause *gradual degradation* of the device. Such mechanisms require that the device state be maintained for a prolonged period in order for the error to manifest itself. Hot carrier effects and negative bias temperature instability (NBTI) are examples of gradual reliability degradation mechanisms. In the presence of a strong electric field, the electrons and holes present in the semiconductor material tend to be accelerated. These high-energy carriers may jump to the oxide region and form pockets (aka interface traps) in the devices that store charge. Charge traps in the oxide change the device's threshold voltage and transconductance, leading to performance-related failures. Channel traps can be caused by substrate effects and secondary hot carrier effects, and they also depend on the temperature of the device under operation. Hot carrier effects have a greater impact on nMOS devices than on other components.

Negative bias temperature instability affects pMOS devices. When a pMOS transistor is negatively biased (i.e., negative gate-to-source voltage) and under high temperature, its threshold voltage increases; this increase affects transistor ON current and device performance. Interface traps that manifest in surfaces of the device lead to performance changes during circuit operation. Trapped charges are contributing factors in both hot carrier injection and NBTI. The physics of NBTI has been hypothesized by several researchers,[1-7] and NBTI effects (unlike those of hot carrier injection) may be reversed under nonnegative bias conditions and lower temperatures.

The reliability effects mentioned thus far include *permanent* failures in circuit logic and timing that arise with aging. In contrast, reliability effects that cause *intermittent* failures in circuit operation are known as soft errors, which are usually not associated with aging. Nonetheless, soft errors are interesting from the perspective of design for manufacturability. For example, increased load capacitance relative to driver strength can increase resilience to soft errors but may also cause performance problems. In addition, large diffusion areas may lead to more soft error–related issues. Thus, soft errors can be addressed by attending to DFM considerations.

Modeling and simulation of reliability effects is key to improved design for yield. With increasing device and interconnect variability, reliability problems become more pronounced. Reliability simulators attempt to predict device lifetime as a function of design parameters. Such analysis facilitates better design for reliability, which is the focus of this chapter. We shall also discuss reliability test techniques.

7.2 Electromigration

Electromigration (EM) is the most widely known reliability failure mechanism, and it has generated much research interest over the past four decades. Electromigration failures are related to current density. Thus, EM failures are more likely for thin power supply lines in a chip, and they may occur in signal interconnects when a thin wire is driven by a relatively large driver. Electromigration problems are also related to properties of conductors—for example, EM problems for aluminum interconnects are worse than for copper interconnects of similar dimensions. Aluminum was the preferred metal for interconnects before the 250-nm technology node. This preference was based primarily on its conductivity, cost, and manufacturability. However, aluminum is highly susceptible to EM failures. The introduction of copper interconnects was motivated both by higher conductivity and reduced EM problems. Yet as feature widths scale and the current density increases, EM for copper interconnects also become a significant concern.

Electromigration is defined as the migration or displacement of atoms due to the movement of electrons through a conducting medium in the presence of an electric field.[8,9] The flow of current through a conductor creates an equal "wind" of electrons in the opposite direction, which causes the metallic (aluminum or copper) atoms to be displaced toward the positive end. When atoms are displaced, vacancies are created that move toward the negative end of the conductor. These vacancies join to form voids, which reduce the conductor's cross section. Any reduction in conductor width will increase current density, which leads to localized heating. Prolonged localized heating in such regions of reduced width causes EM failures such as full and resistive opens. The atoms that migrate to the positive end may increase the cross section of the conductor by forming hillocks, which lead to bridges between adjacent conductors. Figure 7.2 shows the formation of opens and shorts due to EM failures.

Electromigration failures can be observed in metal interconnects, contacts, and gate polysilicon lines. Electromigration manifests itself as either an open defect or an increase in line resistance. With polysilicon lines, voids are caused by the dissipation of phosphorus atoms. In the presence of high temperature and a steady flow of current, the process of EM-induced conductor failure is accelerated. All EM-induced failures are a function of mass transport, temperature gradient, current density, metal dimension, and contact cross section.

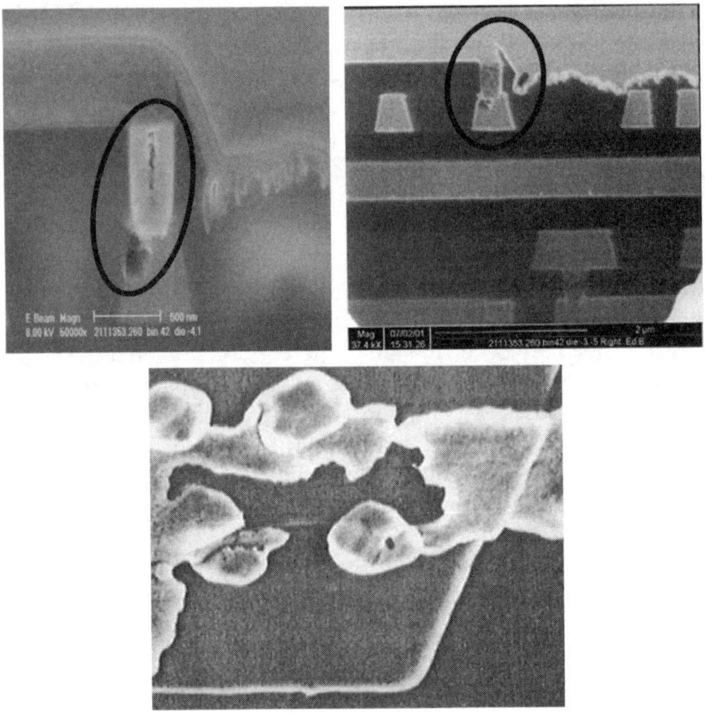

FIGURE 7.2 Electromigration (EM) failure seen in interconnect metal lines. (*Courtesy of Intel.*)

These parameters are used to project the mean time to failure of a device as follows:[10]

$$\text{MTTF} = AJ^{-n} \exp\left(\frac{E_a}{k_B T}\right) \tag{7.1}$$

Here J is the current density, A is a constant that depends on the manufacturing technology, E_a is the activation energy of the conductor material, k_B is the Boltzman constant, and T is the conductor temperature (in Kelvin). The value of T is a function of the reference temperature T_{ref} and the self-heating temperature T_{self}. Self-heating actually forms only a small component of the conductor temperature, so T almost always lies between 100 and 200°C. The value of n depends on the cause of EM failure (e.g., temperature or structural deformity) and also on the degree of coalescence of atoms (vacancies) due to mass transport within the metal under the influence of electric field. For copper metal wires, the value of n is set to 1.2.

Current density J is a function of conductor dimensions and wire capacitance as well as of the supply voltage applied and its frequency and switching probability:[10]

$$J \propto \frac{C \times V}{W \times H} \cdot f \cdot s_p \qquad (7.2)$$

Metal wires and vias of smaller width have higher current density and thus greater susceptibility to EM failure, which leads to a lower MTTF value. A typical scenario for an EM failure is a large gate driving a steady current through a thin interconnect line.

Improper contact cross section leads to "current crowding" that causes localized heating and temperature gradients. Another reliability issue that has become a major concern is junction spiking. (See Figure 7.3, which is derived from Sabnis.[11]) The contact material (metal) punches through the semiconductor and forms spikes. This is observed most often in contacts with low junction depth. Junction spiking and the formation of voids due to silicon migration are the chief EM failure modes in contacts and vias. The mean time to failure for contact and vias is given by the following relation:[11]

$$\text{MTTF} \propto X_d^2 \left(\frac{I}{W}\right)^{-n} \exp\left(\frac{E_a}{k_B T}\right) \qquad (7.3)$$

Here X_d is the junction depth of the contact, I is the current through the contact or via, and W is the width of the region. Because the junction formed is not homogeneous, the MTTF relation here uses current instead of current density. According to the literature, the value of n in equation (7.3) depends on the current passed through the contacts. This value is high ($n = 6$ to 8) for large currents and low ($n = 1$ to 3) for smaller currents that flow through the contact or via.[11]

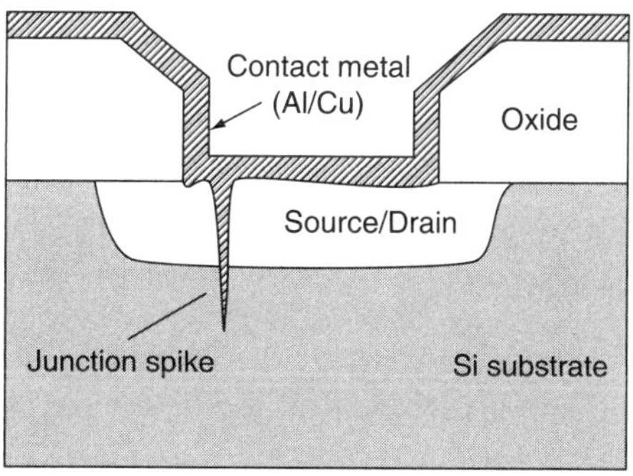

FIGURE 7.3 Junction spikes due to EM failures in metal contacts (lateral view).

The predominant technique used to reduce electromigration failures is increasing the width of the metal line and the contact or via dimensions. Increasing the dimensions of the region reduces the current density and hence the localized heating, thereby reducing EM failures. Increasing via dimensions (even doubling them) incurs very little overhead. However, the potential randomness of failure locations requires that the width of the entire line or via should be increased, which would entail considerable overhead. The increase in width of interconnect lines means that designs must be larger to enable same amount of routing. Increased die area leads to designs that do not conform to performance requirements.

7.3 Hot Carrier Effects

Electrons and holes in the channel region gain kinetic energy in presence of an electric field. When such carriers gain enough kinetic energy to overcome the energy barrier between the oxide and the channel region, it may jump into the gate oxide. Depending on the energy level and the thickness of the oxide, such charges may be trapped in the oxide; this changes the device threshold voltage and hence the drive current.

The typical mechanism for hot carrier injection (HCI) involves the presence of an electric field that induces carriers with high kinetic energy (see Figure 7.4). Hot carrier injection can be caused by scattering

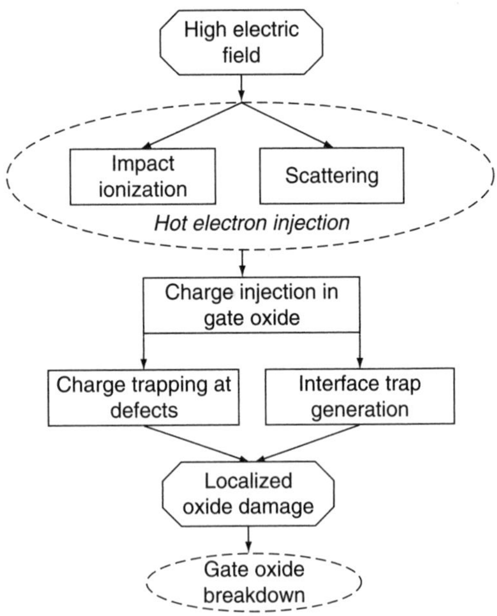

FIGURE 7.4 Hot carrier injection (HCI) mechanisms.

(in the channel) or by impact ionization (avalanche at the drain). The increased energy of the hot carriers exceeds the potential at the oxide-channel boundary, allowing the carriers to tunnel through it. These tunneling carriers form interface traps that, over time, increase in number. Once a critical number of trap states is reached, the result may be an electrical path through the oxide, leading to device breakdown. There are multiple injection mechanisms, which are described in what follows. Sections 7.3.1 to 7.3.4 provide a description of hot carrier generation, trap generation, device degradation, and mitigation strategies.

7.3.1 Hot Carrier Injection Mechanisms

Hot carriers are holes or electrons whose kinetic energy increases under the influence of an electric field. High energy carriers tunnel into semiconductor and oxide materials, forming interface traps. Hot carriers can be injected in one of four injection methods: (1) drain, (2) channel, (3) substrate, or (4) secondary.[12,13]

Drain injection occurs when the device is operating in the active region where the drain voltage is greater than the gate voltage but less than its saturation value. Under the presence of a high drain voltage, channel carriers acquire kinetic energy and collide with silicon crystal lattice at the drain depletion region, forming electron-hole pairs. This occurs because the lateral electric field reaches its maximum at the drain end of the channel. Under normal conditions, the process is similar to impact ionization. But when these electron-holes pairs gain enough energy and potential to tunnel through the boundary between the channel and oxide gate, the result is trapped charges or gate leakage current; see Figure 7.5. As more of these traps are formed in the oxide region, a degradation of device operation is observed. Hot carrier injection induced by drain voltage is the most common type of HCI; it affects the V_T and transconductance of the device.

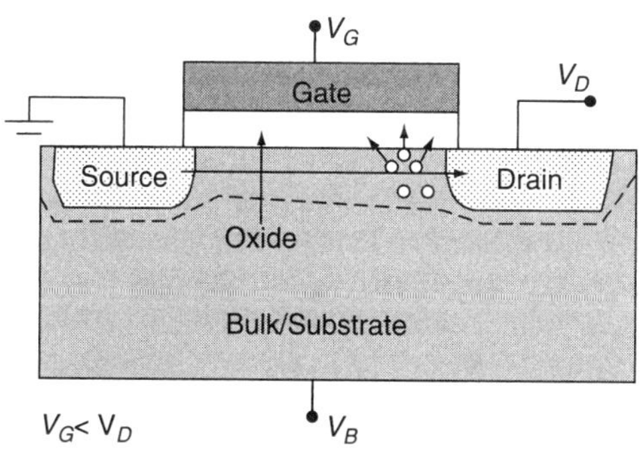

Figure 7.5 Drain injection type of HCI.

A variant of hot carrier injection is *channel injection*, which is illustrated in Figure 7.6. With this injection mechanism, it's not an avalanche of high-energy carriers but rather the scattering of electrons in the channel that causes tunneling into the oxide. When the gate voltage is equal to the drain voltage and the source voltage is at its minimum value, current flows across the channel from the source to the drain. The resulting electric field can induce enough energy in the scattered channel carriers for some of them to penetrate the gate oxide before reaching the drain.

Substrate back-biasing is typically employed to adjust the amount of drain current flowing through a device. The back-bias voltage applied to the substrate can be either positive or negative, depending on whether the device is p-channel or n-channel. If the substrate bias voltage is higher (on the negative or positive side) than the required voltage, then part of the carriers in the substrate are pushed toward the channel. These carriers acquire kinetic energy and tunnel into the gate oxide, forming interface traps. The resulting *substrate injection* is depicted in Figure 7.7.

Secondary injection happens in the same phase as drain-induced carrier injection (i.e., when $V_D > V_G$). As mentioned previously, some carriers with high kinetic energy do not penetrate the oxide and instead move in the opposite direction, toward the substrate, causing

FIGURE 7.6 Channel injection type of HCI.

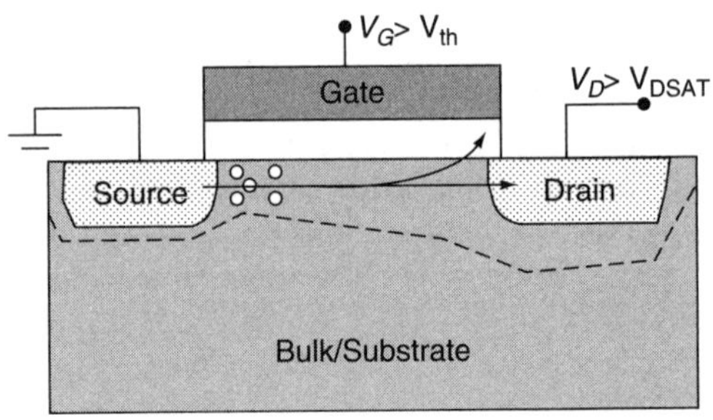

FIGURE 7.7 Substrate injection type of HCI.

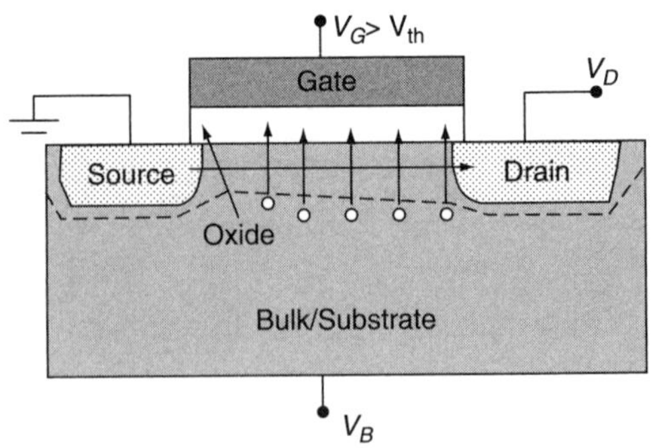

bulk current drifts. When a bulk bias voltage is applied, some of these secondary electron-hole pairs are reflected back toward the channel, penetrating into the oxide.

7.3.2 Device Damage Characteristics

Hot carrier degradation affects n-channel and p-channel devices in much the same way. The effect is somewhat less pronounced in p-channel devices because the drain voltage required for hot carriers to gain kinetic energy is high compared to their n-channel counterparts. Interface traps and trapped oxide induce two major changes in a device. First, trapped charges in the oxide affect the surface potential, which changes the device's flat-band voltage. Any change in flat-band voltage alters the threshold voltage of the device and hence affects its performance. Second, interface traps that are present at the $Si\text{-}SiO_2$ interface affect the mobility of majority carriers through the channel. In turn, reduction in mobility affects the drain current and device performance.

The I-V characteristics for a device before and after HCI stress are plotted in Figure 7.8.[14] Observe that device degradation due to carrier scattering is most pronounced in the linear region of transistor operation, not in the saturation region. This is because, when the device enters saturation, the drain current becomes independent of the voltage at the drain. In this state, the chief cause of device degradation is impact ionization in the channel.

The existence of interface traps does not immediately degrade the device—unlike the case of electromigration. Instead, the device degrades gradually as more traps are formed within the oxide and in the channel-oxide barrier. The mean time to failure of a device subject to hot carrier degradation depends on the device channel width and operating conditions (i.e., bias voltages and temperature). Temperature is a crucial parameter because it reduces detrapping within the oxide

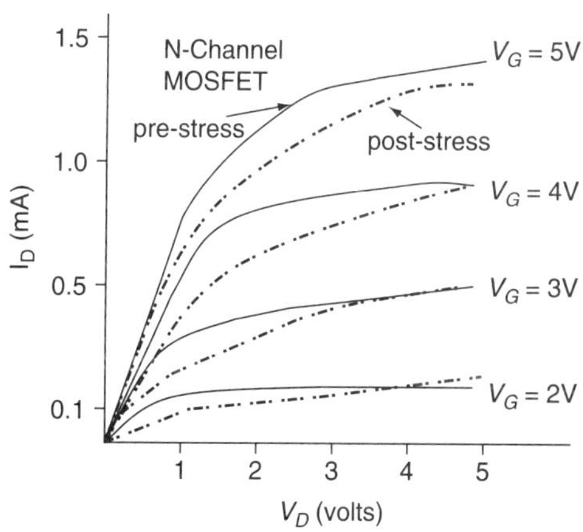

FIGURE 7.8 Drain current versus drain voltage for devices before and after HCI stress.

region. The mean time to failure caused by HCI is modeled by the following expression:

$$\mathrm{MTTF} = \int_{t=0}^{t=T} \frac{I_{\mathrm{DS}}}{W \cdot x} \left(\frac{I_{\mathrm{sub}}}{I_{\mathrm{DS}}} \right)^{m} dt \qquad (7.4)$$

where MTTF is expressed in terms of the degradation age of the device under stress. The resulting value depends on the channel width W as well as the drain and substrate currents I_{DS} and I_{sub}. It can be seen that degradation increases with the time T over which the device is under stress.

7.3.3 Time-Dependent Dielectric Breakdown

Gate oxide failure is sometimes caused by a carrier injection mechanism known as time-dependent dielectric breakdown (TDDB), which results in a multistage destruction of the gate oxide. The first stage is formation of interface traps by hot carrier injection. These interface traps overlap each other to form *defect streaking*, which in turn leads to a path between the gate and the channel or substrate. The injection stages result in the formation of a conduction path between the gate and the source-drain region. The next stage creates an environment that exacerbates this effect in the presence of heat, thereby creating thermal damage. More traps will be generated under such conditions, forming a positive feedback loop that leads to dielectric breakdown. Thus, TDDB consists of a cycle of high-energy carrier injection and trap generation that leads to oxide breakdown.

Thin dielectrics and dielectrics with process deformities can hasten the formation of interface traps and reduce the time to degradation. The presence of a potential difference between source and drain also increases the generation of traps. Experiments have shown that, for a transistor with 4-nm-thick gate dielectric in 45-nm process technology, dielectric breakdown occurs in the presence of an electric field of 5 MV/cm.[15,16] This high electric field is typically caused by a voltage spike greater than the supply voltage V_{DD}. Voltage spikes are usually the result of inductance in power supply lines or other signal integrity issues. The TDDB of gate dielectric can be characterized by using the MTTF as follows:

$$t_{\mathrm{TDDB}} = A \exp\left(-\frac{E_a}{kT_{\mathrm{ref}}} + B_{\mathrm{ox}} V \right) \qquad (7.5)$$

Here A is a constant based on experimentation and process parameters, V is the voltage applied at the gate terminal, and B_{ox} is a voltage acceleration constant that depends on oxide characteristics. It can be seen that t_{TDDB} is a function of temperature.

7.3.4 Mitigating HCI-Induced Degradation

Hot carrier degradation can be mitigated by using device and circuit techniques. Device techniques employed today include using lightly doped drain regions, creating an offset gate, and using buried p+ channels. As explained previously, the drain injection form of HCI results from impact ionization at the drain end of the channel. Therefore, drain regions near and beneath the channel are doped more lightly than elsewhere so there will be fewer electron-hole pairs that can tunnel into the oxide by gaining kinetic energy. The lightly doped regions are illustrated in Figure 7.9.[17]

Circuit techniques that have been incorporated into designs as good practices are also effective at mitigating HCI. Because hot carrier degradation predominantly affects nMOS devices, series connections (such as those in NAND and other complex gates) will be subjected to less degradation. The nMOS closest to the output in a stack is most vulnerable to device degradation that will affect the output. A device can be protected by using an additional MOSFET (in series) that reduces the voltage at the drain of the affected nMOS; see Figure 7.10.[18]

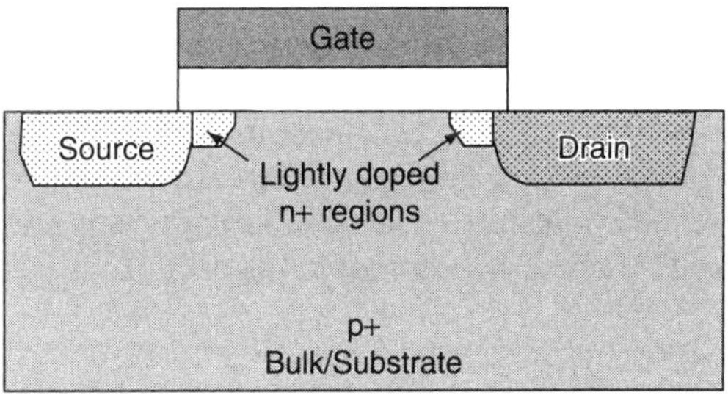

FIGURE 7.9 Device-based HCI mitigation: lightly doped regions reduce drain injection.

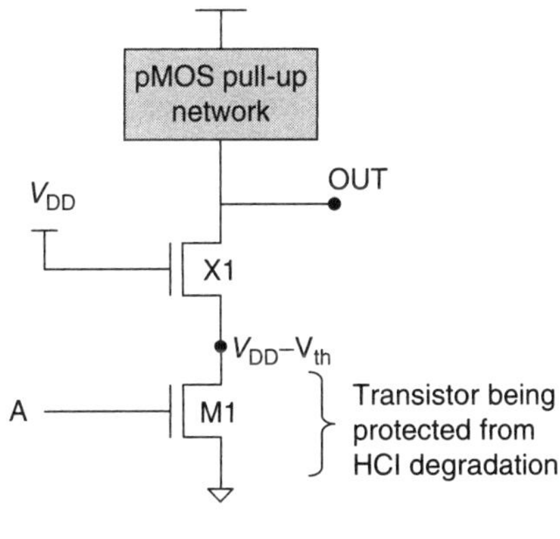

FIGURE 7.10 Circuit-based HCI mitigation technique.

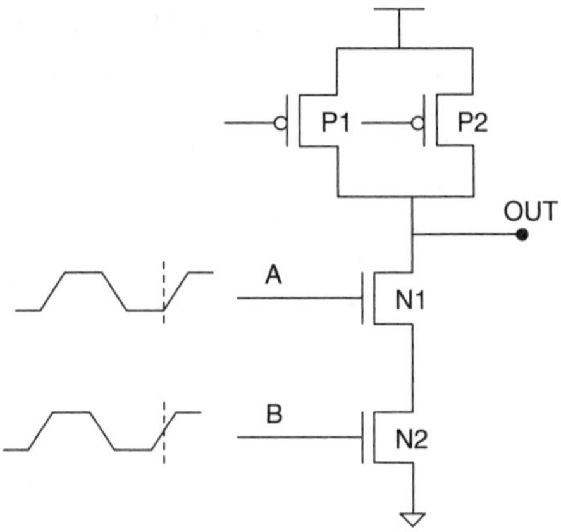

FIGURE 7.11 Scheduling of input signals to mitigate HCI effects.

Gate sizing and input signal scheduling are additional examples of circuit design techniques to mitigate HCI-induced reliability degradation. Gate sizing is a well-known technique for accommodating uncertainties in parameters. In general, increasing the size of a transistor will reduce its susceptibility to HCI. Gate sizing changes the input signal slope, which means the transistor leaves the linear region sooner. This reduces the impact of HCI-induced reliability issues. Gates in the preceding and fan-out stages must be properly sized to ensure overall design reliability. The arrival of input signals can affect the amount of stress and hot carrier resistance in transistors.[19] Hot carrier injection effects are strongly suppressed when input to the nMOS transistor nearest to the output arrives earlier; see Figure 7.11.

7.4 Negative Bias Temperature Instability

Negative bias temperature instability is a circuit aging mechanism that affects pMOS devices. When a pMOS transistor is negatively biased (negative gate-to-source voltage) and under high temperature, the threshold voltage of the device increases, which affects the transistor's ON current and its performance. Concern about NBTI has grown significantly for technologies below 65 nm. Threshold voltage degradation due to NBTI is similar to hot carrier injection in some cases. But unlike HCI, where degradation occurs primarily during active switching of the transistor, NBTI occurs under static stress conditions, when the device is not switching. In today's low-power designs, portions of the circuit are typically power gated to conserve power; static stress under such standby conditions contributes to degradation in those parts of the circuit.

NBTI degradation in pMOS transistors is caused by generation of interface traps at the silicon-oxide barrier.[1] Crystal mismatches between silicon and the oxide results in many free silicon atoms near the interface. The amount of free silicon can be suppressed by *hydrogen annealing,* which bonds the silicon atoms with hydrogen atoms. Yet with continued feature scaling and increased stress on the gate, these Si-H bonds are susceptible to breakage. This breakage leads to interface traps and to the generation of free silicon, causing changes in the device's threshold voltage and drive current.

7.4.1 Reaction-Diffusion Model

Interface trap generation and Si-H bond breakage is well explained by the reaction-diffusion model.[5,6,20,21] This model is divided into two primary phases, the reaction dominated phase and the diffusion dominated phase. Holes in the channel dislodge hydrogen atoms from the Si-H bonds at the interface, creating traps (see Figure 7.12). This process is described by the following expression:

$$\text{Si-H} + h^+ \rightleftarrows \text{Si}^+ + H^0 \tag{7.6}$$

where h^+ denotes the holes in the channel region. This is the *reaction dominated* phase of the NBTI. In this phase, the number N_{it} of interface traps generated is modeled in terms of the rate of bond breaking, the rate of trap generation, and the number of hydrogen atoms diffused toward the gate. The rate of change in the number of interface traps is a function of several factors:

$$\frac{dN_{\text{it}}}{dt} \propto k_r, k_h, N^0, N_{\text{it}}, N^0_{\text{H}_2} \tag{7.7}$$

Here k_r and k_h are (respectively) the reaction rates for reverse reaction (also known as the hydrogen annealing rate) and the rate at which the hydrogen atoms separate from Si-H to form hydrogen molecules (also known as bond-breaking rate), N^0 is the initial Si-H density, and $N^0_{\text{H}_2}$ is the initial hydrogen density at the Si-SiO$_2$ interface.

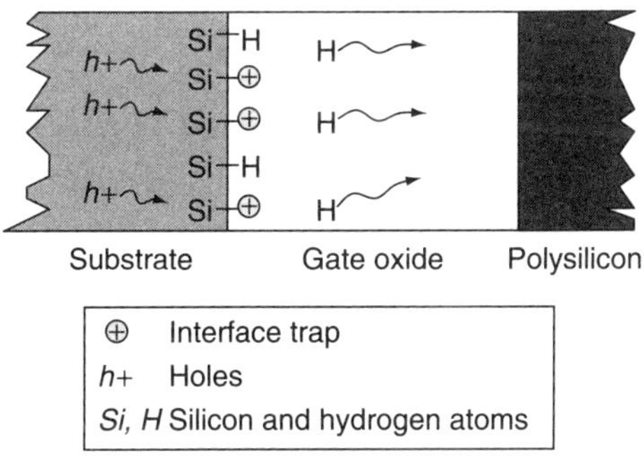

FIGURE 7.12 Si-H bond breakage and interface trap formation during the NBTI stress phase (see Sec. 7.4.2).

After certain period of stress, the $N^0_{H_2}$ density saturates as the device enters an equilibrium state; in this state, then, the formation of interface traps is reduced. The process now moves into the *diffusion dominated* state in which the hydrogen atoms present in the oxide diffuse toward the gate terminal (see Figure 7.12). The change in hydrogen density over time depends on the material diffusion rate, which is given by

$$\frac{dN_{\mathrm{H2}}}{dt} \propto D^2 \frac{dN_{\mathrm{H2}}}{dx^2} \tag{7.8}$$

The diffusion process also slows the formation of more interface traps, because diffusion is much slower than reaction. Hence, in this state the final N_{it} rate is not dependent on $dN_{\mathrm{H_2}}$ in this state. The reaction-diffusion model thus yields the following equation for total interface trap formation due to NBTI stress:[1]

$$N_{\mathrm{it}}(t) = X\sqrt{E_{\mathrm{ox}} \exp\{E_{\mathrm{ox}}/\varepsilon_0\}} \cdot {}^{0.25}\!\!\sqrt{t} \tag{7.9}$$

where X and ε_0 are technology-dependent parameters, E_{ox} is the oxide electric field, and t is the stress time. The generation of interface traps changes the threshold voltage of the pMOS. This change in pMOS threshold voltage is modeled as follows:[7,20-23]

$$\Delta V_{\mathrm{T\text{-}p}}(t) = (\mu_{\mathrm{mob}} + 1)\frac{q\Delta N_{\mathrm{it}}(t)}{C_{\mathrm{ox}}} \tag{7.10}$$

where q is the charge in the channel and C_{ox} is the oxide capacitance. The μ_{mob} term is included because traps in the Si-SiO_2 interface also change the mobility of the device.

7.4.2 Static and Dynamic NBTI

Threshold voltage variation due to NBTI-induced stress occurs in response to trap generation in the Si-SiO_2 interface under negative bias conditions. Unlike HCI-based degradation, however, NBTI is reversible. That is, V_T can recover under nonnegative bias and lower temperature. Thus NBTI has two phases, the stress phase and the recovery phase.[6]

During the stress phase (which is modeled by the reaction-diffusion equations), the source-drain and substrate are at the same potential and the gate is in negative bias. (i.e., $V_{\mathrm{GS}} = -V_{\mathrm{DD}}$). In this phase, interface traps are formed that drive hydrogen atoms toward the gate.

During the recovery phase ($V_{\mathrm{GS}} = 0$), no holes are present in the channel to form interface traps. Some of the diffused hydrogen atoms do not have enough energy to move across to the gate region, and these atoms return to the Si-SiO_2 interface where they rebond with the dangling silicon atoms. In this phase, NBTI degradation is reversed in that the device's threshold voltage returns (partially) to the normal condition. The two phases are illustrated in Figure 7.13.

Static NBTI stress is observed when a device is in the stress phase throughout its lifetime. Devices that are turned off for a prolonged period to minimize static power consumption are the most likely to endure static NBTI stress. The threshold voltage of static-stressed devices tends to change with time; as a result, when they need to be turned back on, their V_T is different and so alters circuit timing. An example of static stress in CMOS gates is illustrated in Figure 7.14(a),

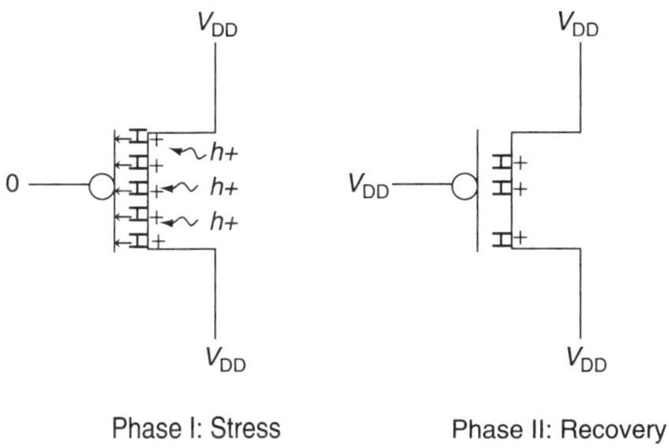

FIGURE 7.13 Two phases of NBTI: stress and recovery.

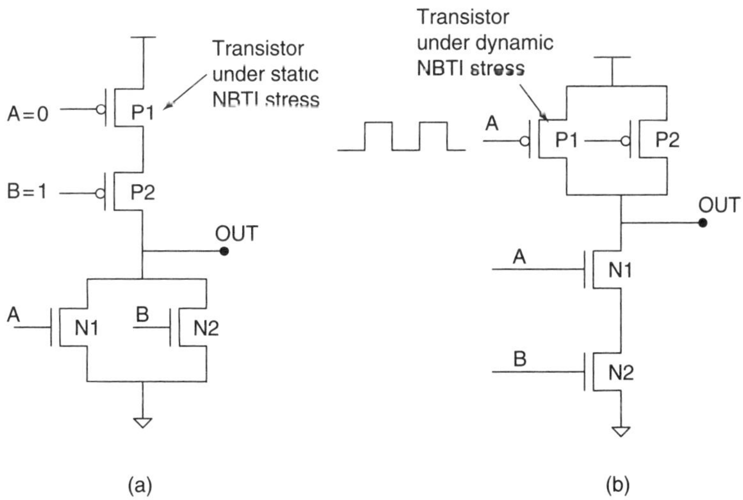

FIGURE 7.14 NBTI stress types: (a) static; (b) dynamic.

where it is evident that the pMOS transistor is under stress when its input voltage is 0. In active circuits, the gate voltage changes between 0 and V_{DD} during proper circuit operation. For a pMOS device, NBTI degradation occurs when gate voltage V_G is 0; the recovery occurs when $V_G = V_{DD}$. This means that dynamic circuit operation alternates between stress and recovery phases, as shown in Figure 7.15.[6] Figure 7.14(b) illustrates the case of *dynamic* NBTI stress applied to a transistor. The gate delay due to V_T variation is found to be highest when P1 switches after being under stress for a prolonged period.

7.4.3 Design Techniques

Design techniques for mitigating NBTI-induced degradation in threshold voltage and drive current include gate sizing, duty-cycle tuning, and V_{DD} and V_T tuning. All these techniques are used in today's designs in order to mitigate process variations.

Duty cycle is defined as the percentage of time over which a particular signal state is active or high. Dynamic NBTI alternates between stress and recovery phases, and the duty cycle determines how much time the device spends in each phase. Using device sizing to properly tune the circuit's duty cycle will mitigate changes in threshold voltage. The longer the device stays in the recovery state, the lower the value of $\Delta V_{T\text{-}p}$.

Variation in $\Delta V_{T\text{-}p}$ is highly dependent on V_{DD} and V_T. The plotted curves in Figure 7.16[24] indicate that V_{DD} tuning is preferable in terms of its effect on $\Delta V_{T\text{-}p}$ and generic ease of control. The extent of tuning

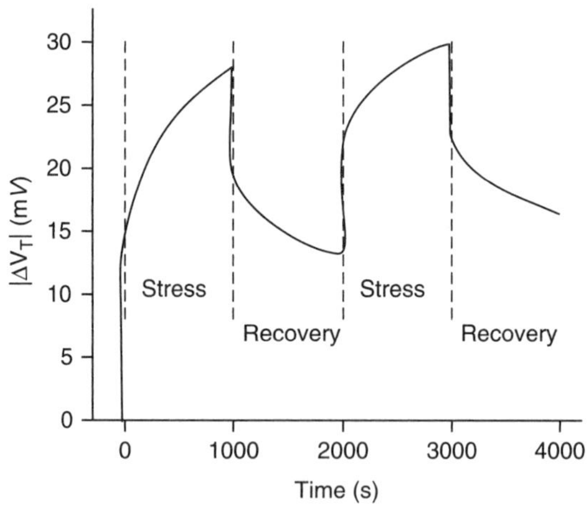

FIGURE 7.15 Variation in threshold voltage V_T during stress and recovery phases of NBTI.

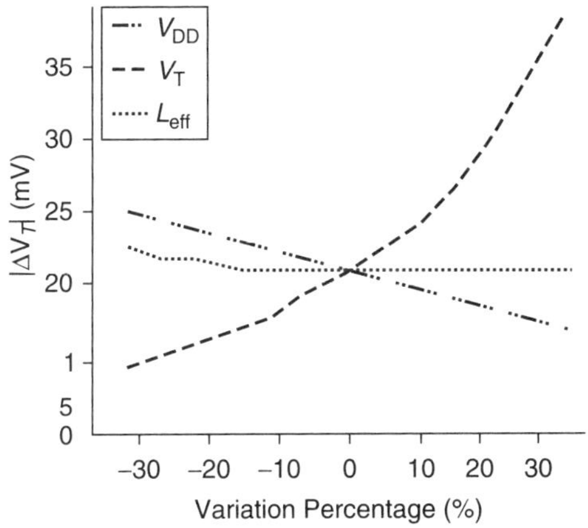

FIGURE 7.16 Effect of variation in V_{DD}, V_T, and effective channel length (L_{eff}) on ΔV_T.

required is based on the process, the type of stress being applied (i.e., static or dynamic), and the range of $\Delta V_{T\text{-}p}$ observed in the device. For a given amount of time under stress, a V_{DD} value can be tuned to minimize $\Delta V_{T\text{-}p}$ variation and performance degradation.

7.5 Electrostatic Discharge

Electrostatic discharge (ESD) is a well-known failure mechanism that occurs when a component is subject to a sudden excessive discharge of static electricity. Such discharges can damage semiconductor components in many ways.[25-27] A MOSFET device is prone to ESD failures because of its high input impedance. Electrostatic discharge failure can be attributed to static electricity generated by the *triboelectric effect*. In other words, when two objects come in contact, their surfaces ionize and inject a charge greater than the work function of the material. The excess charge removes the electrons from one material and attaches them to the other, thereby forming oppositely charged surfaces. When such charged surfaces come in contact with a MOSFET, the ESD can cause gate dielectric breakdown. Since contemporary MOSFETs feature oxides that are less than 40 Å thick, the breakdown voltage is correspondingly lower.

ESD-induced catastrophic failures occur at dielectrics, conductors, and junctions. In gate dielectrics, a high gate voltage can break down the gate oxide. At semiconductor junctions, a high source-drain voltage may cause "punch through" and heating at the source-drain junction, which can lead to silicide cracking and junction failures. When metal conductors are heated by high current levels, the result may be EM- and ESD-induced dielectric defects that cause permanently high current.

As illustrated in Figure 7.17, ESD can cause metal interconnects to heat up, melt, and form bridges or opens. These effects are due to joule heating in the presence of a sudden spike in current across the wire. In MOSFET devices, current flows through a narrow path or filament. When the metal contacts heat up, they melt and fall into the narrow current path, connecting source and drain regions permanently. This phenomenon is known as electrothermomigration. A transfer of mass in the presence of a weak electric field can also cause shorts between MOSFET terminals; in fact, filaments from the polysilicon gate can short all three terminals of the device (see Figure 7.18). Junction breakdowns are characterized by p-n junction rupture due to ESD events, which cause opens or shorts in the p-n junctions of a bipolar device. Electrostatic discharge leads to joule heating, which changes the characteristics of the underlying silicon—for example, its resistivity is reduced, which further increases its susceptibility to heating. The resulting vicious cycle leads to thermal runaway and hence to complete failure of the device.

The failures that could result from electrostatic discharge are prevented by designs that incorporate protection circuits for critical MOSFETs.[28,29] Examples include gated diodes as well as device protection against punch-through.

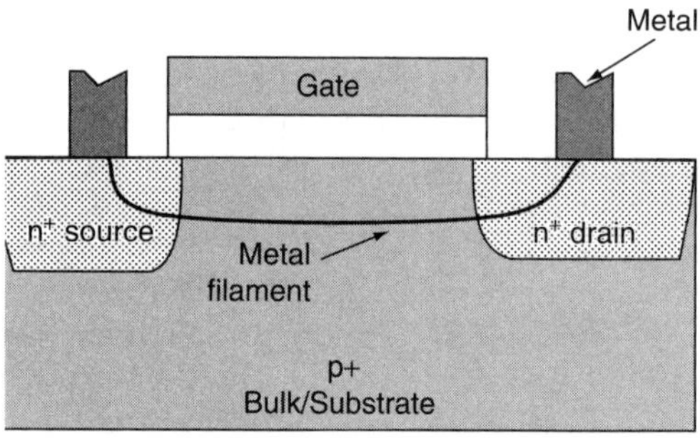

FIGURE 7.17 Short circuit due to a metal filament caused by electrothermomigration.

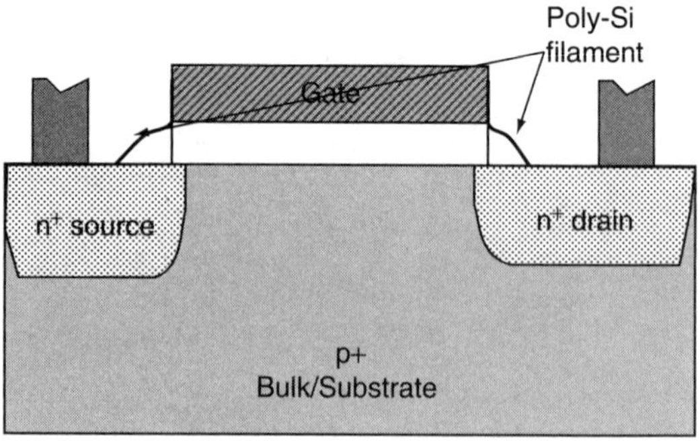

FIGURE 7.18 Short circuits due to polysilicon filaments caused by mass transfer.

7.6 Soft Errors

In order to achieve high density and low power consumption, operating voltages have been scaled for devices whose feature sizes are smaller than 50 nm. However, reduced supply voltage also reduces the noise margin and increases susceptibility to radiation hazards. Soft errors are reliability failure mechanisms that are caused by terrestrial or cosmic radiation. Unlike catastrophic failures and reliability failures due to gradual degradation, soft errors are transient. Soft errors are usually attributed to an error in data or in a circuit signal and do not cause any physical damage to the device. Erroneous bits of data, once detected, can be modified by rewriting. But if these error bits are not detected within a specified time, they can lead to reliability problems in many systems. Therefore, high-reliability systems typically employ mechanisms for detecting and correcting errors in order to avoid sudden system crashes. Memory components are especially vulnerable to soft errors. Techniques for mitigating soft errors include using error-correcting codes, upsizing capacitances, providing spare circuitry, and software techniques such as RMT (see Sec. 6.3.1.2).

7.6.1 Types of Soft Errors

A radiation-induced single event upset (SEU) may or may not affect system operation. If a system is affected, a soft error is said to have occurred. A soft error may propagate across multiple clock cycles without affecting system output. Soft errors that do not propagate to the output (are thus are not found by the user) are not considered to be detectable. Detectable soft errors can be classified as either silent data corruption (SDC) or a detected unrecoverable error (DUE). A DUE will usually cause the system to crash, but SDC errors are also of great concern to users.

The typical duration of an SEU is small (of the order of a few picoseconds), so it may not even propagate to the output of a combinational logic circuit to be detected as a fault. When an SEU becomes latched (into a flip-flop or a latch), however, it may persist or propagate through later cycles. For this reason, soft errors are mostly observed in sequential and memory circuits. It is important to note that, even if an SEU occurs in a circuit, there is only a small likelihood of it causing a system error because logic, timing, and electrical masks prevent its detection.[30] Electrical masks prevent the erroneous value from reaching a detectable logic level, and logic masks prevent faulty values from propagating to the output. Finally, latching time windows may prevent the faulty value from being recorded.

7.6.2 Soft Error Rate

The soft error rate (SER) is the rate at which the system encounters soft errors.[31] The SER can be measured by failure in time (FIT) or by mean time to failure (MTTF). The FIT metric specifies the number of soft errors encountered per billion hours of operation, whereas MTTF reports the SER in terms of the number of years before an IC fails due to soft error. As mentioned previously, SER is not strongly related to aging but is a good measure of the reliability of a circuit. The soft error rate is not related to chip yield, although it, too, can be addressed through design considerations at the circuit or system level.

7.6.3 SER Mitigation and Correction for Reliability

The SER can be mitigated in circuits by radiation hardening, which can be accomplished in many ways. One common method is to increase the capacitance of circuit nodes.[31,32] Another technique is to fracture transistors into parallel "fingers." Mitigation techniques affect circuit power and timing, so the only nodes that are targeted are those for which a high SER is probable.

Error-correcting codes are used to mitigate soft errors in memory circuits.[33] In this method, extra bits are added to detect and correct erroneous storage values. Soft errors can be detected and corrected by using such familiar fault-tolerant techniques as triple modular redundancy. With the TMR technique (explained more fully in Sec. 6.3.1.1), three copies of the same unit are fed as input and a majority voter is used to test for errors. Even if one unit has a soft error, the two other units provide the correct result. Encoding output values in memory circuits is also used to mitigate soft errors. However, all of these fault-tolerant mechanisms have the drawback of increased design area and/or reduced performance.

7.7 Reliability Screening and Testing

Reliability testing is the process of screening to find the "weak" chips in a lot before they are shipped to a customer. The failure of these weak chips leads to the initial high failure rates indicated by the infant mortality portion of the curve plotted in Figure 7.1. *Reliability screening* involves the use of acceleration mechanisms that trigger the failure mode of vulnerable chips. Screening is important because it increases the slope of that figure's bathtub curve (the early-life failures) and also helps tune the process. Temperature, voltage, and mechanical stress are used to accelerate device failures. Mechanical failures in ICs are accelerated when the circuits are subjected to vibrations with high g-force. The IC centrifugal test also falls into this category. Stress tests incorporating high voltage and high temperature in a burn-in chamber are used to accelerate failures due to electromigration and gate oxide shorts. These tests screen chips with latent defects and bonding problems.

Burn-in is the most popular reliability screening procedure, and it uses a combination of temperature and voltage to accelerate failures.

The IC chip is typically maintained at a higher than normal temperature, and test patterns are applied using heat-resistant probes. Burn-in tests differ from other manufacturing tests because patterns are applied but the response is not observed. This is because the components are not usually designed to function normally under stress conditions. The accelerated failures induced by the burn-in test target circuits with weak and thin oxide layers, thin metal lines, improper contacts or vias, and contamination. Two types of burn-in tests are used to detect different weaknesses in the IC: static tests, which are used to cause junction breakdowns; and dynamic tests, which are used to induce electromigration effects.

Apart from screening, *reliability testing* is another important aspect of product characterization. Unlike reliability screening, reliability testing stresses the targeted ICs until they fail. Accelerated reliability testing of circuits is destructive, so only a selected sample of ICs from each lot is chosen for testing to estimate the lifetime of shipped products. The accelerated reliability testing performed today include temperature, voltage, chemical, mechanical, radiation, and humidity tests. The circuits must go through a battery of tests that incorporate heat, high voltage (spikes), corrosive chemicals, shocks and vibrations, and bombardment by alpha particles or neutrons. A comprehensive failure analysis is performed for each stress mechanism to estimate the level of activation energy and stress that the ICs can withstand in the field.

7.8 Summary

In this chapter we reviewed reliability issues that affect contemporary ICs. The objective was to review important reliability mechanisms that lead to aging and permanent failure of devices. We examined failure mechanisms that are parametric (such as HCI and NBTI), recoverable (such as NBTI), and intermittent (such as soft errors). Device reliability is a product of design and manufacturing robustness. Design for reliability (DFR) provides protection against reliability failures when such failures can be modeled, which underscores the importance of reliability modeling. In this chapter, we reviewed various reliability models that are used to predict aging or to identify circuit vulnerabilities. It was observed that, because DFR techniques impose costs related to area, performance, and power, their use should be targeted selectively.

References

1. M. A. Alam and S. Mahapatra, "A Compreshensive Model of pMOS NBTI Degradation," *Microelectronics Reliability* **45**: 71–81, 2005.
2. V. Reddy et al., "Impact of Negative Bias Temperature Instability on Digital Circuit Reliability," in *Proceedings of IEEE International Reliability Physics Symposium*, IEEE, New York, 2002, pp. 248–254.

3. D. K. Schroder and J. A. Babcock, "Negative Bias Temperature Instability: Road to Cross in Deep Submicron Silicon Semiconductor Manufacturing," *Journal of Applied Physics* **94**(1): 1–17, 2003.
4. A. T. Krishnan, V. Reddy, S. Chakravarthi, J. Rodriguez, S. John, and S. Krishnan, "NBTI Impact on Transistor and Circuit: Models, Mechanisms and Scaling Effects," in *Proceedings of International Electron Devices Meeting*, IEEE, New York, 2003, pp. 14.5.1–14.5.4.
5. S. Chakravarthi, A. T. Krishnan, V. Reddy, C. F. Machala, and S. Krishnan, "A Comprehensive Framework for Predictive Modeling of Negative Bias Temperature Instability," in *Proceedings of IEEE International Reliability Physics Symposium*, IEEE, New York, 2004, pp. 273–282.
6. G. Chen et al., "Dynamic NBTI of pMOS Transistors and Its Impact on Device Lifetime," in *Proceedings of IEEE International Reliability Physics Symposium*, IEEE, New York, 2003, pp. 196–202.
7. M. A. Alam, "A Critical Examination of the Mechanics of Dynamic NBTI for pMOSFETs," in *Proceedings of International Electron Devices Meeting*, IEEE, New York, 2003, pp. 345–348.
8. R. Doering and Y. Nishi, *Handbook of Semiconductor Manufacturing Technology*, CRC Press, Boca Raton, FL, 2007.
9. J. Srinivasan, S. V. Adve, P. Bose, and J. A. Rivers, "The Impact of Technology Scaling on Lifetime Reliability," in *Proceedings of International Conference on Dependable Systems and Networks*, IEEE Press, New York, 2004, pp. 177–186.
10. J. Black, "Mass Transport of Aluminum by Momentum Exchange with Conducting Electrons," in *Proceedings of International Reliability Physics Symposium*, IEEE, New York, 1967, pp. 148–159.
11. A. G. Sabnis, *VLSI Reliability*, Academic Press, New York, 1990.
12. T. H. Ning, P. W. Cook, R. H. Dennard, C. M. Osburn, S. E. Schuster, and H. N. Yu, "1 μm MOSFET VLSI Technology: Part IV—Hot-Electron Design Constraints," *IEEE Transactions on Electron Devices* **26**: 346–353, 1979.
13. P. E. Cottrell, R. R. Troutman, and T. H. Ning, "Hot-Electron Emission in n-Channel IGFET's," *IEEE Electron Devices Letters* **26**: 520–532, 1979.
14. A. Schwerin, W. Hansch, and W. Weber, "The Relationship between Oxide Charge and Device Degradation: A Comparative Study of n- and p-Channel MOSFET," *IEEE Transactions on Electron Devices* **34**: 2493–2499, 1987.
15. D. L. Crook, "Method of Determining Reliability Screens for Time Dependent Dielectric Breakdown," in *Proceedings of International Reliability Physics Symposium*, IEEE, New York, 1979, pp. 1–7.
16. S. I. Raider, "Time-Dependent Breakdown of Silicon Dioxide Films," *Applied Physics Letters* **23**: 34–36, 1973.
17. S. Ogura, P. J. Tsang, W. W. Walker, D. L. Critchlow, and J. F. Shepard, "Elimination of Hot-Electron Gate Current by Lightly Doped Drain-Source Structure," in *Technical Digest of International Electron Devices Meeting*, IEEE, New York, 1981, pp. 651–654.
18. H. C. Kirsch, D. G. Clemons, S. Davar, J. E. Harmon, C. H. Holder, Jr., W. F. Hunsicker, F. J. Procyk, et al., "1 Mb CMOS DRAM," in *Technical Digest of International Solid State Circuits Conference*, IEEE, New York, 1985, pp. 256–257.
19. T. Sakurai, M. Kakumu, and T. Iizuka, "Hot-Carrier Suppressed VLSI with Submicron Geometry," in *Proceedings of IEEE Internation Solid–State Circuits Conference*, IEEE, New York, 1985, pp. 272–273.
20. H. Kufluoglu and M. A. Alam, "A Geometrical Unification of the Theories of NBTI and HCI Time-Exponents and Its Implications for Ultra-Scaled Planar and Surround-Gate MOSFETs," in *Proceedings of IEEE Electrons Devices Meeting*, IEEE, New York, 2004, pp. 113–116.
21. K. O. Jeppson and C. M. Svenssen, "Negative Bias Stress of MOS Devices at High Electric Field and Degradation of MNOS Devices," *Journal of Applied Physics* **48**: 2004–2014, 1997.
22. S. C. Sun and J. D. Plummer, "Electron Mobility in Inversion and Accumulation Layers on Thermally Oxidized Silicon Surfaces," *IEEE Journal of Solid-State Circuits* 15(4): 1497–1508, 1980.

23. J. E. Chung, P.-K. Ko, and C. Hu, "A Model for Hot-Electron-Induced MOSFET Linear Current Degradation Based on Mobility Reduction Due to Interface-State Generation," *IEEE Transactions on Electron Devices* **38**(6): 1362–1370, 1991.
24. R. Vattikonda, W. Wang, and Y. Cao, "Modeling and Minimization of pMOS NBTI Effect for Robust Nanometer Design," in *Proceedings of Design Automation Conference*, ACM/IEEE, New York, 2006, pp. 1047–1052.
25. D. P. Renaud and H. W. Hill, "ESD in Semiconductor Wafer Processing—An Example," in *Proceedings of EOS/ESD Symposium*, ESD Association, Rome, NY, 1985, vol. EOS-7, pp. 6–9.
26. W. B. Smith, D. H. Pontius, and P. P. Budenstein, "Second Breakdown and Damage in Junction Devices," *IEEE Transactions on Electron Devices* **20**: 731–744, 1973.
27. L. F. DeChiaro, "Electro-Thermomigration in nMOS LSI Devices," in *Proceedings of International Reliability Physics Symposium*, ESD Association, New York, 1981, pp. 223–229.
28. R. N. Rountree and C. L. Hutchins, "nMOS Protection Circuitry," *IEEE Transactions on Electron Devices* **32**(5): 910–917, 1985.
29. C. Duvvury, R. A. McPhee, D. A. Baglee, and R. N. Rountree, "ESD Protection Reliability in 1 μm CMOS Technologies," in *Proceedings of International Reliability Physics Symposium*, IEEE, New York, 1986, pp. 199–205.
30. E. Normand, "Single Event Upset at Ground Level," *IEEE Transactions in Nuclear Science* **43**(6): 2742–2750, 1996.
31. D. G. Mavis and P. H. Eaton, "Soft Error Rate Mitigation Techniques for Modern Microcircuits," in *Proceedings of International Reliability Physics Symposium*, IEEE, New York, 2002, pp. 216–225.
32. M. P. Baze, S. P. Buchner, and D. McMorrow, "A Digital CMOS Design Technique for SEU Hardening," *IEEE Transactions on Nuclear Science* **47**(6): 2603–2608, 2000.
33. S. Mukherjee, "Architecture Design for Soft Errors," Morgan Kaufmann, San Mateo, CA, 2008.

CHAPTER 8

Design for Manufacturability: Tools and Methodologies

8.1 Introduction

The previous chapters in this text have established that design for manufacturability (DFM) is not just a back-end concern. Optical proximity correction, double patterning, phase shift masking, and other resolution enhancement techniques (RETs) for defect avoidance cannot be decoupled from the physical design process, because some designs cannot simply be "cleaned up" and therefore require redesign. As designers are brought onboard into DFM iterations, design productivity becomes a concern. This productivity must be seen from two aspects: information and tools. The information package must contain process parameter variations (i.e., printed shape representations at various process corners) or the distribution of these variations. Typically, not all this information can be deciphered by the designer because it requires additional knowledge of the manufacturing process and parameters that relate to analysis. Electronic design automation companies and in-house CAD tools seek to provide a bridge between manufacturing specifications, process variabilities, and corresponding design parameter variation by encapsulating this knowledge in technology libraries. Computer-aided design tools are an integral part of the semiconductor design process. During each stage of design, CAD tools perform design transformations guided by analysis and/or empirical and encapsulated knowledge geared to improving the design realization process. Traditionally, designs were guided by the triad of area, performance, and power metrics. However, because of functional and parametric yield concerns as well as the complexity of DFM compliance, the goals of manufacturability, variability, and reliability have become increasingly important in the move to smaller fabrication geometries.

The design realization process consists of adhering to a set of design principles to help improve design productivity, which is defined by ease of the design process and the number of design iterations required. The guidelines are a function of design targets, technology, and capabilities of the tools used. The entire process is commonly referred to as *design methodology.* Adhering to a set of clocking rules, to rules for power supply distribution on chips, or to the prescribed physical dimensions of library cells will simplify partitioning of the design process; this allows the various partitions to be designed concurrently and independently, which increases designer productivity and reduces time to market. Similarly, library cell planning and a set of physical design constraints often obviate the need for multiple iterations to arrive at a DFM-compliant design. These considerations underscore the importance of design discipline to reduce the burden on tools in terms of required types and volume of data. Conversely, suitably sophisticated and capable tools are required because discipline alone cannot accomplish all design and manufacturability goals. Thus, design methodology cannot be addressed separately from the tools involved.

The design realization process involves a series of steps. These steps did not originally include DFM, which was viewed as a one-way, back-end process to improve manufacturability. With the advent of new complexities associated with RET, lithographic variability, and defect avoidance techniques, design methodologies have been forced to accommodate not only DFM but also DFY (design for yield) and DFR (design for reliability). These three design concepts are often lumped together, and referred to as "DFx," because they are often handled in the same manner: analysis, compliance checking, and iteration to optimize the design. The drive to incorporate DFx has increased the need to understand basic manufacturing process, manufacturability models, and process variability. The resulting expanded role for designers is geared to achieving a better design that conforms with goals of power, performance, and manufacturability.

In the future, two principal factors will drive the semiconductor industry toward cost-effective manufacturing: (1) CAD tools and methodologies aimed at enhancing manufacturability, yield, and reliability; and (2) innovative manufacturing ideas that target specific products.

8.2 DFx in IC Design Flow

CAD tools are a major component of DFM and DFR strategy. A typical design flow in semiconductor manufacturing involves the following elements: device and process modeling, standard cell design, library characterization, design analysis, synthesis, placement and routing, layout verification, and mask engineering. The balance of this section, which briefly describes many of these elements, also illustrate how DFM- and DFR-based methodologies have been applied to improve design manufacturability. Semiconductor manufacturing steps incorporate CAD tools that perform analysis, modeling, design modification, or optimization. These CAD tools typically rely on models and parameters. *Models* embody general principles, whereas *parameters* are specific to a technology and so vary from generation to generation. Decoupling parameters from models allows the use of tools over multiple technology generations, thus providing a semblance of repeatability. Parameters are often the most-overlooked part of the design process. Uncalibrated parameters—together with unrealistic assumptions about such environmental conditions as voltage, temperature, and package parameters—lead to design failures.

8.2.1 Standard Cell Design

DFx-compliant standard cell design is vital to improving the manufacturing process and also to achieving a steeper yield curve. For example, designing standard cells in the context of matching FO4 metrics does not, in itself, ensure a timing-compliant design, but it does provide a basis for one. Similarly, addressing DFx issues at the cell level is necessary to build robust designs. At this level, manufacturability and reliability issues arise most frequently in poly-gate masks. Therefore, DFM compliance checks and related modifications must be performed early and often during the design of these cells. Layout issues can arise with respect to polysilicon gate width and length over the diffusion region, minimum pitch spacing between polysilicon lines, contact placement over active areas, within-cell via placement, diffusion rounding, gate length and width biasing, and stressed channel regions. These issues affect the performance (and performance variability) of the standard cells. Reliability issues involving negative bias temperature instability (NBTI), hot carrier injection, and soft errors also require modifications to the standard cell if early-life and intermittent failures are to be prevented. Thus, CAD-based methodology for library cells must include early attention paid to issues that traditionally dominate the back-end design process.

Figure 8.1 reproduces some layout modifications, discussed previously in the text, that are performed by DFx flows. Polysilicon widths outside the diffusion region are increased, and polysilicon lines are spaced to remove forbidden pitches. Dummy fills and subresolution assist features (SRAFs) are employed to reduce the impact of gate length variation in silicon. The edges of diffusion regions are located far away from poly lines in order to preclude variation in gate width and length due to diffusion rounding. Doping

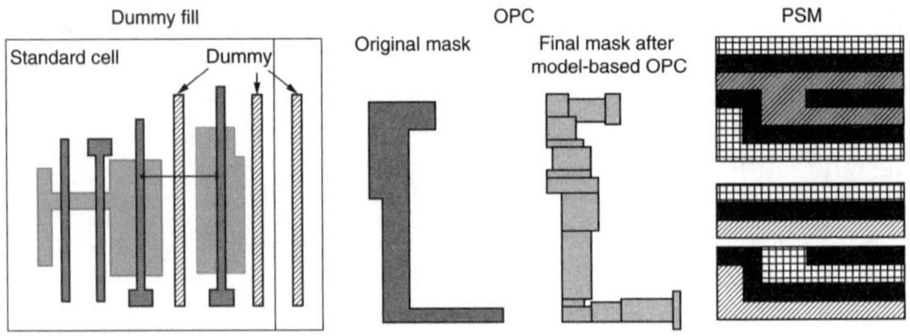

FIGURE 8.1 Layout modifications performed by DFx flows.

with germanium or silicon germanium and using nitride liners both serve to strain the channel; strained silicon increases carrier mobility, which enhances the performance of standard cells. In transistors, carrier mobility is also affected by contact distance to poly and by the spacing between N-active and P-active regions. Layout rules have been proposed to achieve higher carrier mobility without excessive strain, which can lead to crystal dislocation.

Issues related to NBTI were discussed in Sec. 7.4. Negative bias temperature instability places both static and dynamic stress on pMOS transistors, leading to V_T increases that cause gradual degradation of the device. Radiation impact on devices in the integrated circuit can lead to intermittent failures such as soft errors. Such errors are related to diffusion area and to the ratio of channel area to diffusion area. "Hardening" circuits to reduce the soft error rate (see Sec. 6.5) may involve reducing the diffusion area as well as sizing devices optimally to achieve a balance between susceptibility and performance. Resolution enhancement techniques such as optical proximity correction (OPC) and phase shift masking (PSM) are also applied during the final stages of library creation to enhance printability of the gate.

All these methods have become essential for design work under the current state of process parameter variations. Each of the tools involved in the DFx-compliant design process relies heavily on information from the foundry to produce manufacturable designs that meet area and performance targets.

8.2.2 Library Characterization

Library characterization methodology plays a crucial role in design convergence. There are multiple perspectives associated with a given cell; the most common of these are logic, schematic, timing, power, and layout. In the previous subsection we discussed the importance of the physical layout for DFM. The other perspectives also play a role that is related to parametric yield. The standard cell library consists of gates that are tuned for multiple V_T values, multiple drive strengths, and different process corners. However, variations induced by the manufacturing process also have an impact on library cells. Library characterization for timing and static power must therefore take these variations into account.

A cell may be characterized in terms of its performance in slow, nominal, and fast process corners. Pessimistic delay assumptions (slow corners) during design optimization lead to oversizing of gates and thus to increased power consumption. Conversely, optimistic assumptions about cell performance (fast corners) lead to reduced manufacturing yield. For this reason, CMOS circuits are usually sized based on nominal process corners. If variations are large, then slow and fast corners become important for design. For example, violations in hold time or in minimum delay are often analyzed based on fast process corners. If a cell's performance is characterized as being very fast, this may lead to "delay padding" and thus impede design convergence on the target cycle time. Therefore, during cell library characterization, the slow and fast process corners are typically chosen to be ±1.5 standard deviations (rather than $\pm 3\sigma$) in order to strike a working balance between design convergence, area, power, and parametric yield. We remark that, even though most designers take little notice of cell library characterization, the process has a large impact on parametric yield.

Important process parameters are variation in gate length and gate width, random dopant fluctuation (RDF), line edge roughness (LER), gate length biasing, and channel strain. A number of DFM methodologies seek to analyze process and lithographic parameter variations, and their effect on device parameters, in addition to proposing models that fit actual device behavior.[1-13] Lithography-induced across-chip linewidth variation cause changes in gate length and width of the device. Variation in gate length on silicon results in the formation of a nonrectangular gate. Because traditional SPICE models assume that the gate is rectangular, large discrepancies can arise between presilicon simulations of circuit timing or leakage and the actual postsilicon parameters. For this reason, the SPICE models used in DFM-aware methodologies accommodate nonrectangular transistors, which are represented by rectangular transistors of varying lengths and widths for each region of operation. Research results indicate that this technique can be effected by matching the drain current of the transistor during different regions of operation.[3,4,7] Both RDF and LER can be mapped to variation in transistor threshold voltage V_T. Hence, SPICE modeling now accommodates drain current variation due to RDF and LER.

Because post OPC gate length biasing modifies noncritical gates within the standard cell, library characterization in the presence of such biasing must be incorporated when a design is being estimated for timing and leakage. What-if analyses are the most popular approach for such *virtual* models.[14] Process-aware library characterization also enables designers to perform effective logic simulation for the analysis and testing of signal integrity.

8.2.3 Placement, Routing, and Dummy Fills

DFM-aware CAD methodologies have been incorporated into physical design tools for placement and routing. The main purpose of these methods is to produce RET-compliant placement and routing. There are guidelines regarding the *placement* of standard cells that take into account the presence or absence of particular neighboring cells and their orientation. In addition, a new placement approach that incorporates variation in circuit timing based on process parameters has been proposed by Kahng and colleagues.[15] The aim is to model timing variation due to lens aberration and then take this information into account during cell placement. (See Sec. 4.4.5 for more details on this work.)

RET-compliant *routing* algorithms incorporate knowledge of lithography-related issues to stipulate the routing of wires that connect various design blocks. One such approach is to perform initial routing followed by a fast lithography simulation to estimate the edge placement error (EPE) of each metal line in the mask. Hotspots are then marked based on the estimated EPE, after which a series of wire spreading and wire sizing steps are taken. If the hotspots still remain after these steps, the detailed route is ripped and rerouted. This process is repeated until a hotspot-free layout is obtained. Wire spreading is possible only in regions that have sufficient extra space available. Wire spreading, as shown in Figure 8.2(a), reduces critical area and improves printability. Wire widening, as shown in Figure 8.2(b), reduces the critical area for open defects; it also reduces the probability of linewidth reduction due to proximity effects.

Dummy fills are postlayout routing techniques whose purpose is to improve the planarity of oxides and other materials after chemical-mechanical polishing (CMP), whose effects depend on the pattern density of the mask. Dummy fills are added between mask features to minimize postlithography variation in metal and interlayer dielectric thickness. The CMP-induced dishing (erosion) of wide

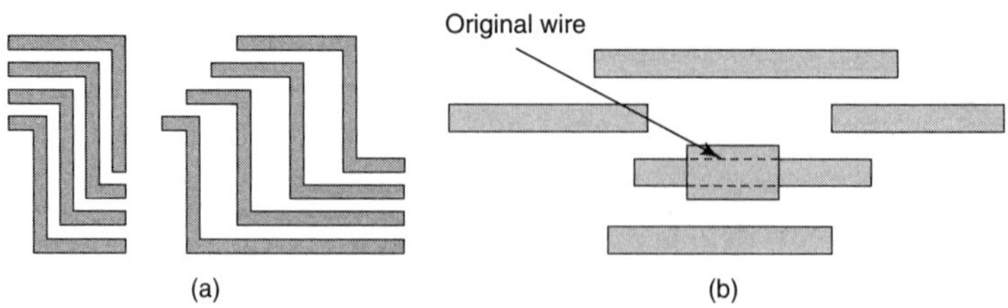

FIGURE 8.2 Lithography-informed routing: (a) wire spreading; (b) wire widening.

metal lines is reduced by *slotting*. This procedure removes certain regions (e.g., power rails and fill oxide) of a metal line to improve planarity (see Figure 5.17).

8.2.4 Verification, Mask Synthesis, and Inspection

Physical verification has become increasingly complex because of the increased pattern density in today's masks. *Verification* is the process of checking the layout for compliance with generic design rules check (DRC), restricted DRC, and lithography rules. For decades, DRC has constituted the final stage before design handoff. However, with increases in layout pattern density, field interactions extend beyond the adjacent polygon. The immense number of interactions that must now be considered has increased the DRC rule count exponentially. Lithography rules check (LRC) is a model-based approach that aims to resolve printability issues in the layout. In particular, LRC makes changes in the design layout so that OPC algorithms can arrive at a suitable solution for hotspots. The LRC algorithms are based on pattern matching of polygons present in the layout to a precompiled, lithography-simulated library of shapes. A library of patterns is created, and lithography simulation is performed to analyze their printability. Pattern matching techniques are used for identifying and fixing hotspots in order to maximize the effectiveness of the final OPC.

Mask synthesis and inspection steps are performed in the design house before mask manufacturing. Effective *mask synthesis* techniques involve DFM strategies that seek to produce an RET-compliant mask. Both OPC and PSM are performed on the final mask to enhance pattern printability. The LRC relies on layout verification to generate a layout that is amenable to optical proximity correction and phase shift masking. Unless a layout is hotspot-free and can be assigned phases, it is ripped and rerouted. (Refer to Sec. 4.3 for more details on PSM and OPC.) *Mask inspection* is performed to analyze the number of shots required for a given layout, where the "shot count" is the number of fractured polygons in the layout. The cost of mask writing is a direct function of the shot count. The inspection step also includes critical area analysis in order to predict the yield of the mask based on a particular defect size. Process yield can be improved by making repairs, but catastrophic faults can only be removed through re-spin.

8.2.5 Process and Device Simulation

The interaction of process and device simulation tools with DFx flows is critical for improving the overall manufacturing process. Device and process simulation tools are categorized as "technology CAD" (TCAD) tools. The purpose of TCAD tools is to produce DFx models that in turn will be incorporated into design tools for both the front-end and back-end design stages.

Technology CAD tools perform process and device simulation in order to assess the impact of process variation on design parameters. Process simulations mimic actual process scenarios to model the stages of oxidation, diffusion, and etching as well as other material deposition stages. The effect of varying process parameters on the final device or interconnect is modeled to facilitate a transistor-level analysis that can improve the overall design. Device simulation characterizes the behavior of a device during different modes of operation. Newer devices, such as FinFETs and tri-gates, are being modeled as effective replacements for the traditional MOSFET in technology nodes below 45 nm. These latest devices require a three-dimensional TCAD formulation incorporating both process and device information in order to analyze all possible variants of the device effectively.

8.3 Electrical DFM

Most of the DFM techniques described in the previous chapters provide strategies for mitigating catastrophic yield. These DFM techniques help improve functional yield but often ignore parametric yield considerations. Parametric yield has been left out of the designer's hand because there are no tools for relating process parameters to their design counterparts. The yield limiters in today's designs are parametric failures: design parameters that vary beyond the prescribed specifications. Electrical-DFM (E-DFM) techniques have been proposed to target such failures and improve design parametric yield. Electrical DFM has become an important part of DFM strategy within design and manufacturing companies. Much as global DFM methodologies, which have infiltrated various steps of the design process, E-DFM may find its way to improving electrical characteristics of the design.

The goal of E-DFM is to improve the parametric yield by fostering communication between manufacturing and design. Electrical DFM analysis techniques addresses the full spectrum of manufacturing stages to obtain comprehensive information about process variability that may be used in electrical optimization. The variations are incorporated into the design tools used to asses overall impact on circuit performance and power. The E-DFM approach improves a design's parametric yield by focusing on leakage power, dynamic power consumption, design timing, electrical variability, and so forth. Examples of E-DFM techniques include leakage-aware standard cell library design in the presence of stress, timing-aware placement to accommodate lithographic variation, electrical-OPC for polysilicon masks, and gate length biasing.[7,15-17] Inserting fill and adjusting vias to reduce parametric variation induced by resist opens are also part of the E-DFM framework. However, dummy fills can induce variation in coupling capacitance between metal lines. Hence, fill insertion techniques have been proposed that optimize both timing and post-CMP material planarity.[18]

8.4 Statistical Design and Return on Investment

There has always been a fuzzy line that separates design for manufacturability and statistical design. The DFM techniques, which usually target catastrophic or parametric yield, follow a series of steps to predict, model, analyze, and compensate for variations. The target of DFM methods are variations that exhibit predictable behavior (or that can be simply generalized as systematic variation). A variation is *predictable* if it can be modeled by a mathematical function. A simple example is how variation in mask feature linewidth on wafer depends on—that is, varies as a function of—the spacing between adjacent metal lines. Another example is the dependence of variation in transistor gate length on standard cell orientation. These relations are predictable and are effectively modeled as simple functions. With some variations, however, quantification is difficult or the affecting parameters cannot be verified or the occurrence is simply random. No single mathematical function can be used to model such behavior, because neither the constituent parameters nor their variations is known. This is where statistical design comes into the picture: it describes phenomena in terms of probability distributions and other statistical attributes.

The objective of statistical design is to address the problem of random parameter variation by assigning a distribution model that can be used to predict the behavior of the device or interconnect. One example of statistical design is the modeling of random dopant fluctuation and its impact on device threshold voltage. The curves plotted in Figure 8.3 reveal the difference in design timing when deterministic versus statistical characterizations are employed. Statistical timing approaches estimate a maximum delay that is less than the worst-case delay ($+3\sigma$) predicted by the deterministic approach. Similarly, Figure 8.4 shows that a design's mean leakage power is predicted to be lower when statistical rather than deterministic estimation approaches are employed. This means that circuits can be designed less conservatively for high-performance applications.

Statistical design approaches come with certain caveats. If the statistical design is based on parameter distributions characteristic of the process's early life, then the results may not be optimal. This is because the variance of process parameters changes as the manufacturing process matures. The implication is that models and CAD tools must be updated on a continuous basis, which is clearly not feasible. Therefore, statistical-based design approaches rely on *performance guard bands* to meet final product specification. Yet given the increasing number of variabilities, even guard banding may prove

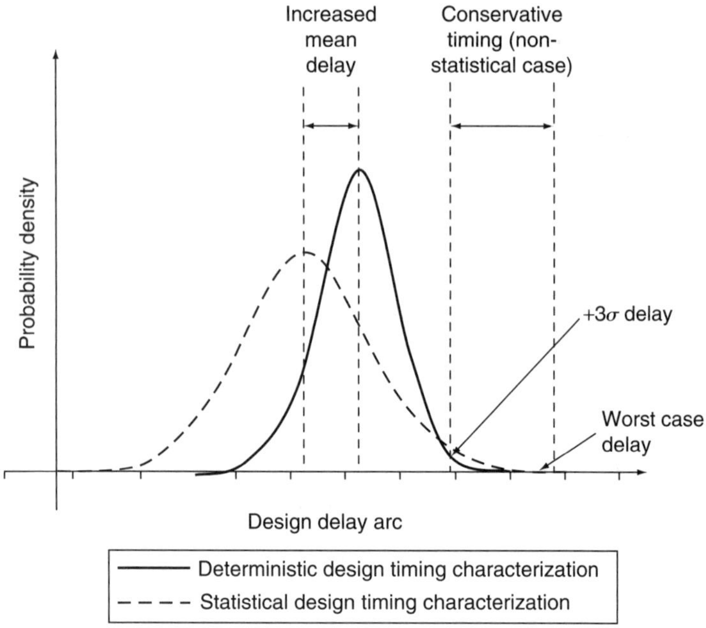

FIGURE 8.3 Difference in the delay estimated by deterministic and statistical algorithms.

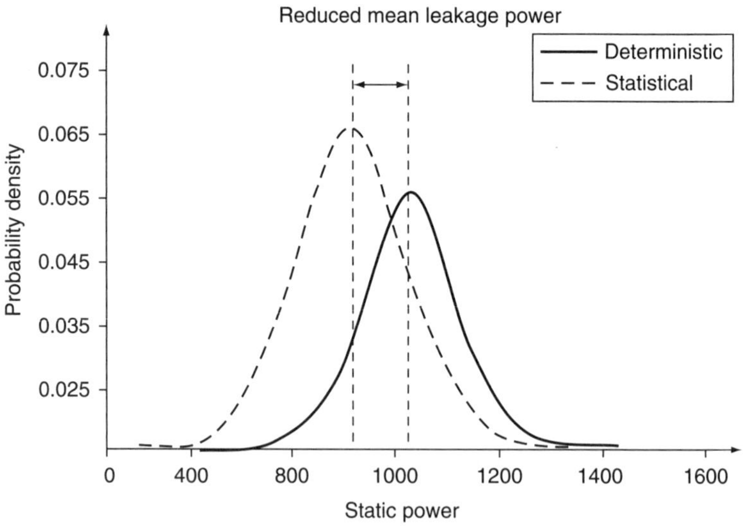

FIGURE 8.4 Difference in the leakage power estimated by deterministic methods and statistical methods using Monte Carlo simulation.

too costly in terms of power. What's needed is analysis of the proper trade-off between defect prevention and parametric yield.[19]

Variation in general can be classified based on different attributes. The distinction between random and systematic variation is well known. Variation can also be classified in terms of location and area of influence as lot-to-lot, wafer-to-wafer, die-to-die, or within-die variation. Each of these types of variations has random and systematic components. The traditional approach is to assume the worst-case variability of a parameter when it cannot be effectively quantified and modeled. However, with the increasing number of possible causes for such random variations and with a broadened scope of parameter interaction, this method is too pessimistic. Statistical distributions provide a region of probable parameter variation that can be used to analyze the effects on design timing and power. A well-known approach that uses statistical design is the simultaneous optimization of timing and leakage power. The recursive approach of optimizing circuit delay and leakage with the aid of threshold voltage distribution is highly effective.[20-23] This method has proven to be reliable and is also being used in circuit synthesis and technology mapping.

The discussion so far has concerned models and analyses. Before venturing into statistical design, a key question must be asked about the value of this approach. In particular, the designer must be aware of the return on investment (ROI) when statistical modeling of specific parameters is performed. It has been demonstrated that the ROI is positive for only a limited number of parameter variations (e.g., temperature, threshold voltage, effective gate length) and their effect on leakage and timing. In contrast, the economic return from the statistical modeling of peak power optimization is quite limited.[24] Therefore, it is important to investigate the ROI before using statistical models, which are both computationally intensive and time-consuming, to derive results comparable to those obtained using straightforward mathematical functions.

Today, there is no settled answer to the ROI question. In general, though, selecting the appropriate modeling approach depends vitally on the application as well as parameter variability and its magnitude and area of influence. The best advice is to keep an open mind to any type of modeling variations that may arise in future technology generations and device structures.

8.5 DFM for Optimization Tools

Since the dawn of deep submicron technology, optimizing power and performance has been the most important goal of IC design. The generation of nano-CMOS VLSI design adds more options to the existing optimization space by manipulating factors related to strain, gate biasing (using SRAFs) and dummy filling. The latter items are mostly associated with DFM, but they may also be used as design parameters when optimizing for power, leakage, and performance.

Circuit designers have a number of choices for optimizing a circuit. Selecting threshhold voltage, transistor sizing, gate biasing, and strain have already been mentioned as factors that affect device optimization. Similarly, layout optimization relies on interconnect modifications, buffer insertion, and logic changes such as negate-invert. Final design yield and performance is a function of all these choices. Figure 8.5 portrays a grand vision of optimization flow. It includes a circuit netlist and/or the design layout as input to the engine, which incorporates information on circuit parameter variability and the required parameter and yield goals. The optimization engine modifies various attributes of the design to generate a final optimized design that satisfies the required goals. There are many engines that addresses some aspect of optimization, and they are still evolving.

Optimization engines have been in existence for the past decade or so. The DFM technique takes V_T variation and the strain factor into account when optimizing timing and leaking power. In contrast, optimizing yield based on layout critical areas has evolved into more lithography-aware techniques for subwavelength patterning. All optimization techniques aim to minimize or maximize a particular function whose limits are defined by the constituent parameter specification. This function is typically referred to as the tool's *cost function*. After each iteration, a new cost is calculated and compared to the existing cost; further iteration is typically not allowed if the change in cost does not support the final goal. Optimization is complete when the final cost has been minimized and all the target parameters are within specifications.

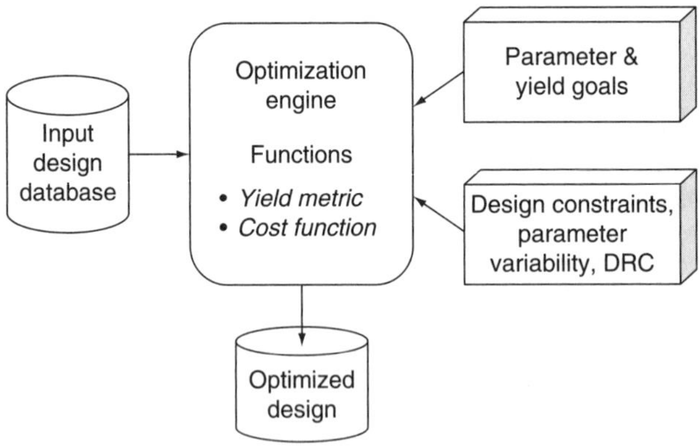

Figure 8.5 Idealized setup for optimization engine.

Standard cell layouts with no change in dimensions now have additional control parameters that can produce higher performance, so there are more varieties of standard library cells available for the designer to use. A leakage and timing optimization flow based on strain and threshold voltage has been proposed by Joshi and colleagues.[16] This design flow uses a combination of stressed and modified V_T-based gates to achieve overall optimization for leakage and performance. The flow is similar to the "dual" V_T-based optimization flow suggested previously by Sirichotiyakul and colleagues.[22] In this dual approach, high and low variants of the parameters are augmented by additional combinations: high and low V_T combined with high and low stress. The benefit of these additional options on leakage (I_{OFF}) and timing (propagation delay) is shown in Figure 8.6.[16]

Yield estimation based on critical area (CA) has also been used in layout optimization. With the advent of DFM, a new kind of yield optimization is required—one that involves lithography simulation. Yield optimization based solely on CA analysis is inadequate because interactions between the drawn lines extend beyond their nearest neighbors. The effect of lithographic parameter variation on linewidth, which is useful in predicting yield, was described in Sec. 3.2. If a line is in danger of disappearing (open) or of shorting with a neighbor under a possible combination of input conditions (e.g., focus, exposure, resist thickness), then that probability counts against the yield. In this case, the layout optimization goal for each mask layer is a lithographic yield metric.

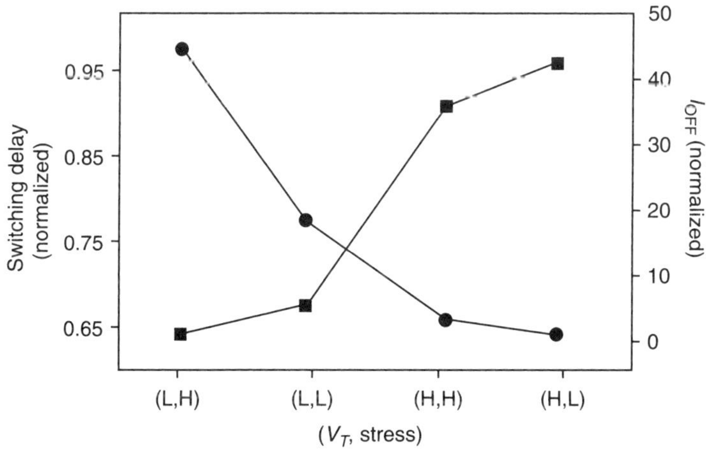

FIGURE 8.6 Leakage and switching delays for various combinations of V_T- and stress-based optimization for a three-input NOR gate.

The overall methodology consists of lithography simulation under multiple process corners to compute the probability of a line being open or short. If this probability is above a certain threshold, the resulting potential hotspots are marked. These hotspots are ranked in terms of severity, and candidates for adjustment are chosen one at a time. The selected candidate may or may not be modified, a decision that is based on its neighborhood. Unlike conventional hotspot-based layout optimization techniques, some lines may be permitted to stay with this methodology. The procedure is akin to the "waiver" process that sometimes accompanies rule-based DRC, in which a few exceptions are sometimes tolerated. This enables a more comprehensive of the overall design objective, namely, meeting area, performance, power, and manufacturability targets.

In sum, DFM-based optimizations have gone well beyond their original intent of improving manufacturability. This is because the same set of manipulations are useful in improving parametric yield and other design metrics, such as performance and leakage power.

8.6 DFM-Aware Reliability Analysis

Computer-aided design methodologies based on design for reliability target electromigration (EM) to extract design geometries and perform drive current analysis; the current density information so obtained is used to estimate a devices's mean time to failure (MTTF) due to EM. For this analysis, drawn layouts are typically used before any OPC is performed. The assumption is that OPC will help preserve the shape in manufactured silicon.

Typical modifications that improve MTTF include changing the interconnect width and reducing the driver size to satisfy current density requirements. As interconnect densities increase with each technology generation, the lithographic processes required to print all features with an acceptable number of irregularities have become highly complex. Resolution enhancement techniques such as OPC and PSM implement changes to the drawn layout. Post-OPC layouts contain modifications to the drawn mask with extra features such as SRAFs, jogs, hammerheads, and serifs that change parameter values for resistance, capacitance, and current density. Other changes to the design include dummy fills for lower metal layers and slotting for higher metal layers such as power rails. This means that reliability verification checks based on drawn layouts are of limited value. More accurate are DFM-aware reliability techniques that perform analysis on predicted linewidth and via width. This is an emerging area of research. Various other reliability tools that operate on drawn geometries must likewise adapt by considering postlithography changes to the device and interconnects. Given the increasing variability across all stages of the manufacturing process, DFM-aware reliability techniques are needed for early identification of probable failure sites.

8.7 DFx for Future Technology Nodes

Scaling of planar bulk CMOS has become increasingly difficult. Apart from the traditional problems of short channel effects (e.g., band-to-band-tunneling) and leakage (oxide and gate-induced drain leakage), manufacturability problems with ultrahigh retrograde channel doping and V_T control are expected to be significant. The need to manage parasitics—such as series source-drain resistance and fringing capacitance—may drive manufacturing toward ultrathin-body, fully depleted, silicon-on-insulator, and/or multiple-gate MOSFET structures.[25] The most challenging issues, in addition to managing the parasitics, are controlling the thickness and variability of ultrathin MOSFETs. Primary design concerns will be control of variability and leakage for better power and performance.

Parameter variation and power issues become paramount in the context of an increased number of devices, higher-density layout patterns, and reduced supply voltage. Any reduction in supply voltage will necessitate reduced threshold voltage in order to maintain the required noise margin for transistors. In turn, reduced threshold voltage leads to increased standby current and/or high leakage power consumption. Variability in temperature and supply voltage also affects leakage, and with the rise in dynamic power management techniques, variations will extend beyond manufacturing parameters to conditions related to workload and environment. The vicious cycle of parameter variation and leakage must be controlled through advanced optimization flows and feedback mechanisms. Consequently, new methods are required that can provide a quick turnaround in variation analysis, modeling, and design modification. Within this requirement lies the need for better SPICE models, whose effectiveness largely determines a design's parametric yield. Frequent updates—based on correlations between models and postsilicon parameter variations—must be incorporated into models for effective control of variability. An OPC-based algorithm for predicting such postlithographic correlation has recently been proposed by Cho and colleagues.[26]

The stress induced by strained silicon (SiGe and nitride liners) and shallow trench isolation (STI) improves transistor mobility by more than 30 percent in today's designs, with minimal impact on leakage. Thus, transistor and field oxide stress factors will play an important role as the size of the standard cell shrinks. Strain due to process-induced factors decreases with scaling, yet strain is critical for maintaining and improving carrier mobility. Hence a detailed understanding is needed of how strain may be induced through physical layout. Placement of STI oxides may then enter the picture, since currently allowed rule-based layout changes to standard cells to control leakage and induce mobility through strained silicon will not be sufficient. CAD tools that can model movement of stress across different standard cell boundaries will be instrumental in revising placement techniques that are driven by timing considerations. In the face of increasing variability in device features and environmental factors, reliability-aware CAD tools must be updated in order to extend the useful lifetime of manufactured products.

At the sub-22-nm technology node, traditional MOSFETs will most likely be replaced by new device structures. These new structures will bring new set of design and manufacturability issues. For example, FinFETs have quantized gate sizing because the gate length depends on the height of the fin. Hence synthesis techniques must be modified to derive a better gate-level description from the register transfer language. Standard cell designs need to incorporate changes based on printability, area, and intracell routing issues. Line edge roughness will remain a problem in the manufacture of FinFETs, and the V_T variation so induced may also have systematic components that depend on the layout. The library characterization of planar double gates, FinFETs, and tri-gates may involve more than parameter changes to the model: they may require different device models to model variations in different ranges. Because these new gate shapes are completely different, strategies for contact placement and metal interconnect routing may need to be adjusted so that performance remains within given constraints. An overall methodology—incorporating CAD tools, DFM, and industry-proven frameworks—must be devised for the devices of future technology generations.

Many of the DFM methodologies have not yet been shown to be effective for large-scale designs, and their applicability to future technology generations is far from assured. Statistical design has been touted as a successor to systematic DFM-based techniques. However, since process parameters (and their standard deviation) changes as the manufacturing process matures, the analysis and optimization solutions that are based on initial parameter distributions may not result in ideal design solutions. Statistical design techniques must be carefully analyzed for their value, in terms of return on investment, before being implemented in the design cycle. In fact, a designer should approach every design methodology with an eye on its ROI value. Design for value is the overarching issue of concern with DFM, DFR, and statistical design methods.

8.8 Concluding Remarks

We hope that this book has provided the reader with valuable insights into the driving forces in nano-CMOS VLSI design—in particular, subwavelength lithography and design for manufacturability. The subject of this text has been the current state of DFM and DFR as understood by academia and practiced by industry. We trust that the comprehensive overview of DFx techniques and issues provided here will motivate readers to explore these topics in more depth. A clear understanding of design for manufacturability and reliability, in the context of semiconductor design and manufacturing, provides an excellent start to a successful career in nano-CMOS VLSI design.

References

1. Artur Balasinki, "A Methodology to Analyze Circuit Impact of Process Related MOSFET Geometry," *Proceedings of SPIE* **5378**: 85–92, 2004.
2. S. D. Kim, H. Wada, and J. C. S. Woo, "TCAD-Based Statistical Analysis and Modeling of Gate Line-Edge Roughness: Effect on Nanoscale MOS Transistor Performance and Scaling," *Transactions on Semiconductor Manufacturing* **17**: 192–200, 2004.
3. Wojtek J. Poppe, L. Capodieci, J. Wu, and A. Neureuther, "From Poly Line to Transistor: Building BSIM Models for Non-Rectangular Transistors," *Proceedings of SPIE* **6156**: 61560P.1–61560P.999, 2006.
4. Ke Cao, Sorin Dobre, and Jiang Hu, "Standard Cell Characterization Considering Lithography Induced Variations," in *Proceedings of Design Automation Conference*, IEEE/ACM, New York, 2006.
5. Sean X. Shi, Peng Yu, and David Z.Pan, "A Unified Non-Rectangular Device and Circuit Simulation Model for Timing and Power", in *Proceedings of International Conference on Computer Aided Design*, IEEE/ACM, New York, 2006, pp. 423–428.
6. A. Sreedhar and S. Kundu, "On Modeling Impact of Sub-Wavelength Lithography," in *Proceedings of International Conference on Computer Design*, IEEE, New York, 2007, pp. 84–90.
7. A. Sreedhar and S. Kundu, "Modeling and Analysis of Non-Rectangular Transistors Caused by Lithographic Distortions," in *Proceedings of International Conference on Computer Design*, IEEE, New York, 2008, pp. 444–449.
8. Ritu Singhal et al., "Modeling and Analysis of Non-Rectangular Gate for Post-Lithography Circuit Simulation," *Proceedings of Design Automation Conference*, IEEE/ACM, New York, 2007, pp. 823–828.
9. Puneet Gupta, Andrew Kahng, Youngmin Kim, Saumil Shah, and Dennis Sylvester, "Modeling of Non-Uniform Device Geometries for Post-Lithography Circuit Analysis," *Proceedings of SPIE* **6156**: 61560U.1–61560U.10, 2006.
10. Puneet Gupta, Andrew B. Kahng, Youngmin Kim, Saumil Shah, and Dennis Sylvester, "Investigation of Diffusion Rounding for Post-Lithography Analysis," in *Proceedings of Asia and South-Pacific Design Automation Conference* IEEE, New York, 2008, pp. 480–485.
11. Robert Pack, Valery Axelrad, Andrei Shibkov et al., "Physical and Timing Verification of Subwavelength-Scale Designs, Part I: Lithography Impact on MOSFETs," *Proceedings of SPIE* **5042**: 51–62, 2003.
12. Puneet Gupta, Andrew B. Kahng, Sam Nakagawa, Saumil Shah, and Puneet Sharma, "Lithography Simulation-Based Full-Chip Design Analyses," *Proceedings of SPIE* **6156**: 61560T.1–61560T.8, 2006.
13. A. Balasinski, L. Karklin, and V. Axelrad, "Impact of Subwavelength CD Tolerance on Device Performance," *Proceedings of SPIE* **4692**: 361–368, 2002.
14. S. Shah et al., "Standard Cell Library Optimization for Leakage Reduction," in *Proceedings of ACM/IEEE Design Automation Conference*, ACM/IEEE, New York, 2006, pp. 983–986.
15. A. B. Kahng, C.-H. Park, P. Sharma, and Q. Wang, "Lens Aberration Aware Placement for Timing Yield," in *Proceedings of ACM Transactions on Design Automation of Electronic Systems* **14**: 16–26, 2009.
16. V. Joshi, B. Cline, D. Sylvester, D. Blaauw, and K. Agarwal, "Leakage Power Reduction Using Stress-Enhanced Layouts," in *Proceedings of Design Automation Conference*, ACM/IEEE, New York, 2008, pp. 912–917.
17. M. Mani, A. Singh, and M. Orshansky, "Joint Design-Time and Post-Silicon Minimization of Parametric Yield Loss Using Adjustable Robust Optimization," in *Proceedings of IEEE/ACM International Conference on Computer Aided Design*, IEEE/ACM, New York, 2006, pp. 19–26.
18. M. Cho, D. Z. Pan, H. Xiang, and R. Puri, "Wire Density Driven Global Routing for CMP Variation and Timing," *Proceedings of IEEE/ACM International Conference on Computer Aided Design*, IEEE/ACM, New York, 2006, pp. 487–492.
19. K. Jeong et al., "Impact of Guardband Reduction on Design Process Outcomes," in *Proceedings of IEEE International Symposium on Quality Electronic Design*, IEEE, New York, 2008, pp. 790–797.

20. Y. Lu and V. D. Agarwal, "Statistical Leakage and Timing Optimization for Submicron Process Variation," in *Proceedings of IEEE VLSI Design Conference*, IEEE, New York, 2007, pp. 439–444.
21. M. Mani, A. Devgan, and M. Orshansky, "An Efficient Algorithm for Statistical Minimization of Total Power Under Timing Yield Constraints," in *Proceedings of Design Automation Conference*, IEEE/ACM, New York, 2005, pp. 309–314.
22. S. Sirichotiyakul et al., "Duet: An Accurate Leakage Estimation and Optimization Tool for Dual-V_T Circuits," *IEEE Transactions on VLSI Systems* **10**(2): 79–90, 2002.
23. L. Wei et al., "Design and Optimization of Low Voltage High Performance Dual Threshold CMOS Circuits," in *Proceedings of Design Automation Conference*, IEEE/ACM, New York, 1998, pp. 489–494.
24. S. M. Burns et al., "Comparative Analysis of Conventional and Statistical Design Techniques," in *Proceedings of ACM/IEEE Design Automation Conference*, IEEE/ACM, New York, 2007, pp. 238–243.
25. *International Technology Roadmap for Semiconductors Report*, http://www.itrs.net (2007).
26. Minsik Cho, Kun Yuan, Yongchan Ban, and David Z. Pan, "ELIAD: Efficient Lithography Aware Detailed Routing Algorithm with Compact and Macro Post-OPC Printability Prediction," in *IEEE Transactions on Computer-Aided Design of Integrated Circuits and Systems* **28**(7): 1006–1016, 2009.

Index

A

B

E

Q

R

S

X

Y

Z